P9-AGS-521

Verilog Digital System Design

Verilog Digital System Design

Zainalabedin Navabi
Northeastern University
University of Tehran

McGraw-Hill

New York San Francisco Washington, D.C. Auckland Bogotá
Caracas Lisbon London Madrid Mexico City Milan
Montreal New Delhi San Juan Singapore
Sydney Tokyo Toronto

Library of Congress Cataloging-in-Publication Data

Navabi, Zainalabedin.
Verilog digital system design / Zainalabedin Navabi.
p. cm.
Includes bibliographical references and index.
ISBN 0-07-047164-9
1. Verilog (Computer hardware description language) 2. Electronic digital computers—Design and construction—Data processing.
I. Title.
TK7885.7.N36 1999
621.39'2—dc21 98-25563
CIP

McGraw-Hill

A Division of The ***McGraw-Hill*** *Companies*

1 2 3 4 5 6 7 8 9 0 AGM/AGM 9 0 4 3 2 1 0 9

P/N 135222-8
PART OF
ISBN 0-07-047164-9

The sponsoring editor for this book was Stephen S. Chapman, the editing supervisor was David E. Fogarty, and the production supervisor was Sherri Souffrance. It was set in Century Schoolbook per the MHC Computer design by Paul Scozzari and Michele Pridmore of McGraw-Hill's Professional Book Group Hightstown composition unit.

Printed and bound by Quebecor/Martinsburg.

This book was printed on recycled, acid-free paper containing a minimum of 50% recycled de-inked fiber.

About the cover. The cover shows steps towards a hierarchical design. Persepolis, Shiraz Iran, 1998 photo by Sane.

To Sadri K. Navabi,
my mother, my inspiration

Contents

Preface

This book is on the IEEE Standard Hardware Description Language based on the Verilog® Hardware Description Language (IEEE Std. 1364-1995). The intended audiences are engineers involved in various aspects of digital systems design and manufacturing or students with the basic knowledge of digital system design. The emphasis of the book is on using Verilog HDL for the design and synthesis of digital systems. We will first present information on the use of hardware description languages in design and synthesis of digital systems; this is followed by a comprehensive presentation of the Verilog HDL and then followed by material related to use of this language in hierarchical design and implementation of a complete system. Embedded in the presentation of the language, the book provides a review of digital system design and computer architecture concepts. This review is useful for relearning these concepts as demanded by new design methodologies and hardware-description-language-based design tools. For practicing engineers, the flow of the book, which starts from introductory material and advances into complex digital design concepts, provides a self-sufficient learning tool. The material is suitable for an upper division undergraduate or a first-year graduate course. For a one-semester course on the Verilog HDL language and its use in a digital system design environment, the book can be used in its entirety. The book can also be used as a supplement for graduate and undergraduate digital system design and computer organization courses.

This book is packaged with a CD that contains Verilog simulation software, a Verilog graphic editor, and a Verilog Circuit Navigator. The navigator provides electronic indexing for the examples of the book. Examples are sorted by their function and the navigator facilitates access to Verilog code of each example. Verilog code of each example has pointers to language constructs used, other examples using the same constructs, and pointers to sections of the book where examples are discussed.

The book starts with an introductory material presenting design flow and the use and methodologies involved in simulation and synthesis of digital systems. This is followed by a brief history of Verilog and its evolution into an IEEE standard. Following this background material, a minimum set of Verilog language elements necessary for generating basic designs are presented. Parallel with this presentation we will show design steps based on hardware description languages, which takes advantage of simulation, synthesis, and predefined design libraries. This presentation is intended to illustrate the use of hardware description languages in new design environments and with new design aids. Following this material, timing concepts, concurrency, and procedural bodies of Verilog are presented. With this presentation, the reader will be ready to learn the details of the language in Chapters 5 to 9. These chapters present the language from a low-level to high-level fashion. The concepts more closely related to hardware, and which are easier for hardware designers to comprehend, are presented first. Structural at the gate and switch level representations of hardware including timing and various delay specification formats are described immediately after the basic concepts have been covered. This material is followed by description of more advanced concepts for higher-level hardware representations, which is then followed by dataflow and behavioral level of hardware description. In the sections on dataflow description, clocking schemes, bussing structures, sequential circuit modeling, and data-control partitioning will be covered. In the parts describing behavioral descriptions, high-level functional descriptions, handshaking, component level data communication, and test benches will be described. We will discuss language utilities for test bench generation and design diagnostics. After completion of Chap. 9, presentation of the Verilog language concepts and constructs is complete and the book continues with presenting cases of design that utilize Verilog-based tools. We will present a top-down design example for manual discrete design and examples of interfacing and board-level design.

Presentation of Verilog examples begins in Chap. 3. In this chapter, examples are designed to give the reader an overview of the language and to illustrate the use of Verilog-based tools in digital system design. The next set of examples is presented in Chap. 4 and consists of partial code for illustration of certain language issues. These examples do not represent actual hardware structures and components. The examples in the rest of the chapters, however, are complete and represent actual hardware components with various degrees of complexity. These examples have been carefully chosen so that their execution and simulation depend on the knowledge of the material covered prior to the presentation of the example. These examples can, therefore, be executed without having to refer to the later parts of the book. Covering Verilog language features along with the presentation of examples is done with the examples of Chaps. 4 through 9. The examples in these chapters get progressively more complex and more complete and prepare readers for the designs of Chaps. 10 and 11. Each chapter includes problems related to the chapter material. Although the book can be used on its own to learn concepts of Verilog, it will benefit the reader most if used with the simulation program provided on the enclosed CD.

Overview of the Chapters

Chapter overviews are presented below. This material is intended to help a reader concentrate on parts of the book that he or she finds suit his or her needs best. Chapters 1 and 2 are introductory, and contain material with which many readers may already be familiar. It is, however, recommended that these chapters not be completely omitted, even by experienced readers. Overview of the language is given in Chap. 3 without concentrating on the syntax and semantics of the constructs used. The details of language syntax and semantics are described in Chaps. 4 through 9. The last two chapters contain discrete design and system design parts of a complete system, and no new language concepts or constructs are presented.

Chapter 1 gives an overview of digital design process and the use of hardware description languages in this process. Levels of abstractions will be defined here and will be referred to in the rest of the text. The role of design tools such as simulation programs, test generation, and hardware synthesis tools in a design flow will be discussed in this chapter.

Chapter 2 describes evolution of Verilog and identifies the main characteristics and capabilities of this language.

Chapter 3 servers the two purposes of giving an overview of the Verilog language and showing how this language and its related tools can help a hardware designer. We will show a top-down design flow that utilizes simulation, synthesis, and predefined parts. The overview given here is needed for understanding the material in the rest of the book.

Chapter 4 is a key chapter covering language timing and concurrency issues. Two main features of a hardware description language are timing and concurrency, and thorough understanding of these issues is essential for correct use of such a language. This chapter explains how concurrent and procedural bodies of Verilog can be used for accurate representation of hardware structures.

Chapter 5 is on structural description of hardware. We will show how transistors and gates are used for describing functional units, and how such units are wired together for creating a higher-level design. Issues regarding use of predefined primitives, creating user primitives, instantiation of modules or primitives, wiring and iterative networks, pin-to-pin and distributed delay specification are emphasized here.

Chapter 6 is on design organization and parameterization. It discusses the use of system- and user-defined tasks, and shows how a parameterized design can be configured for a specific set of values. Hierarchical names and parameter specification through hierarchies is described here. Use of system tasks for reading external data will be discussed in this chapter.

Chapter 7 offers a description of type declaration and usage in Verilog. One- and-two dimensional memories and indexing these structures are discussed here. This chapter presents Verilog operators and system tasks and utilities.

Chapter 8 discusses various signal assignments and dataflow descriptions. State machines, bussing structures, and register clocking schemes are explained in detail. We will present formation of a complete system consisting of busses, registers, and controllers.

Chapter 9 is on the behavioral description of hardware. It discusses high-level timing issues, handshaking, and behavioral representation for state machines. Use of flow control statements in description of handshaking is discussed here. A bus arbiter and a serial to parallel adapter utilizing behavioral constructs of Verilog will be discussed in this chapter. The last part of the chapter presents use of system tasks for text output to a file or the standard output.

Chapter 10 ties it all together in a manual design based on discrete components. Verilog constructs and coding techniques learned in the previous chapters will be used for describing a CPU. The top-down design methodology presented in Chap. 4 will be used for the design of this hardware. The CPU will be partitioned into registers, busses and logic units, and a controller. Each unit will be separately designed and described. The data section wires these components, and a control section generates signals for control of data.

Chapter 11 shows how a board-level designer can use Verilog for design and description of a board using existing parts and new components to be interfaced with. System-level handshaking and interfacing is emphasized here. We will use the CPU of Chap. 10 and interface to it a cache controller and a serial data receiver. The serial data receiver uses a direct memory access (DMA) controller for interfacing with the CPU. High-level bussing and arbitration will be used in this example design.

Appendix A lists commonly used system tasks and briefly describes each task. Appendix B lists Verilog compiler directives and explains their use. Appendix C presents the standard IEEE Verilog HDL syntax. Language constructs terminals and nonterminals are presented here in a formal grammar representation. Appendix D includes complete behavioral and dataflow descriptions of the processor of Chap. 10. In Appendix E several synthesizable examples are presented, and Appendix F lists and explains the software included on the CD-ROM that comes with the book.

Suggested Reading Flow

The book teaches the Verilog language for design, simulation, and synthesis of digital systems. For a complete comprehension of these issues, or for a complete one-semester graduate course, the book is recommended in its entirety. However, for specific needs and requirements or for an undergraduate course on automated design methodologies, parts of the book can also be used. The following paragraphs present several such uses.

For a hardware designer interested in learning about synthesis, Chaps. 1 through 3 and the synthesizable CPU code of Chap. 10 provide sufficient study material. Such audiences as a reference can use the rest of the book as their design needs require.

Chapters 4 through 9 provide learning material for the complete Verilog language. Designers familiar with hardware description languages and design environments based on these languages should study these chapters to learn Verilog and be able to use it in their designs.

Chapters 10 and 11 can be used for learning computer organization concepts and the use of Verilog in description of these structures. A reader familiar with Verilog can use his or her knowledge of Verilog to learn the interworkings of CPU structures, I/O devices, and interfacing hardware.

Code Examples

Among many tasks involved in the preparation of the manuscript for a book describing a language that is as example oriented as this book, selecting an appropriate set of examples and presenting them to the reader are of special importance.

Examples in this book are all related and their complete set defines the flow of the material in the book. With each example, there is a logic design concept that is used and there are several Verilog constructs and features that are covered. The set of examples is chosen to present the complete Verilog language. The flow of the book starts with simple Verilog constructs and progressively moves into more complex ones. Parallel with the flow of language constructs, the book starts with using simple logic design concepts, such as using basic gates for combinational circuits, and moves into advanced logic design concepts such as cache and memory interfacing.

The examples in Chapts. 5 to 9 cover digital system components ranging from simple gates to functional units and state machines. The Verilog code for these examples can be used as a reference for a hardware designer, or as templates for generating models for various digital circuit parts. For all examples presented in the book a test bench and a simulation report has been generated and are available on the enclosed CD through the Verilog Circuit Navigator. Text files for these examples are also available on the CD.

Acknowledgments

Guidelines, comments, reviews, and support of many people helped the development of this book, and the author wishes to thank them. The style used for presenting the material is based on simple examples that cover a certain topic and discussing the issues that the example covers. Professor Fredrick J. Hill of the University of Arizona, with whom I worked many years as a student and a research associate, used this method of presentation in his digital systems book of 1973, and it is from him that I learned this method of presentation. The idea of developing a Verilog book was initiated by Mr. Steve Chapman of McGraw-Hill. Throughout the development of this book I consulted with him when important decisions were to be made, and he was always helpful in presenting new ideas. The editors and staff of McGraw-Hill were very helpful in providing timely feedback on reviewers comments and making corrections to the text.

My students and colleagues were particularly helpful in the development of this book. In the past twelve years my students at the University of Tehran, Northeastern University, and National Technological University have been very helpful in bringing up ideas for more illustrative examples. I thank my students Lily Ghasemzadeh and Sasan Komeilizadeh for reviewing the original manuscript and page proofs. Their comments and corrections were very helpful. At the start of this writing project, Ms. Fatemeh Asgari, computer operations manager in the CAD laboratory of the ECE department of the University of Tehran, assumed responsibility for managing the preparation of the manuscript. She took charge in organizing the efforts for manuscript preparation, keyed in the manuscript, completed the artwork for over 400 figures in the book, and developed the Verilog Circuit Navigator tool. She was always helpful in incorporating my last-minute changes and meeting deadlines. Her contribution to this work and her management of the project made it possible for me to spend more time writing the book and generating examples.

Most of all, I thank my wife, Irma Navabi, for help, encouragement, and understanding of my working habits. Such an intensive work could not be done if I did not have the support of my wife and my two sons, Arash and Arvand. I thank them for this and other scientific achievements I have had.

ZAINALABEDIN NAVABI
Boston, Massachusetts

Chapter

1

Hardware Design Environments

As the size and complexity of digital systems increase, more computer-aided design tools are being introduced into the hardware design process. The early paper-and-pencil design methods have given way to sophisticated design entry, verification, and automatic hardware generation tools. The newest addition to this design methodology is the introduction of hardware description languages (HDLs). Although the concept of HDLs is not new, their widespread use in digital system design is no more than a decade old. Based on HDLs, new digital system computer-aided design (CAD) tools have been developed and are now being utilized by hardware designers. At the same time, researchers are finding more ways in which HDLs can improve the process of digital system design.

This chapter discusses the concept of hardware description languages and their use in a design environment. We will describe a design process, indicate where HDLs fit in this process, and describe simulation and synthesis, the two most frequent applications of HDLs.

1.1 Digital System Design Process

Figure 1.1 shows design steps that must be carried out in a typical design of a digital system. After the initial design idea, a designer goes through several design steps before a hardware implementation is obtained. At each step, the designer checks the result of the last transformation, adds more information to it, and passes it to the next design step.

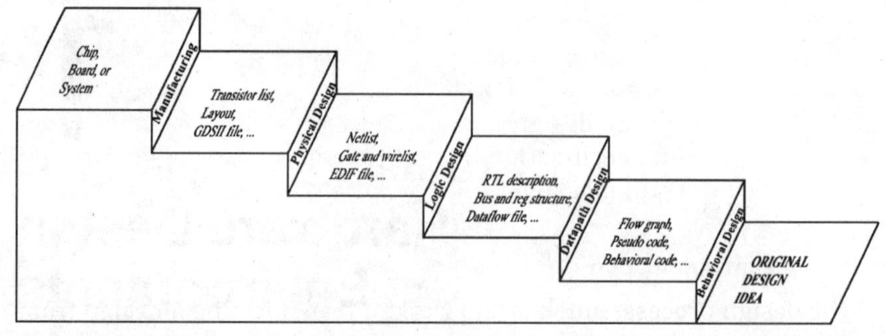

Figure 1.1 Design steps for a digital system.

Initially, a hardware designer starts with a design idea. A more complete definition of the intended hardware must then be developed from the initial design idea. Therefore, it is necessary for the designer to generate a behavioral definition of the system under design. The product of this design step may be a flowchart, a flow graph, or pseudocode. The designer specifies the overall functionality and an input-to-output mapping without giving architectural or hardware details of the system under design.

The next step in the design process is the design of the system data path. In this step, the designer specifies the registers and logic units necessary for implementation of the system. These components may be interconnected using either bidirectional or unidirectional busses. Based on the intended behavior of the system, the procedure for controlling the movement of data between registers and logic units through busses is then developed. Figure 1.2 shows a possible result of the data path design step. Data components in the data part of a circuit communicate via system busses, and the control procedure controls the flow of data between these components. As shown, this design step results in the architectural design of a system with specification of the control flow. No information about the implementation of the controller—hard-wired, encoding technique, or microprogrammed—is given in this step.

Logic design is the next step in the design process; it involves the use of primitive gates and flip-flops for the implementation of data registers, busses, logic units, and their controlling hardware. The result of this design step is a netlist of gates and flip-flops. Components used and their interconnections are specified in this netlist. Gate technology and even gate-level details of flip-flops are not included in this netlist.

The next design step is the transformation of the netlist of the previous level into a transistor list or layout. This involves the replacement of gates and flip-flops with their transistor equivalents or library cells. This step considers loading and timing requirements in its cell or transistor selection process.

The final step in the design is manufacturing, which uses the transistor list or layout specification to burn fuses of a field-programmable device or to generate masks for IC fabrication.

1.1.1 Design automation

In the design process, much of the work of transforming a design from one form to another is tedious and repetitive. From the point of view of a digital system designer, a design is complete when an idea is transformed into an architecture or a data path description. The rest is routine work and involves tasks that a machine can do much faster than a talented engineer. The same can be said about the verification process; that is, a machine can be programmed to verify the functionality or timing of a designed circuit much more easily than any human can.

Activities such as transforming one form of a design into another, verifying a design step output, or generating test data can be done at least in part by computers. This process is referred to as design automation (DA).

Design automation tools can help the designer with design entry, hardware generation, test sequence generation, documentation, verification, and design management. Such tools perform their specific tasks on the output of each of the design steps of Fig. 1.1. For example, to verify the outcome of the data path design step, the bussing and

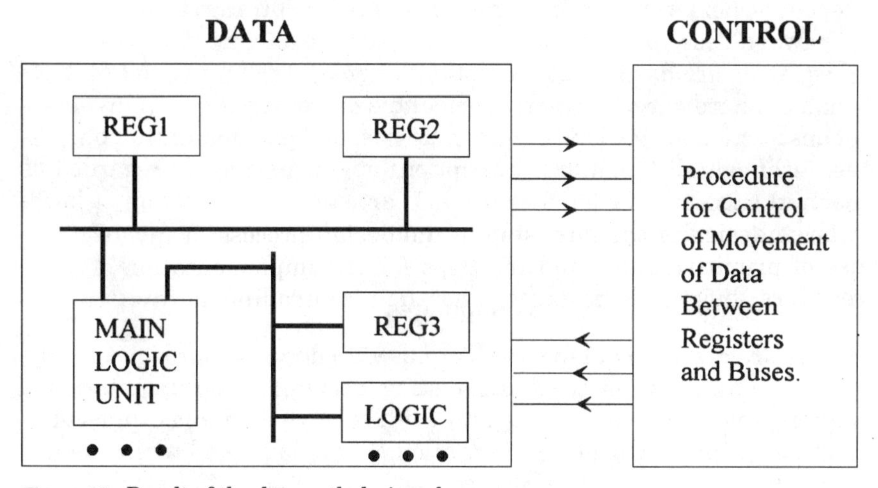

Figure 1.2 Result of the data path design phase.

register structure is fed into a simulation program. Also, to generate tests for register transfer faults, a design automation tool can be used for processing this level of system description and produce tests that can be used by test equipment. Other DA tools include a synthesizer that can automatically generate a netlist from the register and bus structure of the system under design, or one that generates an architectural layout based on the behavioral description of the system.

HDLs provide formats for representing the outputs of various design stages. An HDL-based DA tool for the analysis of a circuit uses this format for its input description, and a synthesis tool transforms its HDL input into an HDL which contains more hardware information. In the sections that follow, we discuss modeling, hardware description languages, digital system simulation, and hardware synthesis.

1.2 Hardware Modeling

A hardware designer models a circuit using available modeling tools. The level of abstraction for this modeling depends on the purpose for which the model is intended. If the model is to be used for documenting the functionality of the circuit, a very abstract behavioral-level model is necessary. On the other hand, if the model is to be used for verification of the timing of the circuit, a more detailed description is needed. A hardware engineer models his or her circuit so that it imitates the actual hardware component as closely and as accurately as possible for its intended purpose. A good modeler uses available hardware modeling tools to produce an elegant model of the hardware part.

Modeling tools available to a hardware engineer include paper and pencil, schematic capture programs, breadboarding facilities, and hardware description languages. The newest and most promising of these tools are hardware description languages. These modeling tools enable a hardware designer to model his or her circuit at many levels of abstraction for various design, analysis, and documentation purposes. Although all hardware description languages may be regarded as such, the level of model elegance and artistic representation of hardware may be different from one language to another.

1.3 Hardware Description Languages

Hardware description languages are used to describe hardware for the purpose of simulation, modeling, testing, design, and documentation of digital systems. These languages provide a convenient and compact format for the hierarchical representation of functional and wiring details of digital systems. Some HDLs consist of a simple set of symbols and notations which replace schematic diagrams of digital circuits, while others are more formally defined and may present the hardware at one or more levels of abstraction. Available software for HDLs includes simulators, test and testability applications, and hardware synthesis programs. A simulation program can be used for design verification, while a synthesizer is used for automatic hardware generation. A test generation program may depend on a hardware description language to provide a netlist format, test application test bench, and fault injection.

In this section, examples of three hardware description languages are presented and discussed. The languages chosen represent the outputs of the first three design steps of Fig. 1.1. In the code examples that follow, uppercase letters are used for keywords and reserved words of the language.

1.3.1 A language for behavioral descriptions

Instruction Set Processor Specification (ISPS) is an HDL for describing the behavior of digital systems. This language was developed at Carnegie Mellon University and is based on the ISP notation, which was first introduced by C. G. Bell in 1971. ISPS was designed for hardware simulation, design automation, and automatic generation of machine-relative software (compiler-compilers). This language is a softwarelike programming language, but it includes constructs for specifying movement of data between registers and busses. CPU-like architectures can be easily and efficiently described in ISPS. The description of the Manchester University Mark-1 computer as it appeared in the ISPS reference manual is given in Fig. 1.3.

The declarative part of this description indicates that the machine has an 8K, 32-bit memory (*m*), a 16-bit instruction register (*pi*), a 13-bit control register (*cr*), and a 32-bit accumulator (*acc*). Also shown in this declarative part is the renaming of bits 15 to 13 of *pi* to *f*, and of its bits 12 to 0 to *s*. The instruction execution of this machine begins by moving a word from the memory into the *pi* register. Following this fetch, a decode language construct decodes the function bits of *pi* (*f* is equivalent to *pi*<15:13>). Based on these bits, one of the seven instructions of *mark1* is executed. For example, if *f* is 3, a store (*sto*) instruction will be executed, which causes the accumulator to be stored at address *s* (bits 12 to 0 of *pi*) of the memory. When an instruction execution is complete, *cr* is incremented by 1, and the next instruction cycle begins.

This example shows that ISPS is easy to read and is close to the way a designer first thinks about the behavior of a hardware component. Referring to Fig. 1.1, ISPS is most appropriate for representing the

```
mark1 :=
  BEGIN
  ** memory.state **
  m[0:8191]<31:0>,
  ** processor.state **
  pi\present.instruction<15:0>'
    f\function<0:2> := pi<15:13>,
    s<0:12> := pi<12:0>,
  cr\control.register<12:0>,
  acc\accumulator<31:0>,
  ** instruction.execution ** {tc}
MAIN i.cycle :=
  BEGIN
  pi = m[cr]<15:0> NEXT
  DECODE f =>
    BEGIN
  0\jmp := cr = [s],
  1\jrp  := cr = cr + m[s],
  2\ldn  := acc = - m[s],
  3\sto  := m[s] = acc,
  4:5\sub:= acc = acc - m[s],
  6\cmp:= IF acc LSS 0 => cr = cr + 1,
  7\stp  := STOP(),
    END NEXT
  cr = cr + 1 NEXT
  RESTART i.cycle
  END
```

Figure 1.3 An ISPS example showing a simple processor. (*From M. R. Barbacci, The ISPS Computer Description Language, Carnegie-Mellon University, 1981, p. 70.*)

output of the behavioral design step in a design process. An ISPS simulator can, therefore, validate the initial plans of a designer for the design of a CPU-like architecture.

1.3.2 A language for describing flow of data

A Hardware Programming Language (AHPL) was developed at the University of Arizona and was used as a tool for teaching computer organization for over two decades. In fact, this language started as a set of notations for representation of hardware in an academic environment. These notations were used instead of spatially inefficient schematic diagrams. The evolution of the initial set of notations led to the development of the AHPL hardware language. The development of a compiler and a simulator established a place for this language in the family of hardware description languages.

Figure 1.4 shows an AHPL example description. This is a 4-bit sequential multiplier that uses the add-and-shift multiplication method. The circuit receives two operands from its *inputbus* and produces the result on its 8-bit *result* output.

The description begins with the declaration of registers and busses. The circuit requires three 4-bit registers for the two operands and the intermediate results, a single flip-flop for the *done* indicator, and a 2-bit counter (*count*) for the number of bits shifted out of the first operand register. The external *dataready* and *inputbus* signals are declared as EXINPUTS and EXBUSES, respectively. The last of the declarations, CLUNITS, indicates the presence of combinational logic networks implementing a 2-bit incrementer and a 4-bit adder.

The circuit sequence part follows the declarations. Step 1 receives the operands and stores them in the *ac1* and *ac2* registers. If there is a 1 on the *dataready* line, control proceeds to step 2; otherwise step 1 remains active. Step 2 sets the *busy* flip-flop to 1 and causes step 3 to be skipped if *ac1[3]*, which is the least-significant bit of the *ac1* register, is zero. In step 3 the addition of the partial products is accomplished. Step 4 right shifts the concatenation of the *extra* and *ac1*

```
AHPLMODULE: multiplier.
        MEMORY: ac1[4]; ac2[4]; count[2]; extra[4]; busy.
        EXINPUTS: dataready.
        EXBUSES: inputbus[8].
        OUTPUTS: result[8]; done.
        CLUNITS: INC[2](count); ADD[5](extra; ac2);
    1       ac1 <= inputbus[0:5];  ac2 <= inputbus[4:7];
            extra <= 4$0;
            => (^dataready)/(1).
    2       busy <= \1\;
            => (^ac1[3])/(4).
    3       extra <= ADD[1:4](extra; ac2).
    4       extra, ac1 <= \0\, extra, ac1[0:2];
            count <= INC(count);
            => (^(&/count))/(2).
    5       result = extra, ac1;  done = \1\;  busy <= \0\;
            => (1).
        ENDSEQUENCE
        CONTROLRESET(1).
    END.
```

Figure 1.4 An AHPL example showing a sequential multiplier.

registers, increments the counter, and activates step 2 if *count* has not reached (1,1). If the *count* register contains (1,1), control will proceed to step 5, where the concatenation of the *extra* and *ac1* registers is placed on the eight *result* lines, a 1 is placed on line *done,* and the *busy* flip-flop is reset to zero. Step 5 returns control to step 1, waiting for another set of operands.

1.3.3 A language for describing netlists

Another way to describe a digital system is by its netlist, which specifies the interconnections of its components. A subset of the VHDL hardware description language can be used for this purpose. Figure 1.5 shows a logic diagram of a full-adder and its corresponding VHDL structural description.

The description contains a part for interface declaration and a part for describing the operation of the unit. The interface description specifies the name of the component and its ports. Following the declaration of the interface, in the architecture of the full-adder, instantiation of components that correspond to the gates of the full-adder is listed. Each component instantiation consists of a generic map which includes rise and fall delays of the corresponding gate, and a listing of its inputs and output connections. The component labeled *g2* is an AND gate with a rise time of 12 nanoseconds (ns) and a fall time of 10 ns. The output of this gate is labeled *w2,* and its inputs are driven by the circuit primary inputs *c* and *b*.

The examples in this section present three very different ways of describing hardware. The information contained in these descriptions varies in the detail of the hardware that they present. Each description is suited for a CAD tool at a different design stage. The ISPS description contains high-level behavioral information and can serve as a modeling tool for a hardware designer or as a CAD system user interface. This level of description is well suited for a manager who wants to learn about the overall functionality of a final product before it is designed. The AHPL description, however, contains more architectural information and is more appropriate for describing a circuit for design and construction. A hardware designer may use this level of hardware description for verifying his or her design of controller and data path. The third description, VHDL, differs from both the ISPS and AHPL descriptions in that it contains information that a CAD tool can use for detailed analysis or manufacturing of a circuit. This level of description is suited for describing predefined cells, or it may be a netlist produced by a CAD tool after an automatic hardware generation phase. New design environments use a single HDL for all levels of abstractions and for all design steps.

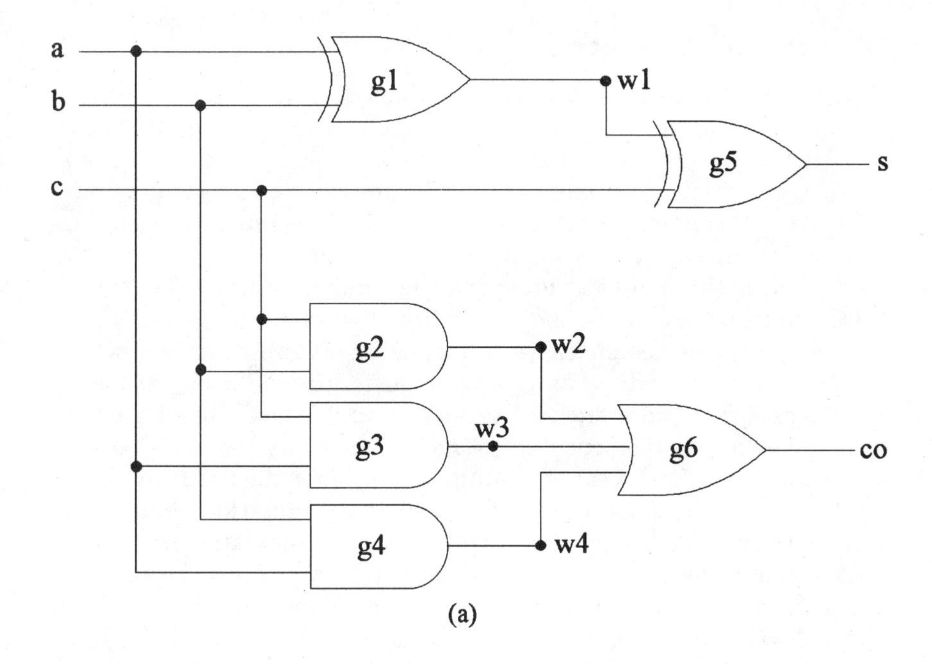

(a)

```
ENTITY full_adder IS
  PORT (a, b, c : IN BIT; s, co : OUT BIT);
END full_adder;
--
ARCHITECTURE gate_level OF full_adder IS
  COMPONENT nand2 PORT (i1, i2: IN BIT; o1: OUT BIT); END COMPONENT;
  COMPONENT xor2 PORT (i1, i2: IN BIT; o1: OUT BIT); END COMPONENT;
  -- Intermediate signals
  SIGNAL im1,im2, im3 : BIT;
BEGIN
  -- sum output
  g1 : xor2 GENERIC MAP (12, 10) PORT MAP (a, b, w1);
  g5 : xor2 GENERIC MAP (12, 10) PORT MAP (c, w1, s);
  -- carry output
  g2 : and2 GENERIC MAP (12, 10) PORT MAP (c, b, w2);
  g3 : and2 GENERIC MAP (12, 10) PORT MAP (a, c, w3);
  g4 : and2 GENERIC MAP (12, 10) PORT MAP (b, a, w4);
  g6 : or3 GENERIC MAP (13, 12) PORT MAP (w2, w3, w4, co);
END gate_level;
```

(b)

Figure 1.5 A full-adder. (*a*) Logical diagram; (*b*) Verilog description.

1.4 Hardware Simulation

Hardware description languages are modeling tools for creating a hardware model. A model may be developed for different applications. One such application is simulation. Simulation is the act of exercising a model of an actual component in order to analyze its conduct under a given set of conditions and/or stimuli. With this definition, a simulation run requires a model of the object being simulated and a set of stimuli for activating or stimulating the conduct of the object that is being studied.

In a hardware design environment, a hardware description uses component models and definitions from a simulation library to form a simulatable hardware model. As shown in Fig. 1.6, the hardware models for library elements, a test bench enclosing the model that is being simulated, and a set of stimuli (test data for the hardware being simulated) are used as inputs of a hardware simulation engine. The simulator produces simulation results that are indicative of the conduct of the hardware component for the given set of test data.

The details of the results obtained from a simulation run depend on the level of detail of a model. A simulation engine may be capable of producing detailed timing and analog voltages, or it may be able to produce high-level function information. The output generated by this simulator depends on the level of detail of the hardware model presented to it. A hardware model for the study of timing and signal voltages must contain this information to be used by a simulator. Also, a hardware description that is being used only for the functional verification of the hardware it is modeling needs to provide information regarding the behavior of the system to the simulator. In either case, the same simulation engine will produce results based on the level of detail of its input model. A simulator based on a hardware description language can process hardware details supported by its input HDL.

In a design automation environment, HDL descriptions of systems can be used for the input of simulation programs. Simulators may be used to verify the results of any of the design steps of Fig. 1.1. A test bench provides stimuli and relative time of data application to a simulation engine. The simulation program applies these data to the input description at the specified times and generates responses of the circuit. Waveforms, timing diagrams, or time-value tabular listings may be used to illustrate the results of a simulation program. These results are interpreted by the designer, who determines whether to repeat a design step if simulation results are not satisfactory.

As shown in Fig. 1.7, simulators can be used at any design step. In the first steps of the design process, simulation provides information regarding the functionality of the system under design. As represented by the hatched areas, simulators for this purpose normally undergo a very quick run on their host computers. Simulation in a later step of the design process—for example, gate-level or device simulation—runs much more slowly but provides more detailed information about the timing and functionality of the circuit. To avoid the high cost of transistor or device-level simulation runs, simulators should be used to detect design flaws as early in the design process as possible.

Regardless of the level of design to which a simulation program is applied, digital system simulators have generally been classified into *oblivious* and *event-driven* simulators. In oblivious simulation, each circuit component is evaluated at fixed time points, while in event driven simulation, a component is evaluated only when one of its inputs changes.

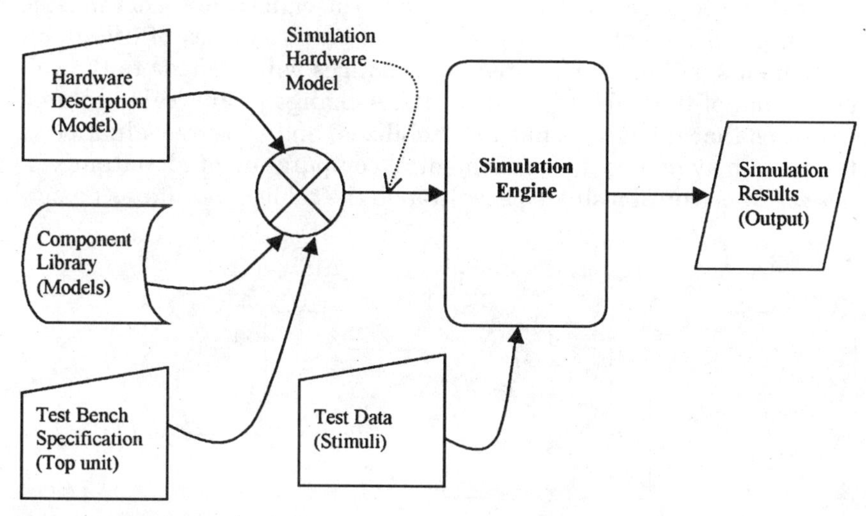

Figure 1.6 Hardware simulation.

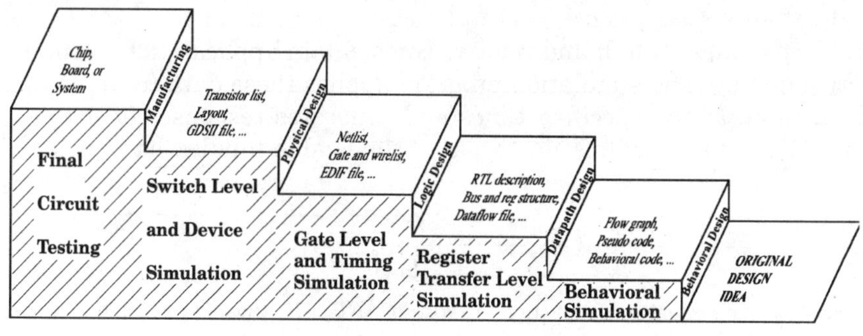

Figure 1.7 Verifying each design step by simulating its output.

1.4.1 Oblivious simulation

As an illustration of the oblivious simulation method, consider the gate network of Fig. 1.8*a*. This is an exclusive-OR circuit that uses AND, OR, and NOT primitive gates and is to be simulated with the data provided in Fig. 1.8*b*.

The first phase of an oblivious simulation program converts the input circuit description to a machine-readable tabular form. A simple example of such a table is shown in Fig. 1.9. This table contains information regarding the circuit components and their interconnections, as well as the initial values for all nodes of the circuit.

After the initialization of the circuit, the simulation phase of an oblivious simulation method reads input values at fixed time intervals and applies them to the internal tabular representation of the circuit. At time t_i, input values of *a* and *b* are read from an input file. These values replace the old values of *a* and *b* in the value column of the table of Fig. 1.9. Using these new values, the output values of *all* circuit components will be reevaluated, and changes will be made to the value column of the affected components. A change in any value column indicates that the circuit has not stabilized, and more reevaluation of the table may be necessary. Sequential computation of all output values continues until a single pass through the table necessitates no new

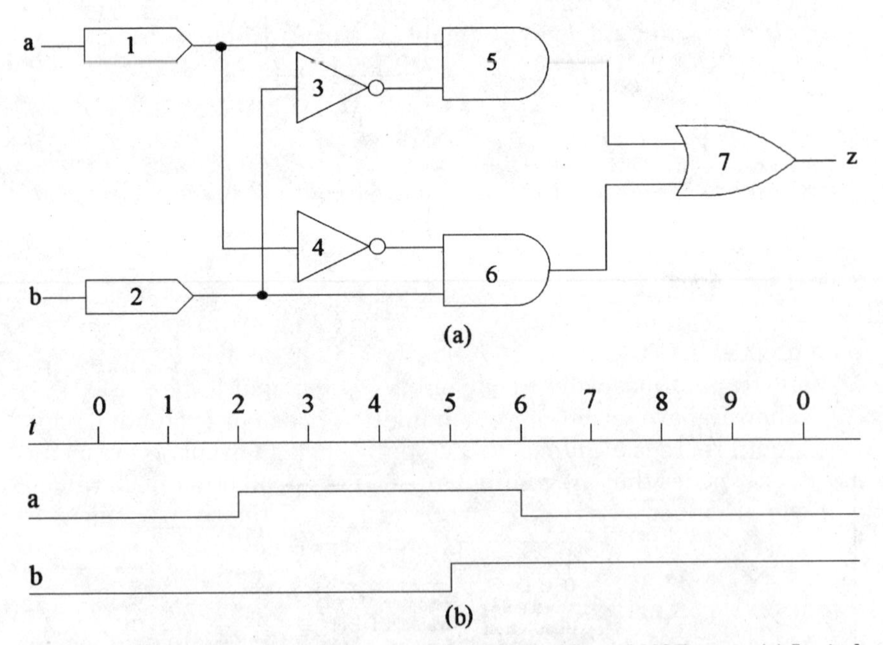

Figure 1.8 An exclusive-OR function in terms of AND, OR, and NOT gates. (*a*) Logical diagram; (*b*) test data.

GATE	FUNCTION	INPUT 1	INPUT 2	VALUE
1	Input	a	--	0
2	Input	b	--	0
3	NOT	2	--	1
4	NOT	1	--	1
5	AND	1	3	0
6	AND	4	2	0
7	OR	5	6	0

Figure 1.9 Tabular representation of exclusive-OR circuit for oblivious simulation.

changes. At this time, all node values for time t_i will be reported, the time indicator will be incremented to t_{i+1}, and new data values will be read from the data file.

1.4.2 Event-driven simulation

Event-driven simulation, while more complex than oblivious simulation, is a more efficient method of digital system simulation. In event-driven simulation, when an input is changed, only those nodes that are affected are reevaluated. A data structure suitable for implementing event driven simulation of our simple gate-level example of Fig. 1.8 is shown in Fig. 1.10.

The first phase of an event-driven simulation program converts the circuit description to a linked-list data structure like that in Fig. 1.10. In the second phase of the simulation, a change on an input triggers only those nodes of the linked list for which an input changes. For example, at time t_2 ($t = 2$) in Fig. 1.8*b*, the transition of *a* from logic level **0** to **1** causes node 1 of the linked list to change its output from **0** to **1.** Since this node feeds nodes 3 and 6, these nodes will also be evaluated, which causes their outputs to change to **0** and **1,** respectively. These changes then propagate to nodes 5 and 7 until the output value is evaluated. No further computations will be done until t_5 when input *b* changes.

As shown above, event-driven simulation does not evaluate circuit nodes until there is a change on an input. When an event occurs on an input, only nodes that are affected are evaluated; all other node values will be unchanged. Since activities occur only on relatively small portions of digital circuits, evaluation of all nodes at all times, as is done in oblivious simulation, is unnecessary. Because of parallelism in hardware structures, event-driven simulation is a more suitable simulation method for digital systems. The speed of this method justifies its more complex data structure and algorithm.

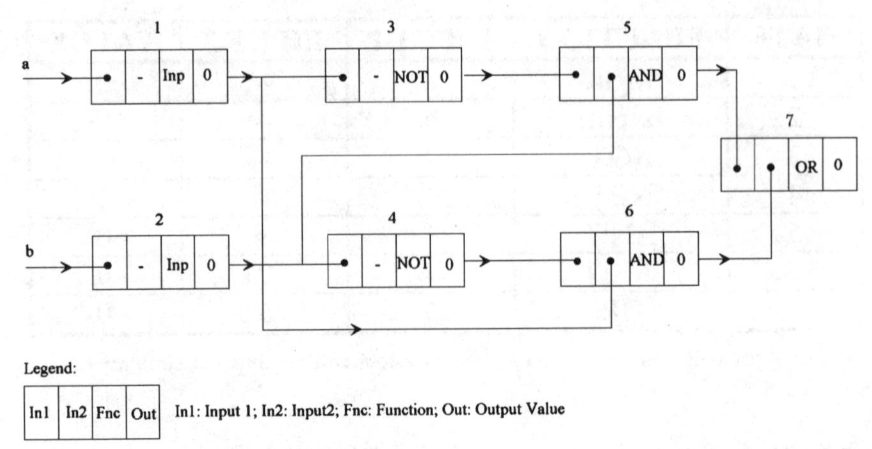

Figure 1.10 Linked-list representation of exclusive-OR circuit for event-driven simulation.

1.5 Hardware Synthesis

A design aid that automatically transforms a design description from one form to another is called a synthesis tool. Such a design aid moves a design one step higher toward its final implementation. Hardware description languages are useful media for input and output of hardware synthesizers. Present synthesis tools replace the designer or provide design guidelines for performing one or more of the design steps of Fig. 1.1. Application of various synthesis tools to the design steps of this figure is shown in Fig. 1.11.

Many commercially available synthesis tools use the output of the data path design stage as input and produce a netlist for the circuit (tool category 2 in Fig. 1.11). Several commercial tools have targeted the field-programmable gate arrays (FPGA) market. FPGA synthesis tools, available from FPGA vendors or general tool vendors, synthesize hardware descriptions in one or more formats to FPGA program file formats that can be fed to a programmer. Such tools concentrate on resource sharing, logic optimization, and binding. Figure 1.12 shows a block diagram of a register transfer level synthesis tool.

Logic units performing register transfer level operations are resources that can be shared between multiple instances of the same operation. If the same operation is used with the same inputs, the sharing is simple and the output of the logic unit will be used in several destinations. An operation used in several instances with different inputs requires the use of multiplexers to provide it with appropriate sets of inputs. Resource sharing may be directed by a synthesis tool user through synthesis directives, or it may be decided by the tool based on the input description. For example, as shown in Fig.

1.13, an add operation on register outputs a and b placing the result on bus c, in one instance, and another operation adding the same inputs and placing the result on bus d, in another instance, can share the same adder unit without any hardware overhead. On the other hand, if the two instances of add operations use different inputs (e.g., a and b in one instance and x and y in another), multiplexers should be used for adder unit input selections.

In the logic optimization pass, boolean expressions are manipulated for better utilization of FPGA cells or chip area. The binding phase in a synthesis tool maps operations of an intermediate data structure,

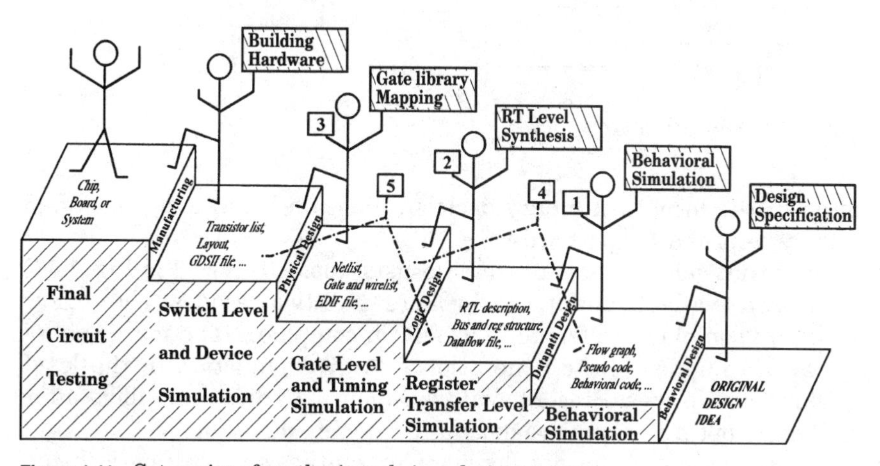

Figure 1.11 Categories of synthesis tools in a design process.

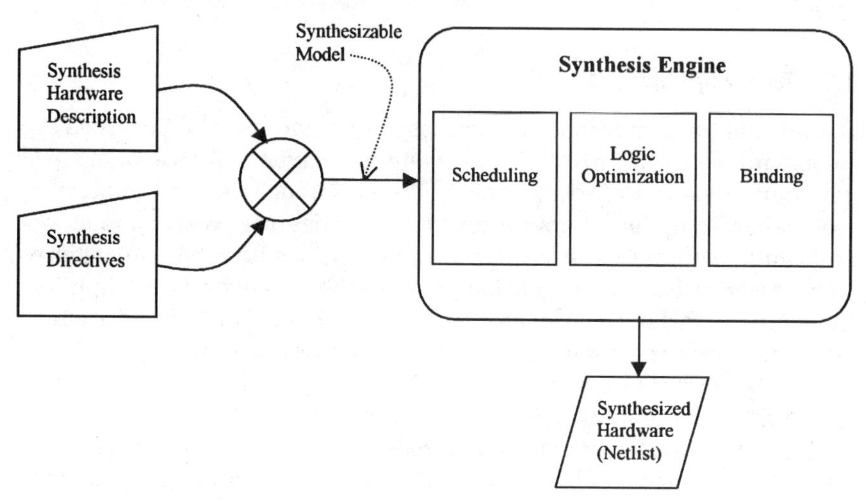

Figure 1.12 Synthesis process.

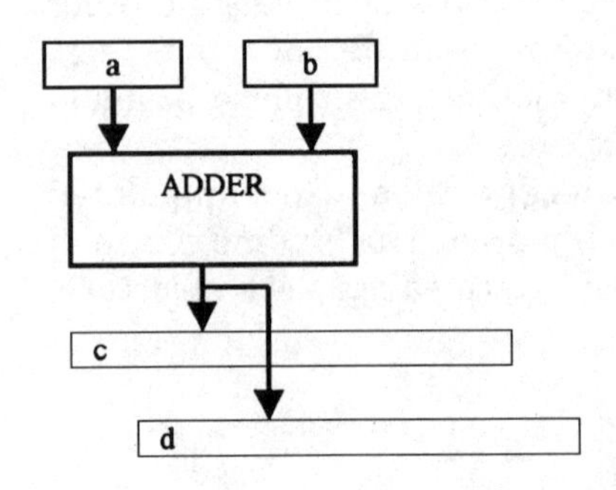

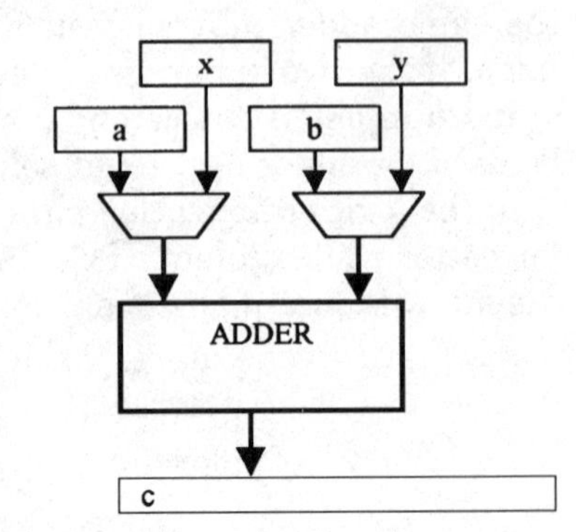

Figure 1.13 Resource sharing.

often in the form of a binary decision diagram (BDD), to predefined library cells of a target hardware.

Tools that generate layout from netlists (tool category 3 in Fig. 1.11) have been available and in use for more than two decades. These tools are more commonly referred to as silicon compilers. The main goal of a silicon compiler is the generation of a given function in the least amount of chip area. Present synthesis tools use silicon compilers after necessary optimizations are done in producing a netlist.

At the present time, development of synthesis programs is being concentrated on ways of producing efficient hardware from general behavioral descriptions of systems (tool categories 1 and 4).

1.6 Test Applications

In addition to simulation and synthesis, test and testability issues are important concerns of a digital system designer. Topics addressed when digital system testing is being discussed include fault collapsing, test generation, fault simulation, test application, test compaction, and fault dictionaries. Although not as advanced in the use of hardware descriptions as simulation and synthesis, several test applications based on HDLs are commercially available, and integration of HDLs into existing commercial fault simulators is under way.

An HDL provides a format for netlist description for the purpose of test generation or fault simulation. Test application test benches for evaluation of fault coverage can be coded in hardware description languages. A waveform subset of a hardware description language can be used for test data representation for test application by test equipment or for stimuli for fault simulation. Ways in which an HDL can benefit a hardware designer in his or her design are still unexplored.

1.7 Levels of Abstraction

The design steps in Fig. 1.1 use hardware descriptions at various levels of abstraction. The difference in the levels of abstraction becomes clear when we compare the three hardware description examples in Sec. 1.3.

The ISPS example presented a readable description of the behavior of the Mark-1 computer. This description does not contain information on the bussing structure of the Mark-1 and does not provide any timing or delay information. The AHPL description of the multiplier described this system based on the flow of data through its registers and busses. This description provides clock-level timing but does not contain gate delay information. The third description in Sec. 1.3 is the structural description of a full-adder in VHDL. This description style provides detailed timing information but becomes very lengthy and unreadable when used for large circuits.

Methods of classifying descriptions of hardware based on the information they contain can be found in the literature. In this book, we take a simple view of this matter and consider abstraction levels of hardware as *behavioral, dataflow,* or *structural* levels. The ISPS description of the Mark-1, the AHPL description of the sequential multiplier, and the VHDL description of the full-adder are examples of these levels of abstraction. The following paragraphs define these abstraction levels.

A behavioral description is the most abstract. It describes the function of the design in a softwarelike procedural form and provides no detail on how the design is to be implemented. The behavioral level of abstraction is most appropriate for fast simulation of complex hardware units, verification and functional simulation of design ideas, modeling standard components, and documentation. A behavioral model is useful for simulation and functional analysis since the details of the hardware, which may not be known to the user of the components, are not required. Such descriptions present an input-to-output mapping according to the data sheet specification provided by the component manufacturer. Descriptions at this level can be accessible to nonengineers as well as to the end users of a hardware component, and can also serve as good documentation media. The operation of large systems can also be modeled at this level for end users and manual writers.

A dataflow description is a concurrent representation of the flow of control and movement of data. Concurrent data components and carriers communicate through busses and interconnections, and control hardware issues signals for the control of this communication. The level of hardware detail involved in dataflow descriptions is great enough that such descriptions cannot serve as an end-user or nontechnical documentation medium. This level of description, however, is abstract enough to be used by a technically oriented designer to describe the components to be synthesized. Dataflow descriptions imply an architecture and a unique hardware. Simulation of these descriptions involves the movement of the data through registers and busses, and therefore is slower than the input-to-output mapping of behavioral descriptions. The function of the hardware is evident from dataflow descriptions.

A structural description is the lowest and most detailed level of description considered, and is the simplest to synthesize into hardware. Structural descriptions include a list of concurrently active components and their interconnections. The corresponding function of the hardware is not evident from such descriptions unless the components used are known. On the other hand, the hardware is clearly implied by these descriptions, and a simple wire router or printed-circuit-board layout generator can easily produce the described hardware. A structural description that describes the wiring of logic gates is said to be the hardware description at gate level. A gate-level description provides input for detailed timing simulation.

1.8 Summary

This chapter presented introductory material that relates to design of digital systems with hardware description languages. The intention was to give the reader an overall understanding of HDLs; the design process based on HDLs; design tools; and simulation, synthesis, and test. Simulation, synthesis, and test, which are the main concerns of a hardware designer, were briefly discussed. The last part of the chapter discussed levels of abstraction that we will reference throughout the book.

Further Reading

Barbacci, M. R., et al., *The ISPS Computer Description Language,* Carnegie-Mellon University, 1981.

Hill, F. J., and G. R. Peterson, *Digital Systems: Hardware Organization and Design,* 3d ed., John Wiley, New York, 1987.

Miczo, A., *Digital Logic Testing and Simulation,* Harper & Row, New York, 1986.

Navabi, Z., and J. Spillane, "Templates for Synthesis from VHDL," *Proc. of the 1990 ASIC Seminar and Exposition,* September 1990.

Palnitkar, S., *Verilog HDL: A Guide to Digital Design and Synthesis,* Prentice-Hall, Upper Saddle River, N.J., 1996.

Wakerly, J. F., *Digital Design Principles and Practices,* 2d ed., Prentice-Hall, Englewood Cliffs, N.J., 1993.

Walker, R. A., and D. E. Thomas, "A Model of Design Representation and Synthesis," *Proc. of 22nd Design Automation Conference,* 1985.

Problems

1.1 Suggest a data path for the Mark-1 computer of Sec. 1.3.1.

1.2 Show the graphical representation of the data path of the 4-bit multiplier of Sec. 1.3.2.

1.3 Show a state diagram for the control part of the multiplier circuit of Sec. 1.3.2.

1.4 Redesign the full-adder circuit of Sec. 1.3.3 using only two-input NAND gates. Show the VHDL description for this design.

1.5 Write pseudocode for implementing the oblivious simulation method of Sec. 1.4.1.

1.6 Write pseudocode (C-like) for implementing the event-driven simulation method of Sec. 1.4.2.

Chapter

2

Verilog HDL Background

In the design of large digital systems, much engineering time is spent in changing formats for using various design aids and simulators. An integrated design environment is useful for better design efficiency in these systems. In an ideal design environment, the high-level description of the system is understandable to the managers and to the designers, and it uniquely and unambiguously defines the hardware. This high-level description can serve as the documentation for the part as well as an entry point into the design process. As the design process advances, additional details are added to the initial description of the part. These details enable the simulation and testing of the system at various levels of abstraction. By the last stage of design, the initial description has evolved into a detailed description which can be used by a program-controlled machine for generation of final hardware in the form of layout, printed circuit board, or gate arrays.

This ideal design process can exist only if there is a language that can describe hardware at various levels so that it can be understood by the managers, users, designers, testers, simulators, and machines. The IEEE standard Verilog hardware description language is such a language. Verilog evolved because a need existed for an integrated design language that could be used to communicate design data between various levels of abstraction.

2.1 Verilog Evolution

Verilog was designed in early 1984 by Gateway Design Automation. Initially the language was used as a simulation and verification tool. After the initial acceptance of this language by the electronics industry, a fault simulator, a timing analyzer, and, in 1987, a synthesis tool based on this language were developed. Gateway Design Automation and its Verilog-based tools were later acquired by Cadence Design System. Since then, Cadence has been a strong force for popularization of the Verilog hardware description language.

In 1987 VHDL became an IEEE standard hardware description language. Because of its DoD support, VHDL was adopted by the U.S. government for related projects and contracts. In an effort to popularize Verilog, in 1990, Open Verilog International (OVI) was formed, and Verilog was placed in the public domain. This created new interest in Verilog on the part of users and EDA vendors.

In 1993, efforts to standardize this language started. Verilog became an IEEE standard, IEEE 1364-1995, in 1995. Since simulation tools, synthesizers, fault simulation programs, timing analyzers, and many other design tools had already developed for Verilog, this standardization helped further acceptance of Verilog in electronic design communities. At the present time Verilog is the most popular HDL in the United States.

2.2 Verilog Attributes

Verilog is a hardware description language that can be used to describe hardware from the transistor to the behavioral level. The language supports timing constructs for switch-level timing simulation, and, at the same time, it has features for describing hardware at the abstract algorithmic level. A Verilog description may consist of a mix of modules at various levels of abstraction with different degrees of detail.

2.2.1 Switch level

Features of the language that make it ideal for switch-level modeling and simulation include primitive unidirectional and bidirectional switches with parameters for delay and charge storage. Circuit delays may be modeled as propagation delay, rise and fall delay, and line delays. The charge storage feature at this level of abstraction in Verilog makes this language capable of describing dynamic CMOS and MOS circuits.

2.2.2 Gate level

Gate-level primitives with predefined parameters provide a convenient platform for netlist representation and gate-level simulation. For more detailed and special-purpose gate simulations, gate components may be defined at the behavioral level. Verilog also provides utilities for defining primitives with special functionality.

A simple four-value logic system is used in Verilog. However, for more accurate logic modeling, the Verilog model for wires also includes 16 levels of strength.

2.2.3 Pin-to-pin delay

A utility for timing specification of components at the input/output level is provided in Verilog. This utility can be used for back annotation of timing information in original predesigned descriptions. Moreover, the pin-to-pin language facility enables modelers to fine-tune the timing behavior of their models based on physical implementations.

2.2.4 Dataflow level

Bus and register modeling utilities are provided in Verilog. For various bus structures, Verilog supports predefined wire and bus resolution functions using its four-value logic system. The combination of bus logic and resolution functions enable modeling of all physical busses. For register modeling, behavioral clock and timing control constructs can be used for representation of registers with various clocking schemes.

2.2.5 Behavioral level

Procedural blocks of Verilog enable algorithmic representations of hardware structures. Constructs similar to those in software programming languages are provided for describing hardware at this level.

2.2.6 System utilities

System tasks in Verilog provide designers with tools for test bench generation, file access, data handling, data generation, and special hardware modeling. Verilog display and I/O tasks can be used to handle all input and output for data application and simulation.

The programming language interface (PLI) of Verilog provides an environment for accessing Verilog data structures using a library of C-language functions.

2.3 The Verilog Language

The Verilog HDL satisfies all requirements for the design and synthesis of digital systems. The language supports hierarchical description of hardware from system to gate or even switch level. Verilog has strong support at all levels for timing specification and violation detection. Timing and concurrency required for hardware modeling are specially emphasized.

In Verilog, a hardware component is described in a **module** language construct. The module description specifies the input and output list as well as internal component busses and registers. Within a **module,** concurrent signal assignments, component instantiations, and procedural blocks can be used to describe a hardware component.

Several modules can be hierarchically instantiated to form other hardware structures. Leaves of a hierarchical design specification may be modules, primitives, or user-defined primitives. For simulating a design, it is expected that all leaves of the hierarchy are individually compiled.

Many Verilog tools and environments provide simulation, fault simulation, and synthesis. Simulation environments provide graphical front-end programs and waveform editing and display tools. Synthesis tools are based on a subset of Verilog. For synthesizing a design, target hardware, e.g., the specific FPGA or application-specific integrated circuits, must be known.

2.4 Summary

This chapter provided the reader with the history of the evolution of Verilog. With this standard HDL, the efforts of tool developers, researchers, and software vendors have become more focused, resulting in better tools and more uniform environments. In the next chapter we will present features of Verilog that will be used in an automated design environment.

Further Reading

IEEE Standard Hardware Description Language Based on the Verilog Hardware Description Language, IEEE Std. 1364-1995, Institute of Electrical and Electronic Engineers, New York, 1996.

Palnitkar, S., *Verilog HDL: A Guide to Digital Design and Synthesis,* Prentice-Hall, Upper Saddle River, N.J., 1996.

Smith, Douglas J., *A Practical Guide for Designing, Synthesizing and Simulating ASICs and FPGAs Using VHDL or Verilog,* Doone Publications, Madison, Ala., June 1996.

Thomas, D. E., and P. R. Moorby, *The Verilog Hardware Description Language,* 3d ed., Kluwer Academic Publishers, Norwell, Mass., 1996.

Chapter

3

Design Methodology Based on Verilog

The intent of this chapter is to present an overview of Verilog and the way it may be used in a real design. Various concepts of a language, be it software or a hardware language, are interdependent. A general knowledge of the entire language is therefore needed before advanced concepts are described in detail. This chapter will provide the necessary Verilog background to facilitate learning the details of the language in the forthcoming chapters. In addition to the overview, this chapter shows how a designer can take advantage of Verilog and tools based on this language to better organize his or her thoughts and resources for managing the implementation of large designs.

In the sections that follow, we will first present an overview of Verilog and elements of this language, and we will then describe the top-down design process. Section 3.3 presents an example using Verilog in a top-down design process. The example is staged to show steps necessary in a complex design using a simple problem. The Verilog example code in this chapter provides the necessary background for presentation of more complex features of Verilog in Chaps. 5 through 9. In the top-down design process, we will show various levels of abstraction (structural, dataflow, and behavioral) in Verilog, and where and how each level of abstraction can be used.

3.1 Elements of Verilog

Constructs of the Verilog language are designed for describing hardware modules and primitives. This section presents language features related to these applications.

3.1.1 Describing hardware modules

The Verilog HDL is used to describe hardware modules of a system and complete systems. Therefore, the main component of the language, which is a *module,* is dedicated for this purpose. As shown in Fig. 3.1, a module description consists of the keyword **module,** the name of the module, a list of ports of the hardware module, the module functionality specification, and the keyword **endmodule.** Following the list of ports of a module, a semicolon separates this part from the next part, in which usually declarations of parameters, ports, and signals of a module are specified. After the declarations, statements in a module specify functionality of the module. This part defines how output ports react to changes on input ports.

As in software languages, there is usually more than one way in which a module can be described in Verilog. Various descriptions of a component may correspond to descriptions at various levels of abstraction or to various levels of detail of the functionality of a module. As shown in Fig. 3.2, one module description may be at the behavioral level of abstraction with no timing details, while another description for the same component may include transistor-level timing details. A module may be part of a library of predesigned library components and include detailed timing and loading information, while a different description of the same module may be at the behavioral level for input to a synthesis tool. It must be noted that descriptions of the same module need not all behave in exactly the same way. Nor is it required that all descriptions describe behavior correctly. In a fault simulation environment, faulty modules may be developed to study various failure forms of a component.

3.1.2 Primitives

Generally hardware components are described as modules. Often, however, in a design environment, bit-level primitives exist that are used as primitive gates or memory elements. Verilog allows definition of bit-level sequential or combinational primitives. Figure 3.3 shows the basic structure of a user-defined primitive in Verilog.

```
module module_name
    list of ports;
    declarations;
    specification of the
    functionality of module;
endmodule
```

Figure 3.1 Module specifications.

```
module module_i
    (ports, ...);
    declarations, ...;
    behavioral
    specification
    of module_i
    with no timing
    specification
endmodule
```

```
module module_i
    (ports, ...);
    declarations, ...;
    structural
    specification
    of module_i
    with transistor
    level timing
    information
endmodule
```

• • •

```
module module_i
    (ports, ...);
    declarations, ...;
    synthesizable
    specification
    of module_i
    for input to
    a synthesis
    tool
endmodule
```

Figure 3.2 Module descriptions with various levels of detail.

```
primitive udp_name
    (bit level ports, ...);
    declaration;
    tabular specification
    of mapping of inputs
    to the single bit output;
endprimitive
```

Figure 3.3 Verilog user-defined primitive.

3.2 Top-Down Design

Instead of trying to implement the design of a large system all at once, a divide-and-conquer strategy is taken in a top-down design process. Top-down design is referred to as recursive partitioning of a system into its subcomponents until all subcomponents become manageable design parts.

Design of a component is manageable if the component is available as part of a library, if it can be implemented by modifying an already existing part, or if it can be described for a synthesis program or an automatic hardware generator. Figure 3.4 outlines recursive partitioning in a top-down design process. This figure shows that the *Partition* procedure is called recursively until the design of all subcomponents is completed by the *HardwareMapping* procedure.

Mapping to hardware depends on target technology, available libraries, and available tools. For example, if the target technology is full custom CMOS VLSI layout, good synthesis tools and routing and placement programs are required for generating the hardware of a system. On the other hand, for a programmable device target, many readily available tools can be used to generate device programming files or fuse layouts. Generally, the unavailability of good tools and/or libraries can be compensated for by further partitioning of a system into simpler components.

When the recursive *Partition* procedure of Fig. 3.4 completes its job, a partition tree and hardware implementation of the terminals of the tree, such as the example shown in Fig. 3.5, become available. After the completion of this top-down design process, the bottom-up implementation phase begins. In this phase, hardware components corresponding to the terminals of the tree (hatched boxes in Fig. 3.5) are recursively wired to form the hierarchical wiring of the complete system.

Figure 3.5 shows that the original design is initially described at the behavioral level. In the first level of partitioning, two of its subcomponents (SSC1 and SSC2) are mapped to hardware. Further partitioning is required for hardware implementation of SSC3 and SSC4. The SSC3 subcomponent is partitioned into n identical subcomponents, and each of these is realized by SSC3i1 and SSC3i2 hardware parts. The SSC4 subcomponent is partitioned into SSC41 and SSC42 components, for which hardware implementations are available.

```
Partition (system)
        IF HardwareMappingOf (system) IS done THEN
            SaveHardwareOf (system)
        ELSE
            FOR EVERY Functionally-Distinct Part_I OF system
                Partition (part_i);
            END FOR;
        END IF;
END Partition;
```

Figure 3.4 Recursive partition procedure.

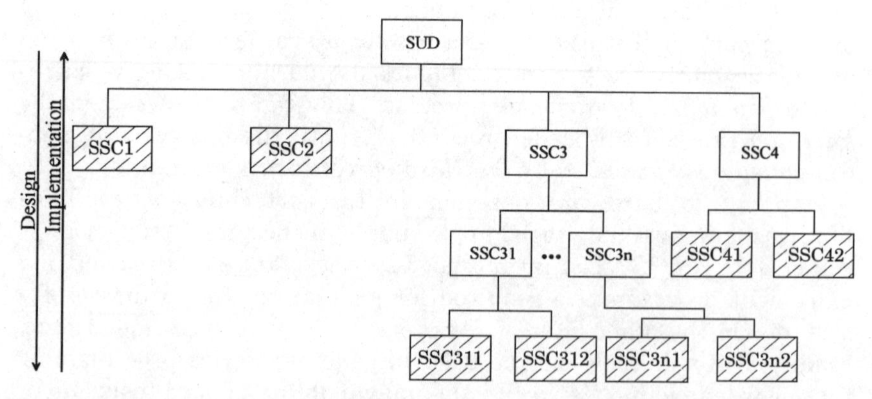

Figure 3.5 Top-down design, bottom-up implementation. (SUD=system under design; SSC=system, subcomponent; hatched areas designate subcomponents with hardware implementation.)

3.2.1 Verification

At each and every step of a top-down design process, a multilevel simulation tool plays an important role in the correct implementation of the design.

Initially a behavioral description of the system under design (SUD) must be simulated to verify the designer's understanding of the problem. After the first level of partitioning, a behavioral description of each of the subcomponents must be developed, and these descriptions must be wired to form a structural hardware model of the SUD. Simulation of this new model and comparison of the results with those of the original SUD description will verify the correctness of the first level of partitioning. Figure 3.6 shows simulation of the first level of partitioning for the top-down design tree of Fig. 3.5. The dotted circles in Fig. 3.6 represent multilevel simulation of the enclosed components. Bold boxes signify that the behavioral descriptions of components are being used in forming the simulation model.

After verification of the first level of partitioning, Fig. 3.6, the hardware implementation of SSC1 and SSC2 must be verified. For this purpose, another simulation run, in which behavioral models of SSC1 and SSC2 are replaced by more detailed hardware-level models, will be performed. Figure 3.7 shows this verification phase. In this figure, shaded boxes signify component models that are functionally equivalent to the actual hardware implementation of a component and have representations for many of the physical characteristics of the hardware. Typical physical characteristics are timing, power consumption, and temperature dependencies. We will refer to such a model as the hardware-level model.

The process of partitioning and verification stated above continues throughout the design process. At the end, a simulation model consisting of the interconnection specification of hardware-level models of

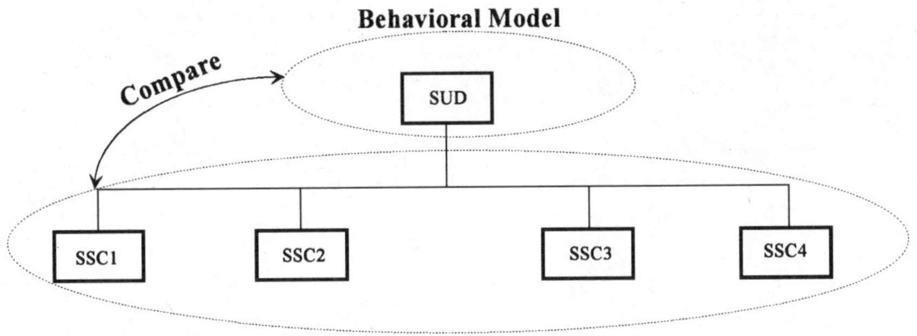

Figure 3.6 Verifying the first level of partitioning.

the terminals of the partition tree will be formed. The simulation of this model, as shown in Fig. 3.8, and comparison of the results with those of the original behavioral description of the SUD verify the correctness of the complete design.

In a large design, where simulation of a complete hardware-level model, such as that shown in Fig. 3.8, is too time-consuming, subsections of the partition tree can be independently verified. Verified behavioral models of such subsections are used in forming the simulation model for final design verification. Figure 3.9 illustrates a simulation and comparison run for verifying the behavioral model of the SSC3 component.

After this verification phase, the behavioral model of SSC3 can be confidently used in all simulation runs for verifying partitioning or

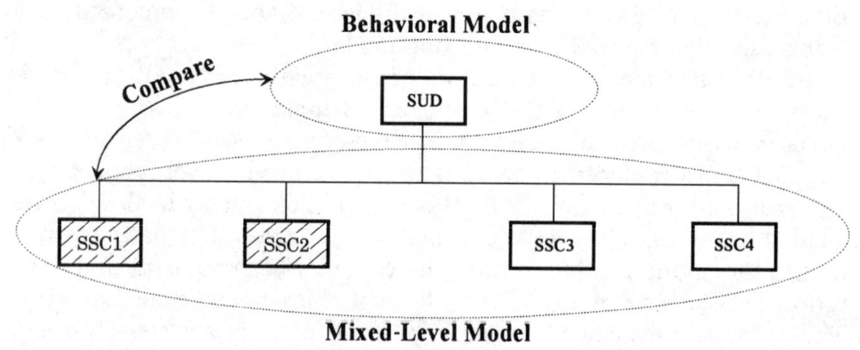

Figure 3.7 Verifying hardware implementation of SSC1 and SSC2.

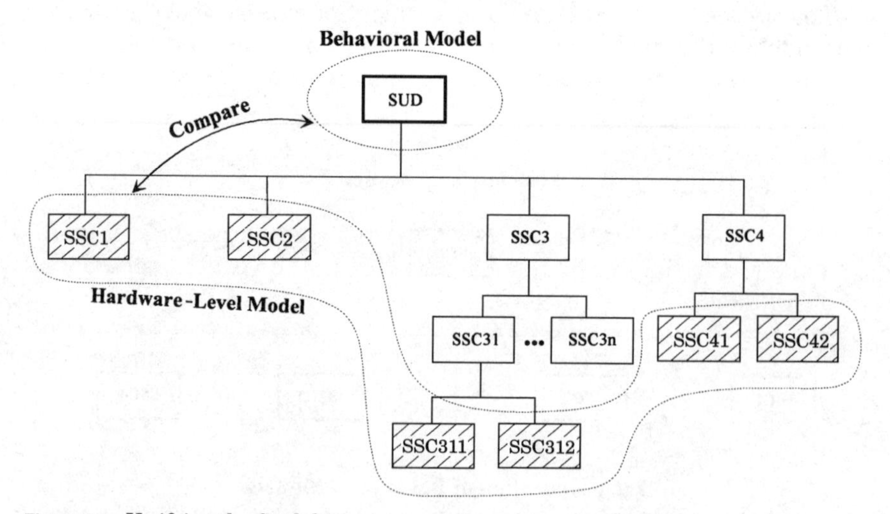

Figure 3.8 Verifying the final design.

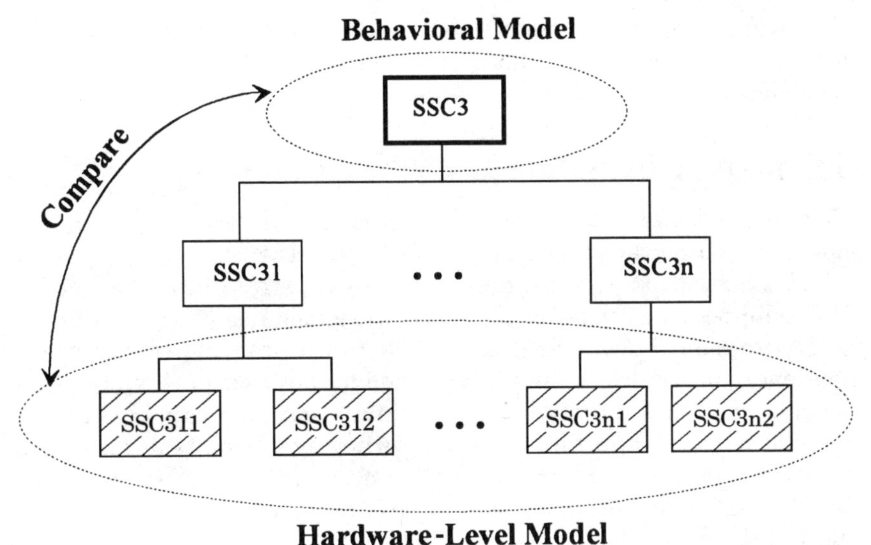

Figure 3.9 Verifying hardware implementation of SSC3.

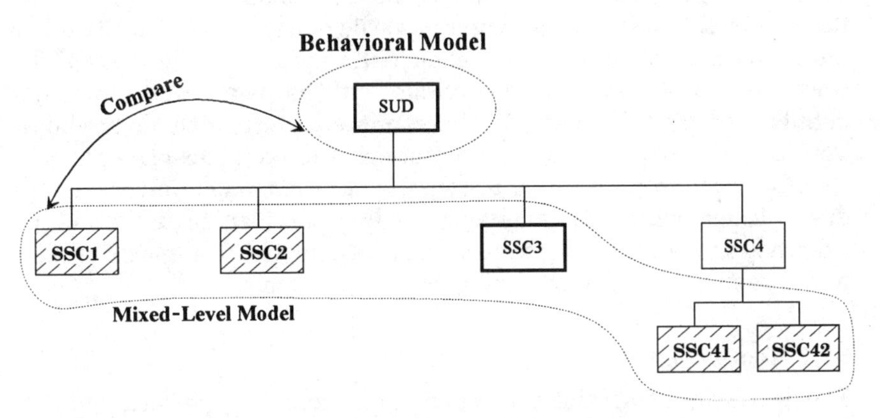

Figure 3.10 Verifying the final design, an alternative to the setup of Fig. 3.8.

hardware implementation of other system subcomponents. Figure 3.10 shows the final design verification using the verified behavioral model of SSC3.

Although the behavioral and hardware-level models of SSC3 may be functionally equivalent, differences may exist in their timing and other physical properties. If such characteristics become necessary in the verification of a design, the behavioral model of a subcomponent can be adjusted to mimic the properties of the hardware-level model. Adjusting upper-level models based on characteristics of actual devices or more detailed models is referred to as back annotation.

Hardware-level models often contain sufficient timing properties of a device to be able to be reliably used for timing back annotation of behavioral models.

3.3 Top-Down Design with Verilog

The previous two sections discussed elements of Verilog and the top-down design methodology. This section puts the two topics together in an example. In this example, we will illustrate how the Verilog HDL can be used for describing hardware modules at various levels of abstraction, how such descriptions can be assembled to form a top-level, mixed-level simulation model; how various simulation models can be simulated; how a simulation run can be verified against another; and finally how a Verilog synthesis tool can help a hardware designer. In short, the example will illustrate that the Verilog HDL provides a convenient environment for top-down design of digital systems.

Selecting an appropriate example for this purpose is a challenging task. On the one hand, the example must be complex enough to show the reader the power of Verilog and Verilog-based tools. On the other hand, the example must be small enough to illustrate the steps of the work instead of burdening the reader with hardware and functional details of the specific example. Unfortunately, our readers do not know Verilog at this point, and so we cannot present a complex example anyway. In order to present an overview of the language and show top-down design with this language, we have decided to use a simple example and act as if it were a real design with complexities that would require a complete Verilog-based set of tools.

3.3.1 Design to perform

The design we have selected for giving an overview of Verilog and various levels of abstraction as well as top-down design is a serial adder. As shown in Fig. 3.11, the circuit has two data inputs a and b and a *start* input. The outputs of the circuit are *ready* and *result.* The circuit is a synchronous sequential circuit that uses the *clock* signal for the synchronization clock.

Serial data are synchronized with *clock* and appear on the a and b inputs. Valid data bits on these inputs begin after a pulse on *start,* and the ordering of these bits is from least to most. After eight clock pulses, the *result* output of the circuit will have the add result of the serial data on a and b. The *ready* output becomes **1** to indicate that one data set has been collected and the add result is ready. This output remains **1** until another synchronous *start* pulse is observed.

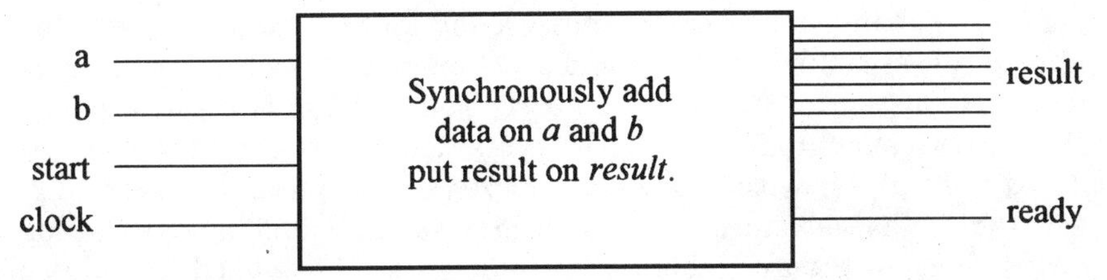

Figure 3.11 Serial adder.

We assume that a word description such as that of the previous paragraphs is given, and the task is to generate a Verilog model for this application. Where we start from and how we achieve this task will be discussed next.

3.3.2 Setting the stage

As discussed in the previous section, the design partitioning process continues until a design is partitioned into "manageable" parts. In a real design environment, synthesis tools, complex library elements, parts from previous designs, and design primitives are available to the designer. In our case, if we were to take advantage of such facilities, the example design would be done in one process and we would not be able to demonstrate the top-down methodology. Therefore, we will set an acting stage in which synthesis tools with limited capabilities and limited libraries and parts are available. This way, even a design the size of our example would require several levels of partitioning.

Available synthesis tools. Present synthesis tools are capable of translating high-level hardware descriptions into an interconnection of logic cells or FPGA layouts. In our example, we assume that we have access to a synthesis tool that can only synthesize logic expressions to generate nonfeedback, memoryless combinatorial circuits. Our synthesis tool for this act uses a Verilog module hardware description of a multi-input multioutput combinatorial circuit and generates a VLSI layout.

Available libraries. Prepackaged libraries are available in most technologies. Libraries consist of predesigned and tested commonly used functional units. At the board level, 7400 series libraries have been used by board-level designers for many years. Because many designers are familiar with these parts, IC, and in particular application-specific integrated circuit (ASIC), manufacturers provide libraries equivalent to 7400 parts for their specific ASIC parts. This way, a board-level designer can make a smooth transition to using ASIC parts for his or her design.

In our setting, however, we have a far more limited library. Our library contains a multiplexer and a D-type flip-flop. Graphic symbols for these parts are shown in Fig. 3.12. The flip-flop is a synchronous D-type with synchronous active high-reset input. The multiplexer is a 2-to-1 multiplexer with active high inputs and outputs. In this setting, we assume that our library manufacturer has provided us with Verilog descriptions of the available library elements. These Verilog modules contain timing parameters that are specific to a given manufacturer and technology.

Figure 3.13 shows the Verilog code for the 2-to-1 multiplexer of Fig. 3.12*a*. Since this is a library element, we have used the border style shown in the figure. This style corresponds to the hatching used for library element terminal nodes of partition trees in Sec. 3.2. This indicates that the multiplexer is a library element and that its Verilog description models its hardware-level characteristics.

The Verilog code in Fig. 3.13 describes the *mux2_1* module. Following the *timescale,* which defines a time unit of 1 ns and time precision of 100 ps, the first line of code specifies the name of the module and its ports. Four input and output ports of *mux2_1* are named *sel, data1, data0,* and *z.* The line following this header line specifies that the first

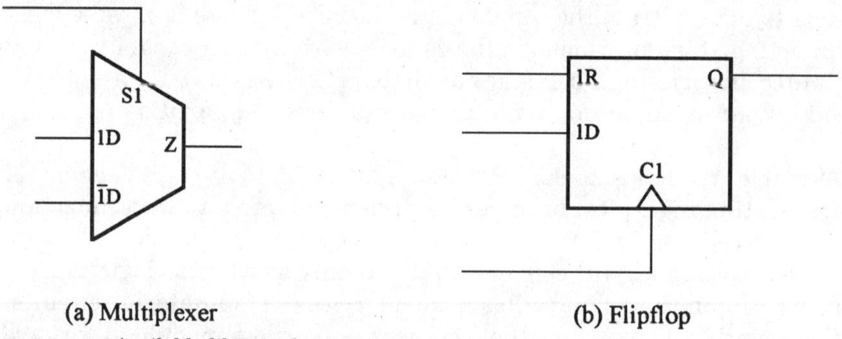

Figure 3.12 Available library elements.

```
`timescale 1ns/100ps

module mux2_1 (sel, data1, data0, z);
input sel, data1, data0;
output z;
  assign #6 z = (sel == 1'b1) ? data1 : data0;
endmodule
```

Figure 3.13 Verilog code for the multiplexer library element.

three ports are inputs, and the following line declares *z* as an output. This ends the declarations of inputs and outputs. If other signals were necessary for describing the operation of *mux2_1,* their declaration would be also included in this part of the code. All declared signals are of type bit.

Following the declarations, the main body of a Verilog module describes the operation of the module. In this part, a module may be described in terms of its subcomponents, its register and bus structure, or its behavior. In the *mux2_1* example, an assign statement is used to specify output values for various input combinations. This statement specifies a 6-ns delay for all values assigned to *z.* The right-hand side of this statement selects *data1* or *data0* depending on whether the *sel* value is binary 1 or not. Signals, such as *z,* to which assigning is done are presumed to be driven by their right-hand side at all times. Such signals are considered wire and do not need to hold any value.

Because this kind of coding describes the flow of data from data inputs to outputs under the control of control inputs, we will refer to it as the dataflow level of abstraction. In a more general case, dataflow or register transfer level (RTL) descriptions may be represented by the schematic in Fig. 3.14. As in the case of the multiplexer, a description corresponding to this hardware structure consists of a flow of data between registers, through busses, under the control of signals coming from the control unit. In later chapters we will see Verilog constructs that enable designers to describe hardware at this level of abstraction.

Back to describing our stage. Figure 3.12*b* shows that a flip-flop is also part of our library. Figure 3.15 shows the Verilog code for this library element. The border style in this figure indicates that this element becomes a terminal node in the partition tree which will be developed for the design of our example serial adder.

As in Fig. 3.13, the first line in Fig. 3.15 specifies the time unit and its precision. Also as in the description of *mux2_1,* the first line in the *flop* code specifies the input and output ports of the flip-flop. Declarations following this header specify which ports are inputs and which are considered outputs of the module. An additional declaration specifies that *qout* is a signal that has the capability of holding its values. This becomes clearer in the following paragraph.

The part of the code in Fig. 3.15 that begins with the **always** keyword specifies the values assigned to *qout* in response to *clk* and input changes. As specified by the statement following the @ sign, the body of this always statement is executed at the negative edge of the *clk* signal. At such times, if *reset* is true, *qout* receives 1'b0 (1-bit binary 0); otherwise, *qout* receives *din.* Value assignments to *qout* take place only on the negative edge of the clock. Therefore, in order for this output to hold its value between clock edges, it has been declared as a **reg.**

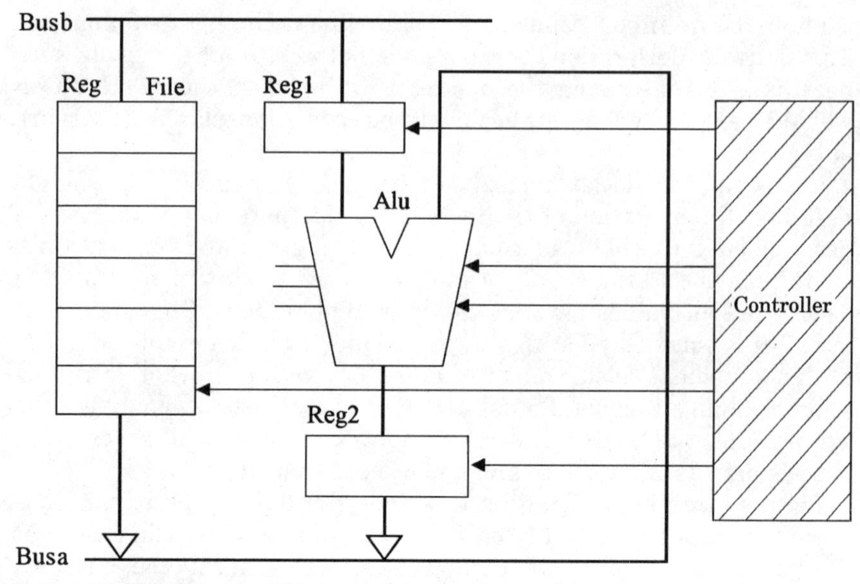

Figure 3.14 Dataflow descriptions.

```
`timescale 1ns/100ps

module flop (reset, din, clk, qout);
input reset, din, clk;
output qout;
reg qout;
  always @(negedge clk) begin
    if (reset) qout = #8 1'b0;
    else qout = #8 din;
  end
endmodule
```

Figure 3.15 Verilog model of the flip-flop library element.

In all Verilog descriptions, a delay value is specified by an integer following a # sign. In Fig. 3.15, the 8-ns delay value specified on the right-hand side of assignments to *qout* specifies the time delay between evaluation of the right-hand side expression and its assignment to *qout*.

A softwarelike, sequential coding style is used for describing the flop model. In this description we are concerned only with assigning appropriate values to circuit outputs. Neither the structure of the circuit nor the details of the hardware in which the data flow are of any concern. Such descriptions will be referred to as *behavioral*. At the behavioral level of abstraction, ports of a model correspond to the actual inputs and outputs of the actual circuit. Input/output mapping and functionality, not hardware details, are of primary concern. Figure 3.16 shows a graphical representation of behavioral descriptions.

> A flowchart, pseudocode, or English-language statements are good tools for describing a system at the behavioral level. However, we will use the Verilog language for this purpose. Behavioral constructs enable designers or modelers to concentrate on *what* the hardware is supposed to do rather than *how* the task is to be done.

Parts from previous designs. A good resource in any design environment is parts that have been designed and tested in previous designs. After doing several designs, a hardware designer forms his or her own library of parts. A new design can take advantage of these predesigned

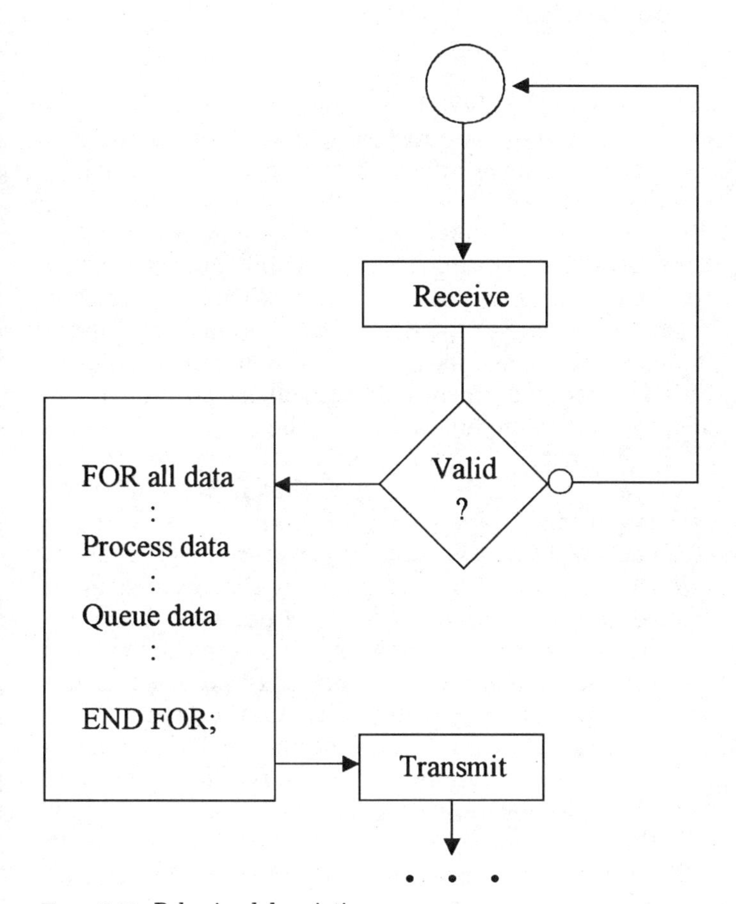

Figure 3.16 Behavioral descriptions.

```
`timescale 1ns/100ps

module counter (reset, clk, counting);
input reset, clk;
output counting;
reg counting;
integer count;
integer limit; initial limit = 8;
  always @(negedge clk) begin
    if (reset) count = 0;
    else begin
      if (count < limit) count = count + 1;
    end
    if (count == limit) #8 counting = 0;
    else #8 counting = 1;
  end
endmodule
```

Figure 3.17 Divide-by-8 counter.

parts either by using them directly or by reconfiguring them for a new design application. In our stage setting, we will take advantage of a counter that we assume we have inherited from one of our previous designs.

Figure 3.17 shows Verilog code for a counter. The counter has a synchronous *reset* input and a *counting* output. While counting falling clock edges to 8, the *counting* output stays high. When the circuit is reset or when 8 clock edges are counted, *counting* becomes **0.**

The Verilog code for the counter begins with the module name and port list. Input and output declarations as well as declaration of *counting* as a **reg** follow the module heading. The signal *counting* is to hold values between activations of assignment of values to this signal, and therefore it is declared as **reg** so that it has the value-holding property. Two integers, *count* and *limit,* are also declared in Fig. 3.17. The former keeps the count of the counter until it reaches the value of *limit.* As in the description of *flop,* an **always** statement that becomes active on the negative edge of *clk* encloses the statements specifying the behavior of the counter. Following the keyword **begin,** an if-then-else statement increments *count* if *reset* is not active and *count* has not reached the limit. Another if-then-else statement sets the *counting* output of the counter to **1** if the count limit of 8 has been reached. Assignments to *counting* are delayed by 8 ns, specified by integers following # signs.

With the presentation of *counter,* our stage setting is complete. As shown in Fig. 3.18, the stage consists of a logic synthesis tool, two library elements, and a predesigned counter.

3.3.3 Design scenario

With the stage setting of the previous section, we are now ready to present our serial adder design. The steps that we follow are those of top-down design as discussed in Sec. 3.2.

Analyzing the requirements. The design begins with analysis of the requirements and development of a simulatable model of system under design. The description shown in Fig. 3.19 has resulted after the following analysis:

> A *count* variable keeps the count of clock edges received. After the circuit receives the *start* pulse, bits of data on *a* and *b* inputs will be added and shifted into the *result* shift register. Because adding data bits requires a carry from previous add results, a *carry* variable is used for this purpose. Adding bits is done by boolean expressions assigning values to *sum* and *carry.* When *counting* reaches 8, the *ready* output is assigned a **1.**

The Verilog code in Fig. 3.19 is at the behavioral level. The heading contains the list of data and control inputs and outputs. An **always** block that becomes active on the falling edge of the clock encloses several sequential if-then-else statements. As data bits are read from *a*

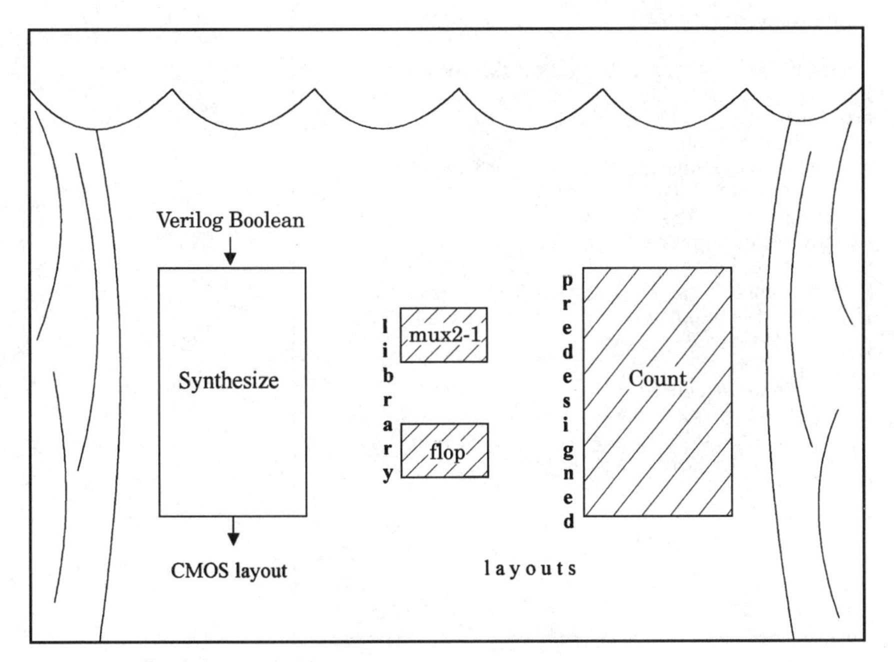

Figure 3.18 Design stage setting.

and *b* inputs, their add result will be shifted into the *result* output. The readability of this code enables designers to communicate their understanding of the design to others, and the simulatability of this code enables a designer to make sure his or her understanding of the design requirements is correct.

Before the next step in the design is taken, this behavioral description must be simulated. Many Verilog simulation environments provide a mechanism for applying test vectors to inputs and observing output waveforms. Figure 3.20 shows a typical simulation-run waveform output. This simulation result shows binary value 10011111 on *result* after 8 falling edges of *clock* following a *start* pulse. Data on *a* and *b* that are being added are 01100111 and 00111000, respectively.

Recursive partitioning. After verification of the behavioral description of *serial_adder* by simulating it and studying the waveform in Fig. 3.20, the code must be studied for generating the general layout of the design. The statements circled in the partial code of Fig. 3.21 each indicate sections of code for the functionality of which an easy hardware correspondence exists.

The functionality of the first circle in this figure can be implemented by a counter or a divide-by-8 circuit. The second circle shows that a

```
`timescale 1ns/100ps

module serial_adder (a, b, start, clock, ready, result);
input a, b, start, clock;
output ready;
output [7:0] result;
reg sum, carry, ready;
reg [7:0] result;
integer count; initial count = 8;
  always @(negedge clock) begin
    if (start) begin
      count = 0; carry = 0; result = 0;
    end else begin
      if (count < 8)
      begin
        count = count + 1;
        sum = a ^ b ^ carry;
        carry = (a & b) | (a & carry) | (a & carry);
        result = {sum, result [7:1]};
      end
    end
    if (count == 8) ready = 1;
    else ready = 0;
  end
endmodule
```

Figure 3.19 Serial adder behavioral description.

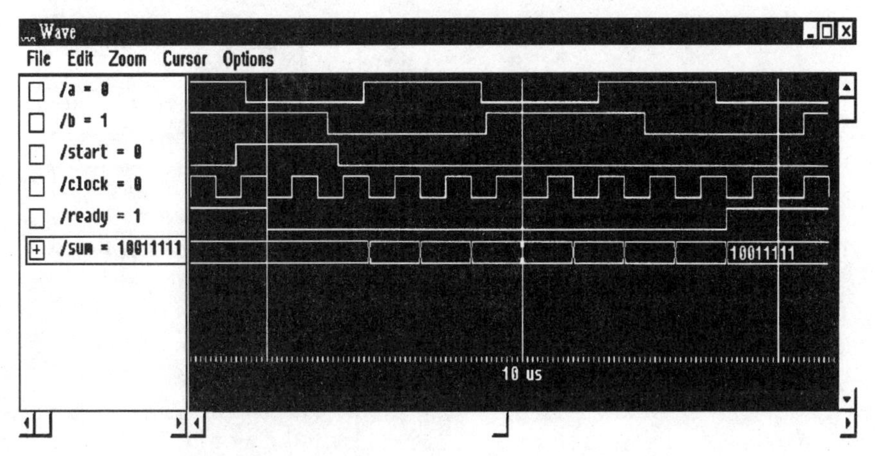

Figure 3.20 Verilog simulation results.

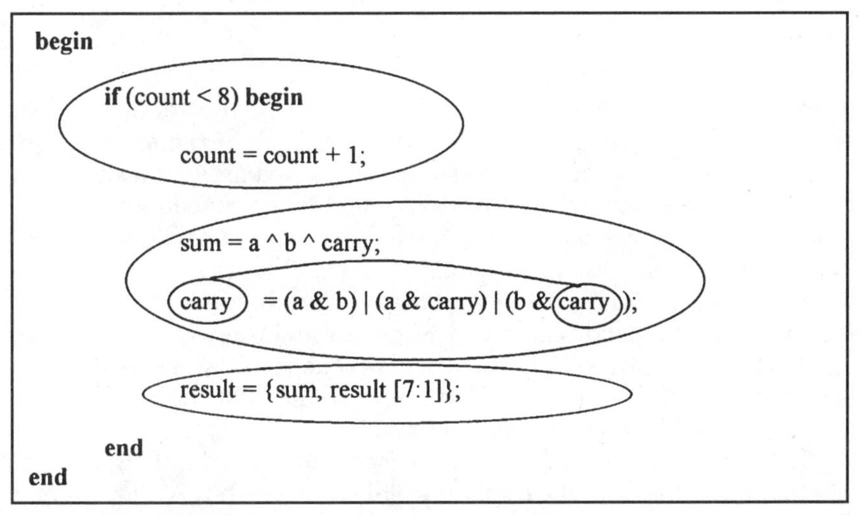

```
begin
    if (count < 8) begin
        count = count + 1;

        sum = a ^ b ^ carry;
        carry = (a & b) | (a & carry) | (b & carry));

        result = {sum, result [7:1]};
    end
end
```

Figure 3.21 Partial code of *serial_adder.*

full-adder is necessary, and the appearance of *carry* on the right- and left-hand sides indicates that a flip-flop is needed to use old values of *carry* for calculating new *sum* and *carry* values. The third circle shows shifting of the result one place to the right. The shifting is achieved by concatenating *sum* with the 7 leftmost bits of *result* and assigning this concatenated vector to the *result* signal. Curly brackets signify **nets** that are being concatenated. Implementation of this function requires a shift register. Figure 3.22 shows the general layout of our design based on the analysis presented in Fig. 3.21.

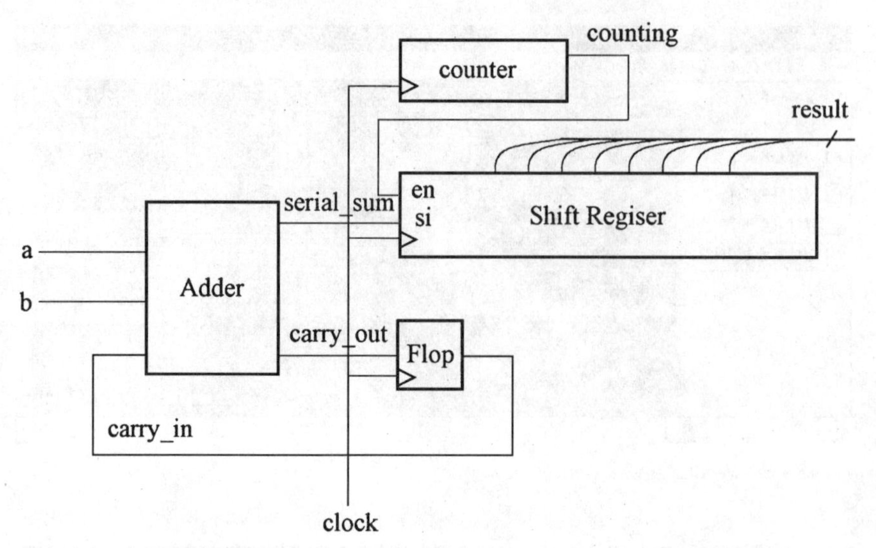

Figure 3.22 General layout of *serial_adder.*

First level of partitioning. The layout of Fig. 3.22 results in the first level of partitioning, as shown in Fig. 3.23. Each of the terminals of the partition tree of this figure should be checked for synthesizability, availability in a library, or availability as a predesigned part.

Figure 3.18 shows our design stage. We find that the flip-flop and counter of our partition tree exactly match *flop* and *counter* of our stage setting. There is no direct correspondence between the other two terminals of the partition tree of Fig. 3.23 and the available components. Therefore, we will develop Verilog code for these components to study ways of designing them. Figures 3.24 and 3.25 show the Verilog code for *fulladder* and *shifter,* respectively.

The full-adder is described at the behavioral or dataflow level of abstraction. This unit is a combinatorial circuit with three inputs and two outputs. Referring back to our stage setting, we find that a synthesis tool available to us can synthesize this description. The synthesis tool generates an EDIF file, and this file is used by layout editors for routing and placement, resulting in CIF or GDSII file formats.

The shift register (Fig. 3.25) is described at the dataflow level. Following the module header and declarations, an **always** block that is sensitive to the negative edge of the clock encloses a sequential assignment to *parout* shifter parallel output. The value assigned to *parout* is determined by nested expressions using the Verilog conditional operator (?:).

On the falling edge of the clock, if *reset* is **1,** an 8-bit decimal 0 is placed on *parout*; if reset is **0,** then a second conditional operator checks for *enable* being **1.** If *enable* is **1,** then *sin* is left concatenated with the previous value of *parout* and will be assigned to *parout* as its new value. This concatenation causes *parout* to be shifted to the right by 1 bit. On the other hand, if *enable* is **0,** the second conditional operator in Fig. 3.25 will place the previous value of *parout* back on itself, causing the shift register state to remain unchanged.

The *shifter* is a sequential circuit that cannot be synthesized with the synthesis tool of our stage setting. Therefore, further partitioning is needed for the design of this unit. Before continuing, we must verify the first level of partitioning, using behavioral descriptions of the terminals

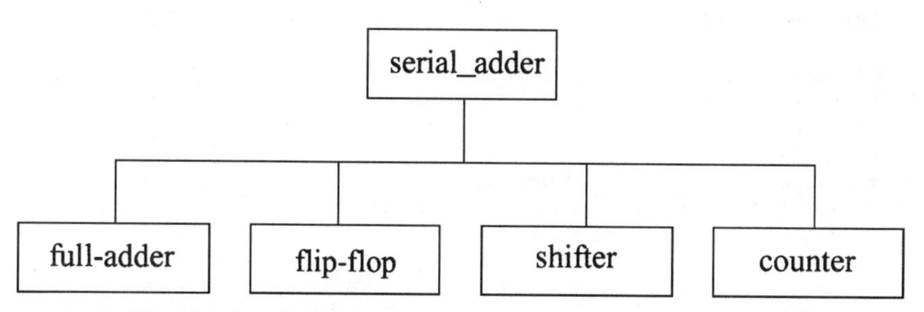

Figure 3.23 First level of partitioning.

```
`timescale 1ns/100ps

module fulladder (a, b, cin, sum, cout);
input a, b, cin;
output sum, cout;

  assign sum = a ^ b ^ cin;
  assign cout = (a & b) | (a & cin) | (b & cin);

endmodule
```

Figure 3.24 *fulladder* description.

```
`timescale 1ns/100ps

module shifter (sin, reset, enable, clk, parout);
input sin, reset, enable, clk;
output [7:0] parout;
reg [7:0] parout;
  always @(negedge clk)
    parout = (reset) ? 8'd0 : ((enable) ? {sin, parout [7:1]} : parout);
endmodule
```

Figure 3.25 Shifter Verilog code.

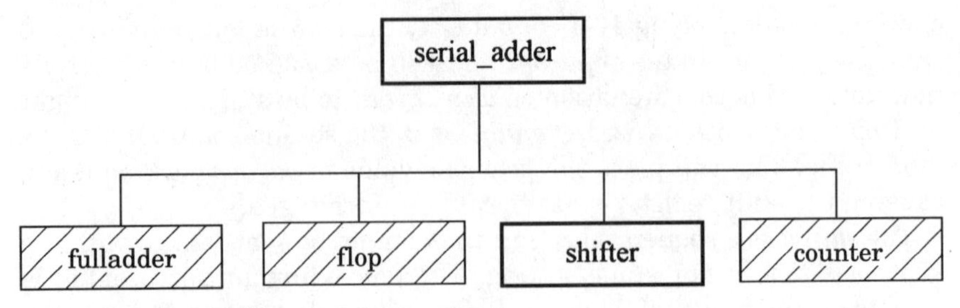

Figure 3.26 Completed parts of first partitioning.

```
`timescale 1ns/100ps

module serial_adder (a, b, start, clock, ready, result);
input a, b, start, clock;
output ready;
output [7:0] result;
wire serial_sum, carry_in, carry_out, counting;
   fulladder u1 (a, b, carry_in, serial_sum, carry_out);
   flop u2 (start, carry_out, clock, carry_in);
   counter u3 (start, clock, counting);
   shifter u4 (serial_sum, start, counting, clock, result);
   assign ready = ~ counting;
endmodule
```

Figure 3.27 Structural description of *serial_adder.*

of the partition tree. Figure 3.26 shows our description status to this point. Hatched areas signify models of libraries or completed parts, and bold boxes indicate that a behavioral description is available for the part.

Verification of the current status of our design is done by forming a model by wiring *fulladder, flop, shifter,* and *counter* and comparing the simulation of this model with that of the behavioral description of the *serial_adder.*

Figure 3.27 shows the structural description of *serial_adder,* consisting of interconnection of its four subcomponents. The module declaration consists of a port list identical to that of Fig. 3.19, which describes *serial_adder* at the behavioral level. Declarations in the structural description of Fig. 3.27 declare inputs and outputs of *serial_adder* as well as wires for the interconnection of the components of the serial adder. Following the declarations, the *serial_adder* module includes instantiation of *fulladder, flop, counter,* and *shifter* modules. Instantiation refers to naming a module within another module for use as a subcomponent. Every instantiation begins with the name of the module that is being instantiated, followed by an arbitrary label and a list of module port connections.

The format used here for module port connections is an ordered list. Signals in the port connections of an instantiated module (e.g., *full-adder* instantiation in Fig. 3.27) connect to actual ports of the module (e.g., the *fulladder* module in Fig. 3.24) in the order in which ports appear in the port list of the module. Within a module, signals in the port list of the module and wires declared in the module are visible signals that can be used for any instantiated module. For example, signals *a, b, start, clock, ready,* and *result,* which appear in the port list of *serial_adder* in Fig. 3.27, and declared wires in this module, i.e., *serial_sum, carry_in, carry_out,* and *counting,* can be used for connections to ports of *fulladder, flop, counter,* and *shifter.* As shown in Fig. 3.28, through instantiation of *fulladder,* wires visible in the *serial_adder* module are connected to ports of the *fulladder.* The *carry_out* signal connects to the *cout* output of *fulladder* at the same time this signal is used in the port connection of *flop* to connect to its second port. As shown in the partial code of Fig. 3.29, the *carry_out* wire causes the *cout* output of *fulladder* to connect to *din* of *flop.*

The complete code of Fig. 3.27 corresponds to the first-level partitioning of Fig. 3.26. This code is simulatable, and by simulation of this structural Verilog code, we will be able to verify our design to this point.

Second level of partitioning. Our design is not complete until a partition tree is generated in which all the terminals have a corresponding hardware. For this to happen, the shifter unit of Fig. 3.26 must be transformed into submodules that can be realized using tools or library parts of our stage setting (Fig. 3.18). The functionality of Fig. 3.25 indicates that eight flip-flops with synchronous reset and clock

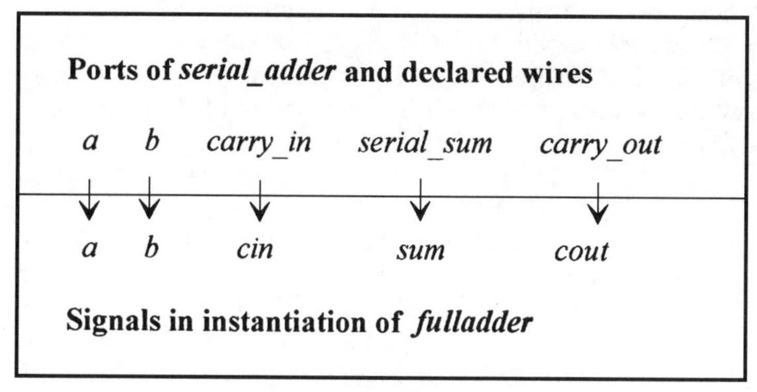

Figure 3.28 Signal mapping for *fulladder* instantiation.

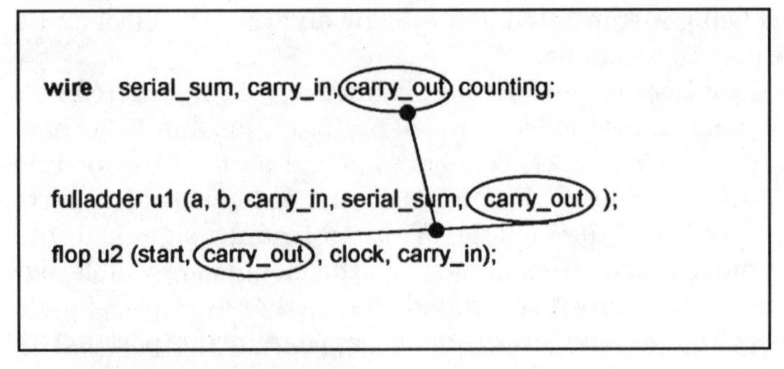

Figure 3.29 Interconnecting ports.

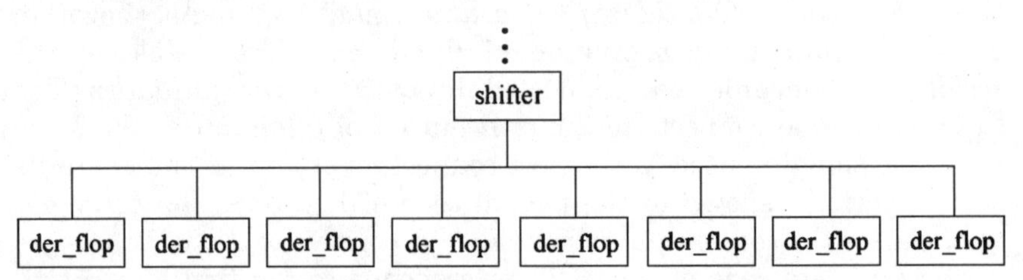

Figure 3.30 Partitioning of *shifter.*

enabling are needed for the hardware realization of the shifter. We will refer to this flip-flop as *der_flop.* Partitioning of *shifter* into eight *der_flop* units is shown in Fig. 3.30.

Because *der_flop* is not directly implementable with the tools and libraries of our stage setting, we will develop a behavioral code for this unit (Fig. 3.31) to serve (1) as a description for verifying the partitioning of Fig. 3.30 and (2) for analyzing it for the purpose of designing it.

The behavioral code of *der_flop* follows the general format of other behavioral descriptions presented in this chapter. The structural description of *shifter* based on *der_flop* is shown in Fig. 3.32. Simulation of this code verifies our understanding of *der_flop* requirements as well as the design of *shifter* based on eight of these units.

After the above-mentioned verification, *der_flop* must be designed. The hardware of this flip-flop, shown in Fig. 3.33, can be deduced from the behavioral code in Fig. 3.31.

The *der_flop* unit is partitioned into a multiplexer and a D-type flip-flop (Fig. 3.34). These units are available in the library of parts of our stage setting. To verify this design, a structural code corresponding to the hardware of Fig. 3.33 is developed. Simulation of this code, Fig. 3.35, indicates that the hardware of Fig. 3.33 satisfies the requirements of the behavioral description of Fig. 3.31.

Putting together the partition trees of Figs. 3.26, 3.30, and 3.34 yields the complete partition tree of Fig. 3.36. All terminal nodes of this tree have hardware correspondence as set in our stage setting of the previous section.

Design implementation. The complete *serial_adder* design is now done. The bottom-up implementation consists of wiring *mux2_1* and *flop* to build *der_flop* (the code is shown in Fig. 3.35), wiring eight *der_flop* units to build *shifter* (the code is shown in Fig. 3.32), and wiring *fulladder, flop, shifter,* and *counter* to build *serial_adder.* The final circuit specification is that shown in Fig. 3.37.

In a final verification phase, the hierarchical description of the *serial_adder* corresponding to Fig. 3.36 can be simulated along with its original behavioral description. There will be timing differences

```
`timescale 1ns/100ps

module der_flop (din, reset, enable, clk, qout);
input reset, din, clk, enable;
output qout;
wire dff_in;
reg qout;
  always @(negedge clk) begin
    if (reset) qout = #8 1'd0;
    else
      if (enable) qout = #8 din;
  end
endmodule
```

Figure 3.31 Behavioral code of *der_flop.*

```
`timescale 1ns/100ps

module shifter (sin, reset, enable, clk, parout);
input sin, reset, enable, clk;
output [7:0] parout;
  der_flop b7 (        sin, reset, enable, clk, parout[7]);
  der_flop b6 (parout[7], reset, enable, clk, parout[6]);
  der_flop b5 (parout[6], reset, enable, clk, parout[5]);
  der_flop b4 (parout[5], reset, enable, clk, parout[4]);
  der_flop b3 (parout[4], reset, enable, clk, parout[3]);
  der_flop b2 (parout[3], reset, enable, clk, parout[2]);
  der_flop b1 (parout[2], reset, enable, clk, parout[1]);
  der_flop b0 (parout[1], reset, enable, clk, parout[0]);
endmodule
```

Figure 3.32 Structural description of *shifter.*

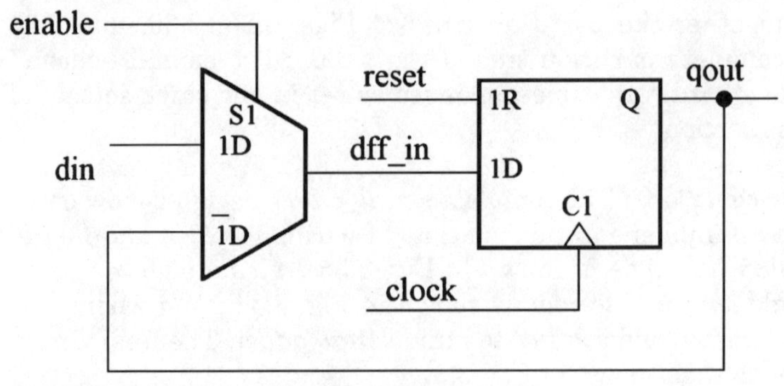

Figure 3.33 Hardware realization of *der_flop*.

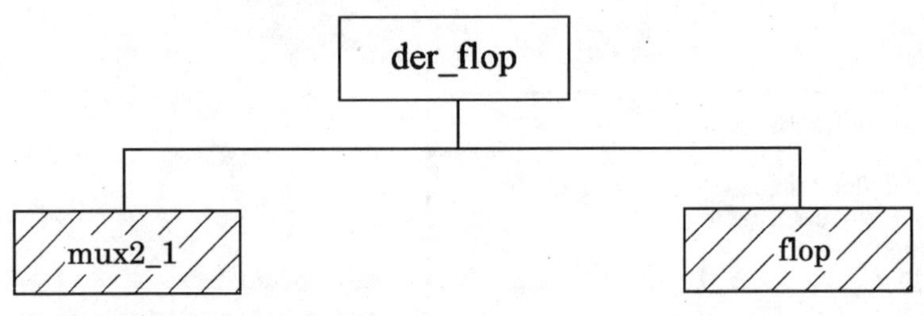

Figure 3.34 Partitioning *der_flop*.

```
`timescale 1ns/100ps

module der_flop (din, reset, enable, clk, qout);
input reset, din, clk, enable;
output qout;
wire dff_in;
  mux2_1 mx (enable, din, qout, dff_in);
  flop ff (reset, dff_in, clk, qout);
endmodule
```

Figure 3.35 Structural code of *der_flop*.

between the two, but their functionality must be the same. In Verilog we can develop a test bench for instantiating both models and reporting functional differences between their output signals.

The behavioral model of *serial_adder* does not have all the timing details of the structural model, but it simulates faster. If it becomes necessary to have a model that runs fast and contains most of the structural details (such as timing), such information can be extracted from the structural model and inserted into the behavioral model. This process is called *back annotation.* Timing and other physical characteristics from models that have a close hardware correspondence can be back annotated into higher-level models.

3.3.4 Final act

The final layout of our example design is shown in Fig. 3.37. The hatched areas indicate elements that have hardware correspondence and those that appear as the terminal nodes of our final partitioning tree (Fig. 3.36).

This design resulted in 19 hardware modules from our libraries or generated by our synthesis tools. The same design methodology would be used in an actual large design. However, the size and complexity of our libraries, tools, and Verilog descriptions would differ. In Chap. 10 a CPU system will be designed by the methodologies presented here. Chapters 4 through 9 present more features of Verilog to set the stage for presentation of the example of Chap. 10.

3.3.5 Real world

Sections 3.3.1 to 3.3.4 presented a design scenario in order to highlight the steps necessary for a real-world design. However, our design is simple enough that real-world tools could be used to complete its implementation much more easily than presented in our scenario. If we were to use commercial synthesis tools, the Verilog code of Fig. 3.38, which is a slight modification of that of Fig. 3.19, could directly be synthesized to a designed target hardware.

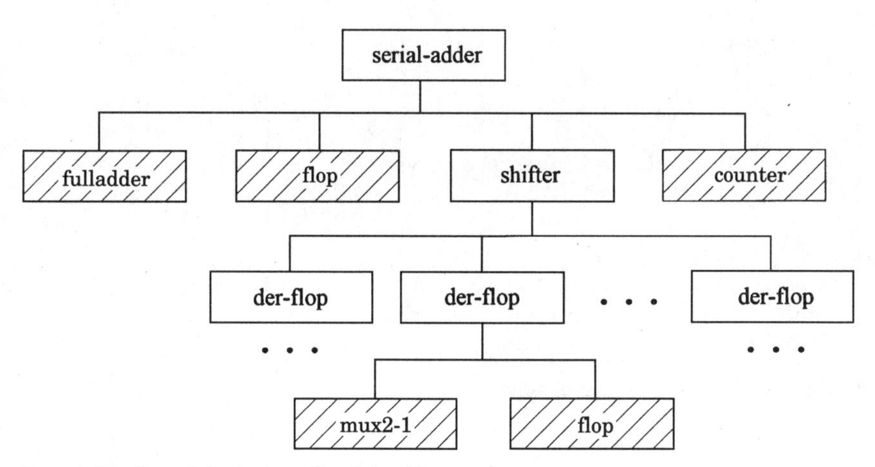

Figure 3.36 Complete design of *serial_adder.*

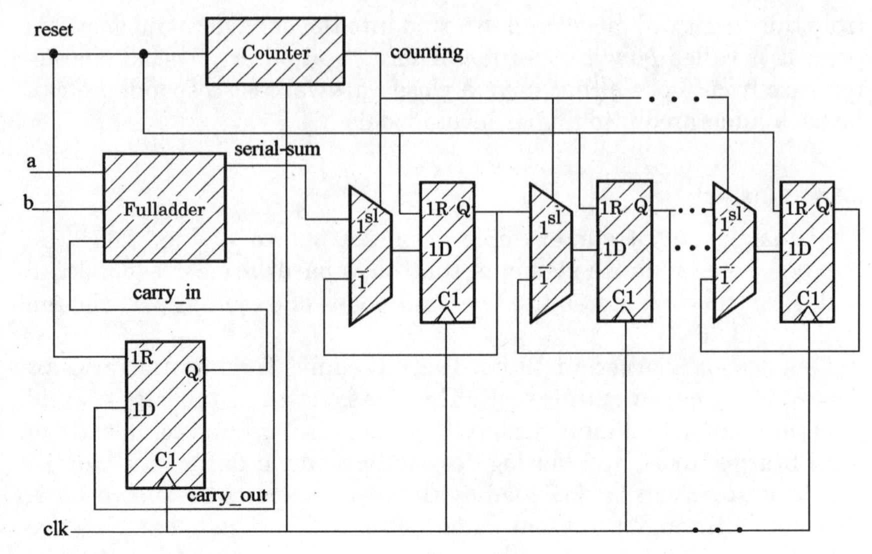

Figure 3.37 Final design.

```
`timescale 1ns/100ps

module serial_adder (a, b, start, clock, ready, result);
input a, b, start, clock;
output ready;
output [7:0] result;
reg sum, carry, ready, result;
reg [3:0] count; initial count = 4'b1000;
  always @(negedge clock) begin
    if (start) begin
      count = 0; carry = 0;
    end else begin
      if (count < 8)
      begin
        count = count + 1;
        sum = a ^ b ^ carry;
        carry = (a & b) | (a & carry) | (a & carry);
        result = {sum, result [7:1]};
      end
    end
    if (count == 8) ready = 1;
    else ready = 0;
  end
endmodule
```

Figure 3.38 Synthesizable serial adder.

To put a limit on the size of a synthesized register for *count,* in the description of Fig. 3.38, it is declared as a 4-bit register. Partial layout of the *serial_adder* synthesized to an FPGA target is shown in Fig. 3.39. The figure shows wiring of FPGA cells for the implementation of serial adder registers and logic. Several cells of the *result* register are shown in this figure. A Verilog synthesis tool automatically generates this layout along with timing and various netlist files.

3.4 Subprograms

Like most high-level languages, Verilog allows the definition and use of functions and procedures. In Verilog, functions are referred to as functions, and a procedure is referred to as a task. Functions and tasks may be used to correspond to hardware entities, or they may be used for writing structured code, in much the same way as they are used in software languages.

Typical applications of functions include representation of boolean functions, data and code conversion, and input and output data formatting. Generally, any time the final value of a process is used on the right-hand side of an expression, a function can be used to simplify the expression.

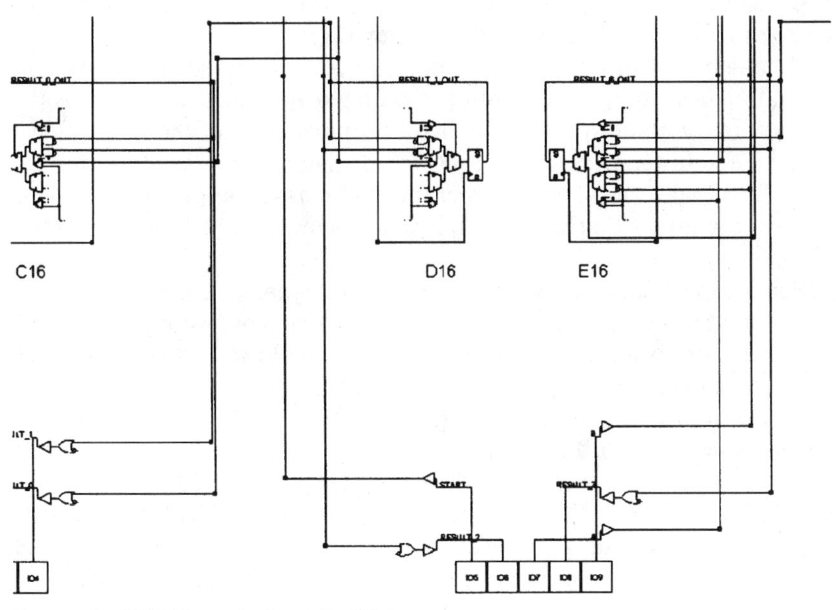

Figure 3.39 FPGA layout of *serial_adder.*

A task can represent a submodule within a Verilog module. A task begins with the **task** keyword and uses a format that is very similar to the format of a module.

As a function example, we replace the boolean expressions in Fig. 3.24 with invocation of a function. In this figure, two assignments to *sum* and *cout* take place. If these expressions were to repeat in other parts of the code, it would be more efficient to write a function and replace these assignments by a function invocation.

Figure 3.40 shows a description for *fulladder* in which the function *fadd* is used for generation of the sum and carry outputs of the full-adder. The function, which appears in the lower part of the code in this figure, has a 2-bit output. This 2-bit function is used on the right-hand side of an assignment, with *sum* and *cout* being concatenated on the left-hand side. Curly brackets signify concatenation of operands that are separated by commas. Concatenation in Fig. 3.40 forms a 2-bit operand on the left-hand side of the assignment to match the function *fadd* invoked on the right-hand side.

3.5 Controller Description

The main purpose of this chapter has been the presentation of design methodology, levels of abstraction, and an overview of Verilog. A complete design usually consists of a data unit and a controller. Because of the size of our top-down design example, we were not able to discuss controllers and Verilog coding styles for them.

By use of a stand-alone example, this section shows Verilog coding style for describing a controller. As shown in Fig. 3.41, a controller circuit has control input and control output signals. The controller is a state machine, and state transition decisions are based on the values of control inputs. State machine states issue appropriate output control signals. All state transitions are synchronized with a system clock.

Sequence detectors are simplified controllers. Instead of many inputs and outputs, sequence detectors have one or two input and output lines, and instead of complex decision makings and input conditions, sequence detectors generally search for a sequence of **1**s and **0**s on their input. We will present a Verilog description for a simple sequence detector in Fig. 3.42.

A Moore machine sequence detector, the pseudocode description of which is shown in Fig. 3.42, searches on its x input for the 110 sequence. When this sequence is detected in three consecutive clock pulses, the output (z) becomes **1** and stays **1** for a complete clock cycle. The state machine for this detector is shown in Fig. 3.43. States are

```
`timescale 1ns/100ps

module fulladder (a, b, cin, sum, cout);
input a, b, cin;
output sum, cout;
  assign {sum , cout} = fadd (a, b, cin);

  function [1:0] fadd;
  input a, b, c;
    fadd = {(a ^ b ^ c) , ((a & b) | (a & c) | (b & c))};
  endfunction
endmodule
```

Figure 3.40 *Fulladder* using *fadd* function.

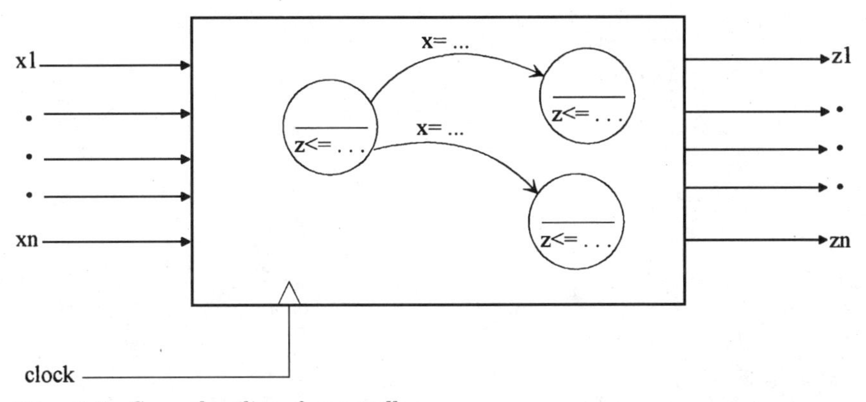

Figure 3.41 General outline of a controller.

named according to bits detected on the *x* input. Starting in the *reset* state, it takes at least three clock pulses for the state machine to get to the *got110* state, in which output becomes **1.**

The Verilog behavioral description for this machine is shown in Fig. 3.44. The list of ports contains *x, clk,* and *z,* which match those in the block diagram of Fig. 3.42. A 2-bit variable is declared as **reg** to hold the current state of the machine. Four 2-bit parameters define state names and their binary assignments. In an **initial** block, the present state is set to the *reset* state. The main flow of the state machine is implemented by an **always** block that is sensitive to the positive edge of the clock. In this statement, **if-else** statements are used to check for each of the four cases of the *current* state. For each state that matches the *current* state, a conditional operator is used to set the *current* state to the next active state of the machine, depending on the value of the *x* input. Figure 3.45 shows state *got1,* its next

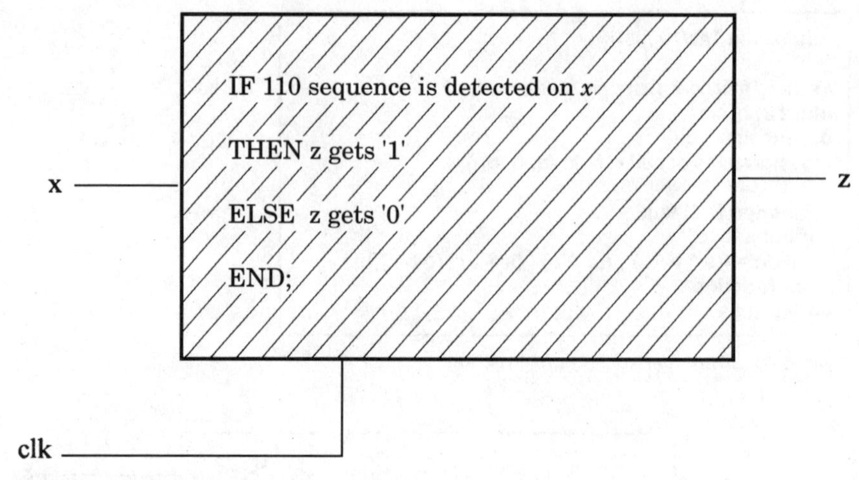

Figure 3.42 Moore machine description.

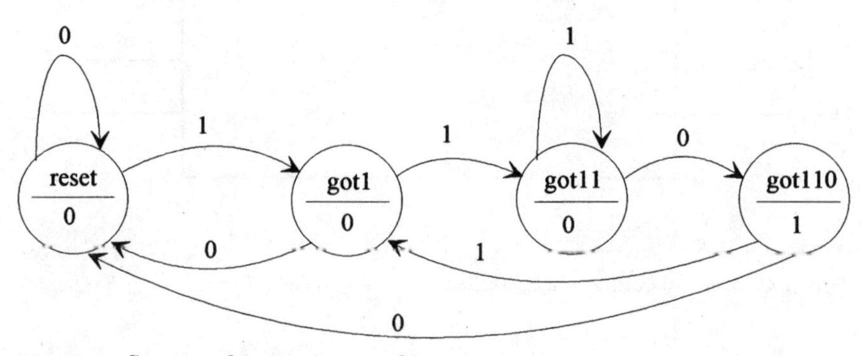

Figure 3.43 Sequence detector state machine.

state, and the corresponding Verilog code in the **always** block of Fig. 3.44.

The style presented here is simple and flexible in terms of its number of inputs, outputs, and states. More state machine coding styles will be presented in Chaps. 8 and 9. A state machine will be used for describing the controller of the CPU of Chap. 10.

3.6 Verilog Operators

Logical operations are the most common type of operations for describing the function of hardware components at the gate level. In addition to these operations, there are operations for the behavioral or functional description of hardware. Most operations found in software languages are also supported by Verilog. Figure 3.46 shows the operations that can be used in Verilog.

Operators in Fig. 3.46 are grouped according to their functionality. Basic arithmetic and relational operators are listed first. Arithmetic operators operate on all data types and have their conventional mathematical meanings. Relational and equality operators return **1** or **0** for true or false results. Case equality operators match **X** and **Z** logic values as well as **0** and **1.** The three groups of logical operators include logic operations for vectors, scalars, and reducing vectors to scalars. The right-hand side of shift operators specifies the number of bits to shift the left-hand-side operand to the right or left. Other operators listed in Fig. 3.46 include concatenation, which is used for forming larger vectors from smaller ones by concatenating them. Concatenation and replication operators can be nested.

```
`timescale 1ns/100ps

module moore_110_detector (x, clk, z);
input x, clk;
output z;
reg [1:0] current;
parameter [1:0] reset = 0, got1 = 1, got11 = 2, got110 = 3;
  initial current = reset;
  always @(posedge clk) begin
    if      (current == reset) current = x ? got1 : reset;
    else if (current == got1) current = x ? got11 : reset;
    else if (current == got11) current = x ? got11 : got110;
    else if (current == got110) current = x ? got1 : reset;
  end
  assign z = (current == got110) ? 1'b1 : 1'b0;
endmodule
```

Figure 3.44 Verilog code for 110 detector.

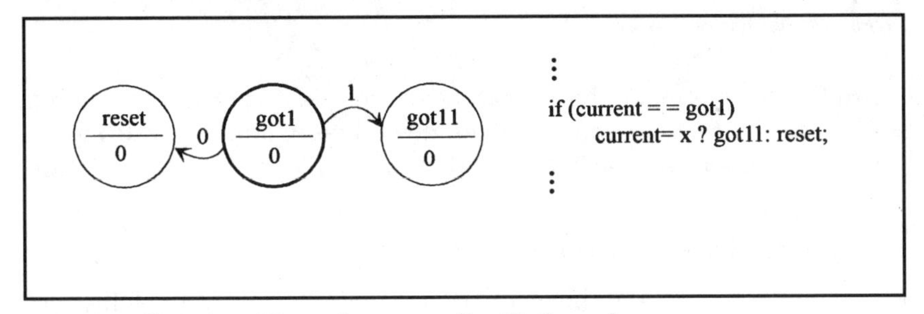

Figure 3.45 State transition and corresponding Verilog code.

BASIC	OPERATION	DESCRIPTION	RESULT
Arithmetic	+ - * /	Basic arithmetic	Multi-bit
Relational	> >= < <=	Compare	One-bit

EQUALITY	OPERATION	DESCRIPTION	RESULT
Logical	== ! =	Equality not including Z, X	one-bit
Case	=== ! ==	Equality including Z, X	one-bit

<table>
<tr><th>LOGICAL</th><th>OPERATION</th><th>DESCRIPTION</th><th>RESULT</th></tr>
<tr><td>Connectives</td><td>&& || !</td><td>Simple logic</td><td>one-bit</td></tr>
<tr><td>Bit-wise</td><td>~ & | ^ ^~
~^</td><td>Vector logic
operations</td><td>one-bit</td></tr>
<tr><td>Reduction</td><td>& ~& | ~| ^
^~ ~^</td><td>Perform operation
on all bits</td><td>one-bit</td></tr>
</table>

SHIFT	OPERATION	DESCRIPTION	RESULT
Right	>> *n*	Zero-fill shift *n* places	multi-bit
Left	<< *n*	Zero-fill shift *n* places	multi-bit

OTHERS	OPERATION	DESCRIPTION	RESULT
Module	%	Remainder	multi-bit
Concatenation	{ }	Join bits	multi-bit
Replication	{{ }}	Join & replicate	multi-bit
? :	Conditional	If-then-else	multi-bit

Figure 3.46 Verilog operators.

Brief descriptions of the operators in the above paragraph should suffice for the code presented in the next three chapters. In Chap. 7, we will discuss Verilog operators and their data types in greater detail.

3.7 Conventions and Syntax

In all the Verilog code in this book, we use indentation on the left-hand side of the code to illustrate nesting levels. In all Verilog code, we use bold type for Verilog keywords, predefined entities, and standards. All other code, including names and labels, is in lightface. When a Verilog keyword or predefined name is used in the text, it is also in bold type. All other parts of code used in the text are in *italics*.

The syntax of Verilog is shown in illustrations such as Fig. 3.47. This format is extracted from a Verilog program, with individual elements isolated to make labeling easy. These "syntax details" are only for the example code that is being discussed when the illustration is presented and do not necessarily present the general syntax of the language. However, they are designed to cover as much of the general case as possible. For example, Fig. 3.47 shows a test bench for testing the *flop* of Fig. 3.15. This illustrates that the body of a *module_declaration* consists of *module_item,* which includes declarations and other constructs. Obviously this does not show all the variations of these language constructs. Where variations are important, we highlight them by showing syntax details in other examples. Appendix C shows the complete syntax of Verilog, from which all the syntax details illustrations of this book are drawn.

As stated earlier, Verilog is not case-sensitive and has a free format. In Verilog, long statements can continue over several lines. For comments, a pair of slashes (//) is used. This pair, anywhere in a line, makes the rest of the line a comment. Also, bracketing a section of code by /* and */, as in the C programming language, makes that section a comment.

3.8 Summary

This chapter introduced the main concepts of the Verilog hardware description language. Entities needed for description of designs were presented. Top-down design methodology and the use of Verilog in this design flow were discussed. We have shown where simulation tools,

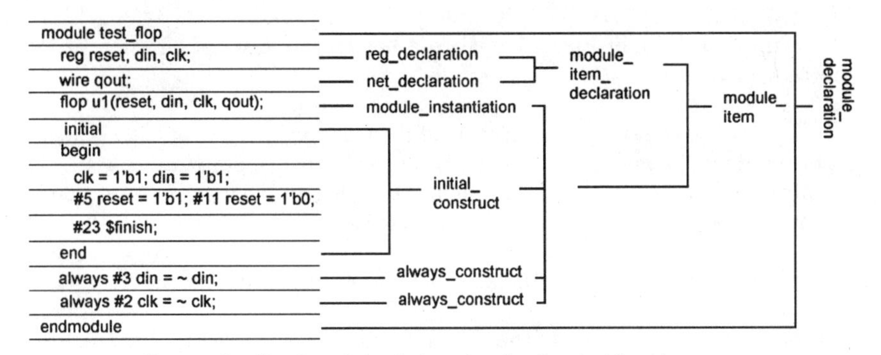

Figure 3.47 Syntax details of *module_declaration* for *flop* test bench.

synthesis tools, and libraries fit in a top-down design flow. In our design scenario, we also discussed reuse of previous designs.

Further Reading

IEEE Standard Hardware Description Language Based on the Verilog Hardware Description Language, IEEE Std. 1364-1995, Institute of Electrical and Electronic Engineers, New York, 1996.

Palnitkar, S., *Verilog HDL: A Guide to Digital Design and Synthesis,* Prentice-Hall, Upper Saddle River, N.J., 1996.

Smith, Douglas J., *A Practical Guide for Designing, Synthesizing and Simulating ASICs and FPGAs Using VHDL or Verilog,* Doone Publications, Madison, Ala., June 1996.

Thomas, D. E., and P. R. Moorby, *The Verilog Hardware Description Language,* 3d ed., Kluwer Academic Publishers, Norwell, Mass., 1996.

Problems

3.1. Assuming that the only available part is a two-input NAND gate, show the partition tree for a 2-bit magnitude comparator. Form Karnaugh maps and generate the complete gate list using the output maps. Students should first review their basic logic design material before doing this problem.

3.2. Assuming that the only available parts are D-type flip-flops and two-input NAND gates, show the partition tree for a Moore 1011 sequence detector. To achieve this, a state diagram must be formed and all design steps must be carried out. Students should first review their basic logic design material before doing this problem.

3.3 Write a dataflow description for a 1-bit comparator. The output of the comparator becomes 1 when its two input bits are equal.

3.4 Write a dataflow description for a 1-bit subtractor. Include a borrow input and a borrow output.

3.5 Modify the description of *counter* of Fig. 3.17 to include a *preset* input. When this input becomes 1, the counter gets set to 1111.

3.6 Using the style of Fig. 3.19, write a behavioral description for a serial subtractor.

3.7 Modify the behavioral description of the shift register of Fig. 3.25 to form a universal shift register with right and left shift modes. An *rl* input controls the direction of shift. For this problem, you need to modify the conditional operation used in the **always** block of Fig. 3.25.

3.8 Wire four full-adders of Fig. 3.24 to build a nibble adder.

3.9 Modify the description of the *der_flop* to include a preset input. The flip-flop output must be set to 1 when a synchronous 1 appears on this input.

3.10 Modify the *shifter* in Fig. 3.32 to include a *preset* input. Use the flip-flop developed in Prob. 3.9.

3.11 Modify the description of the sequence detector in Fig. 3.44 to detect the 1011 sequence. Your circuit should detect overlapping sequences.

3.12 Write a behavioral description for a clock generator generating two nonoverlapping clock phases.

3.13 Write a function for subtracting 2 data bits using a borrow input. Your function should produce a borrow and a result output. This function is similar to that shown in Fig. 3.40.

3.14 Use the function developed in Prob. 3.13 to develop a module for a bit subtractor. This module is similar to that shown in Fig. 3.40.

Chapter

4

Basic Concepts in Verilog

Because Verilog is a language for description of hardware, it has features which are conceptually different from those of software languages. These features are in the language to represent special characteristics of hardware structures and variables. Two main features that characterize hardware languages are timing and concurrency. Timing is associated with values that are assigned to hardware carriers, while concurrency refers to simultaneous operation of various hardware components. This chapter discusses concepts related to timing and concurrency in Verilog. Continuous and procedural assignment of values to variables and concurrent and procedural program flow will be discussed.

4.1 Characterizing Hardware Languages

Timing and concurrency are the main characteristics of hardware description languages. These features are instrumental in the correct description of hardware components at various levels of abstraction.

4.1.1 Timing

Transfer of values between hardware components or within a component is done through wires or busses. Variables in Verilog may be used for representation of actual wires, and because of delays associated with the transfer of values through wires, variable assignments in Verilog can include timing specification.

Figure 4.1 shows a gate output connected to two gate inputs through long capacitive wires. Dotted capacitors represent wire capacitance proportional to the length of the wire segments. Wire capacitance, together with gate pull-up or pull-down resistances, causes propagation delays through wires.

In a software language, assignment of the value of node x to a and b of Fig. 4.1 is done by simple assignments such as those shown below:

```
a := x;
b := x;
```

These assignments consider only value transfer, from x to a and b, and ignore the timing of such transfers, as well as the fact that the value of x is targeted toward a and b at the same time.

In a hardware language, assignments can be timed such that both the value assignment and the time of such assignment can be specified. Furthermore, assignments can be made to become active simultaneously. Transfer of the value of node x to a and b of Fig. 4.1 can be more accurately represented in Verilog by the following concurrent assignments:

```
assign #(4*unit_delay) a = x;
assign #(3*unit_delay) b = x;
```

In these assignments, when the value of x changes, its new value appears on a and b after 4 and 3 *unit_delay* values, respectively.

The time at which a value appears on a signal depends on the time at which the right-hand side is evaluated and the specified delay value. In the previous example, when x changes, the right-hand sides of both assignments are evaluated at the same time. The new value is scheduled for a and b and appears on b one delay unit ahead of its appearing on a.

4.1.2 Concurrency

Like timing, concurrency is an essential feature of any language for description of hardware. When a software programmer develops code for performing a certain task, he or she thinks of this task in a

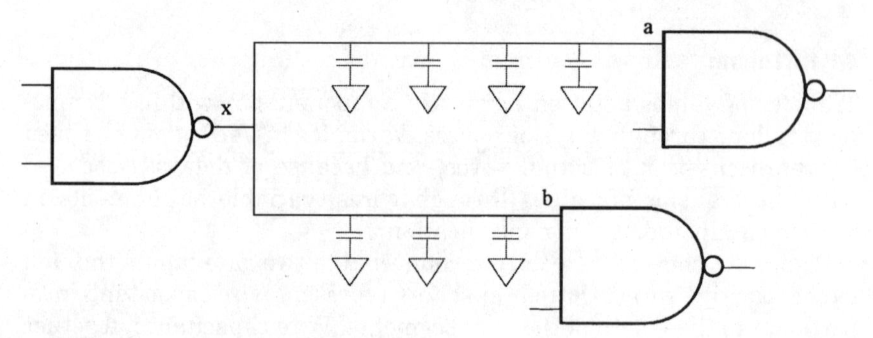

Figure 4.1 Value transfer through wires.

sequential manner. The software developed this way will have a top-down sequential flow. On the other hand, when a hardware designer or modeler is to describe a hardware system, he or she thinks of this hardware as interconnections of components. The functionality of the overall system is achieved by concurrently active components communicating through their input and output ports. The functionality of each component may be described by concurrent subcomponents or described by a program in a sequential manner.

We refer to concurrency as the way the simulation of components or constructs appears to the user. Obviously, Verilog is a language for which simulators have been developed on single-processor platforms, and true concurrency in the execution of thousands of components cannot exist. Through the use of concurrent constructs, timing of interconnecting signals, and order of simulation of constructs or components, a Verilog simulator makes us (the users) *think* that such execution is being done concurrently.

The Verilog language has constructs that allow the creation of a virtually concurrent environment. These constructs satisfy the concurrency required for description of hardware. Figure 4.2 shows a system consisting of three subcomponents. The overall system is a concurrent body consisting of three concurrent substructures. Each component may itself be described by concurrent constructs or by a procedural code specifying the behavior of the construct.

A section of code or a Verilog body in which interconnection of subsystems, such as those in Fig. 4.2, is specified will be referred to as a concurrent body of Verilog. Statements immediately enclosed in a module that are part of the *module_item* construct of Verilog constitute its concurrent body. Module and gate instantiations and wiring, procedural constructs, and continuous assignments can appear in this part. This body may also contain variable and parameter declarations. Unless a statement uses a declared variable, the order of the statements and declarations in a concurrent body does not affect the behavior of a module.

Figure 4.3 shows a block representing a Verilog concurrent body. A Verilog simulator simulating this body makes it appear as if all statements that appear here are simulated concurrently. This is done by executing each statement only when a change occurs on the right-hand side of a continuous assignment or on an input of an instantiated gate or module.

Verilog concurrent bodies also describe blocks A and B of Fig. 4.2. As in S, A is at the structural level of abstraction and describes interconnection of subcomponents. Block B is at the dataflow level of abstraction and may be described by continuous assignments in a Verilog module specifying the flow of data between registers and busses.

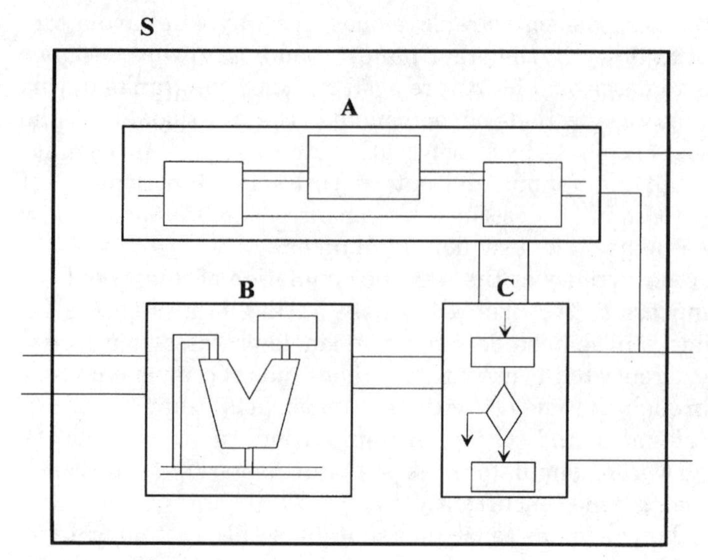

Figure 4.2 Describing subcomponents.

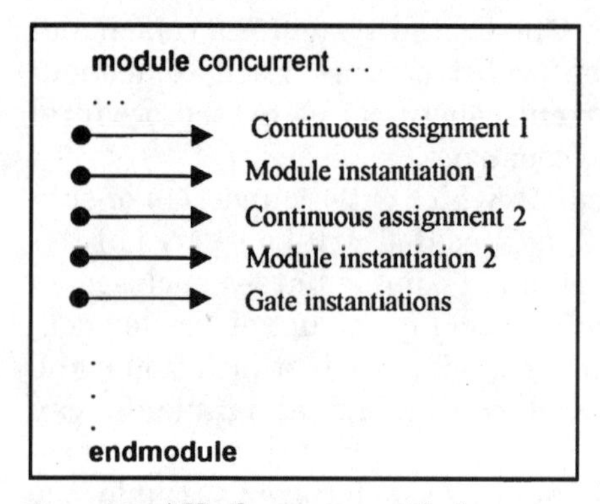

Figure 4.3 A Verilog concurrent body.

Even though the combined operation of various system components and communication of components with each other are concurrent, operation of an individual component may be specified by a top-down procedural flow. Figure 4.2 shows that the internal operation of block C is specified by a flowchart in a procedural manner. This subcomponent is communicating with other subcomponents, whose operations may be specified concurrently or sequentially.

Verilog constructs for the specification of the behavior of a system are referred to as procedural constructs. These constructs allow a designer to express the functionality of hardware under design instead of its structural details.

Figure 4.4 shows a block representing a Verilog procedural body. The flow of the program begins after the keyword **always** in the *always_statement* and continues sequentially. Statements similar to those of programming languages may appear in this Verilog construct.

4.1.3 Modeling hardware

In a simple example, we will illustrate how the concepts of concurrency and timing become useful in modeling actual electronic circuits. Various Verilog representations of this example will be used in later sections of this chapter to demonstrate continuous assignments and timing.

As in all electronic circuits, components of the circuit in Fig. 4.5 are always active, and there is a timing associated with every event in the circuit. We will assume that each gate has a delay of 12 ns, and that at time 100 ns all inputs are high. In this state, if no event occurs on any of the nodes of the circuit for a long time, the output remains at a stable **1** and nodes *w, x,* and *y* assume **0, 1,** and **0** values, respectively. As illustrated in Fig. 4.6, if input *a* switches from **1** to **0,** gates *g1* and *g2* concurrently react to this change. As shown in the timing diagram in Fig. 4.7, the *g1* gate causes node *w* to go to **1** after 12 ns, and *g2* causes node *x* to go to **0** after the same amount of time. At this time, gates *g3* and *g4* see a change at their inputs, and they start reacting to their new input conditions. The change at the *x* input of *g4* turns the output off

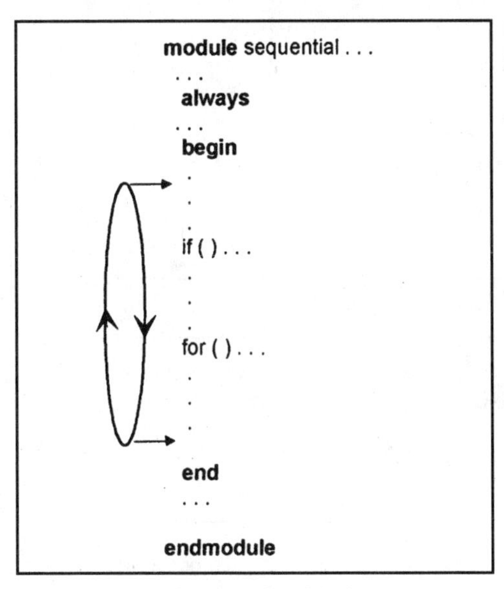

Figure 4.4 A Verilog procedural body.

after 12 ns. At this same time, the change on w has caused node y to become **1**. The OR gate (*g4*) now has a **1** at its y input, which causes it to go back to **1** only 12 ns after it has gone to zero. As shown in Fig. 4.7, this causes a 12-ns-wide zero glitch on the output of the circuit, which must be properly represented in a simulation model of this circuit.

This analysis would be more complex if the gates had unequal delay values, or if other inputs change when the circuit has not stabilized. A Verilog description capable of modeling this circuit at the level of timing discussed here would consist of four concurrent expressions repre-

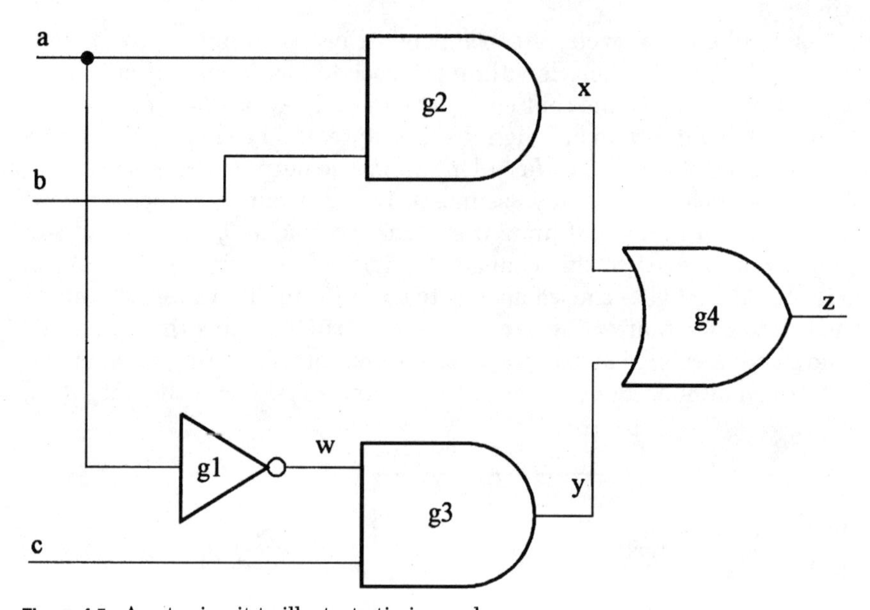

Figure 4.5 A gate circuit to illustrate timing and concurrency.

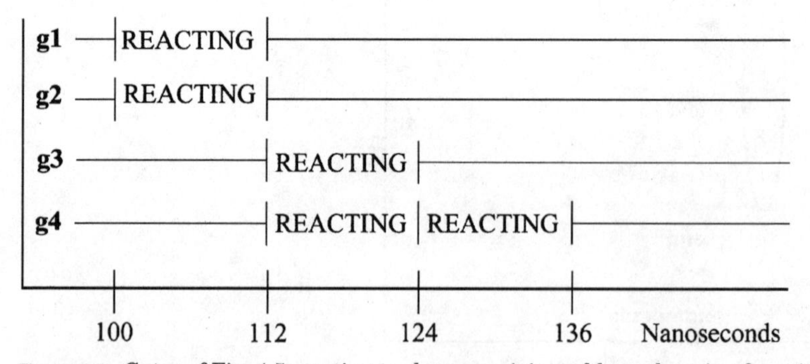

Figure 4.6 Gates of Fig. 4.5 reacting to changes originated by a changing from **1** to **0**.

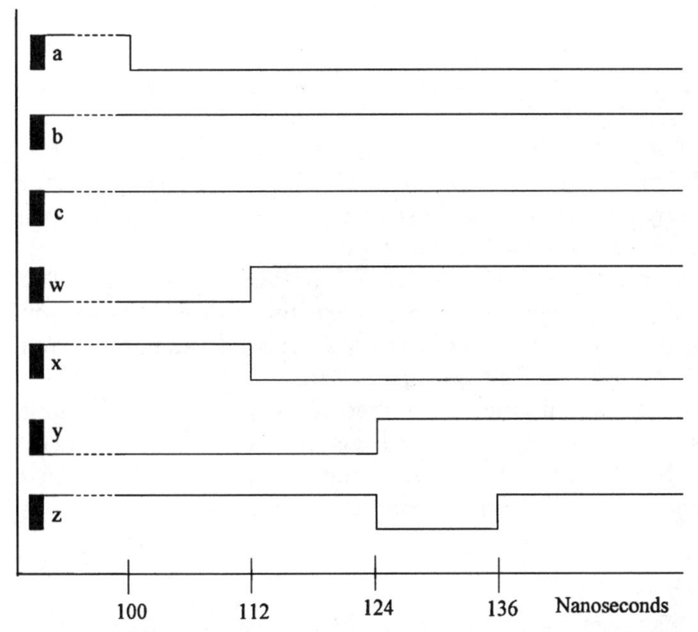

Figure 4.7 Timing diagram resulting from input *a* of the circuit of Fig. 4.5 changing from **1** to **0** at time 100 ns.

senting the gates of the circuit. Each expressions becomes active when a change occurs on an input of the gate it is representing. At this time, a new output value will be evaluated and, after a delay that represents the gate delay, will be assigned to the output variable.

4.2 General Format

This section presents conventions, lexical issues, and code formats in Verilog. The standard IEEE Verilog document has a formal presentation of these topics. In this section we will discuss only the most essential format issues.

4.2.1 Code format

Verilog code is free format, with spaces and new lines serving as separators. Source text is case-sensitive, i.e., identifiers using lowercase or uppercase characters are distinguished from each other. The language uses certain keywords, all of which must use lowercase characters.

Comments may appear anywhere in a Verilog source text. A comment designator starting with **//** makes the rest of the line, up to a new-line character, a comment. The symbols **/*** and ***/** bracket a section of code as a comment, and they go across new-line characters.

4.2.2 Names

A stream of characters starting with a letter or an underscore forms a Verilog identifier. The $ character and the underscore are allowed in an identifier. A stream of special characters may be used as an identifier if preceded by a backslash character. Verilog uses keywords that are all formed by streams of lowercase characters. In our example, we use bold type for Verilog codes for keywords.

System tasks and functions are part of the Verilog standard. The names of these utilities begin with a **$** character. An example system task is **$display,** which is used for formatted output. System tasks and functions will be discussed in Chap. 7.

The Verilog language defines a number of compiler directives that will be discussed in Chap. 7. Compiler directive names are preceded by the ` (back single quote) character. An example is **`timescale,** which defines the time unit for a Verilog code in a source text.

4.2.3 Numbers

The logic value system used in Verilog is the four-value system. In this system, logic values **1** and **0** have their standard boolean meanings. Logic value **X** is used for unknown or conflict, and value **Z** is used for high impedance or the open state.

Integer formats provide various ways for representing bit streams. Integers may be sized or unsized. A sized integer begins with the number of equivalent bits, followed by a single quote character ('), a base specifier, and the digits of the number in the specified base. The base specifier is a single lower- or uppercase character, **b, d, o,** or **h** for binary, decimal, octal, and hexadecimal bases. The general format for integers is

number_of_bits '**base-identifier** *digits*

Digits in the decimal (**d**) system are **0** through **9.** For the hexadecimal, octal, and binary systems, in addition to their standard digits, **X** and **Z** (both upper- and lowercase) characters are also allowed. Hexadecimal and octal **X** and **Z** digits expand to 4 or 3 bits of **X** and **Z,** respectively.

If a number contains fewer digits than are specified by the equivalent *number_of_bits,* zeros will be padded to the left of the number if the leftmost bit is not **X** or **Z.** In the case of a leftmost value of **X** or **Z,** this same value will be expanded to fill all left bits of the number.

If there are more *digits* than are specified by the equivalent number of bits, the extra digits will be ignored. Constants can also include the underscore character for separation of groups of digits. This character is ignored.

Several examples and their binary equivalent are shown here:

```
4'd5   = 0101
8'b101 = 00000101
-8'b101 = 11111011
10'o752 = 0111101010
8'hF   = 00001111
12'hXF = xxxxxxxx1111
```

In the above format, eliminating the number of bits yields an unsized number. The size of an unsized number is whatever is required by the number of actual digits of the number. For example, 'h5F becomes an 8-bit binary number 01011111.

Real numbers are also allowed in Verilog, and for their representation, decimal or scientific notation may be used. As in integers, the underscore character may be used for separation of digits and better readability.

4.3 Data Types

Verilog variables may be of **net** or **reg** data types. The **net** type represents wires or busses that are driven by gates or hardware sources, while the **reg** type represents variables that hold the value they are assigned until they are overwritten. Declaring **net** and **reg** types and their significance in hardware modeling will be discussed here.

4.3.1 net declarations

A **net** represents a hardware wire driven by one or more gates or other types of signal sources. The simplest form of a **net** declaration begins with a keyword specifying the type of **net** followed by a list of identifiers.

Allowed **net** types are shown in Fig. 4.8. Types **wire** and **tri, wand** and **triand,** and **wor** and **trior** are equivalent. Types **supply0** and

	NET TYPES	PROPERTIES	INITIAL
Supply	supply0	Driven 0	0
	supply1	Diven 1	1
Three-state	wire (tri)	tri-state wired logic	Driven: X Not Driven: Z
	wand (triand)	wired-and logic	Driven: X Not Driven: Z
	wor (trior)	wired-or logic	Driven: X Not Driven: Z
Capacitive	trireg	Hold old value	X

Figure 4.8 **net** types and properties.

supply1 are used for declaring signal names for supply voltages. The **trireg net** type declares three-state capacitive signals. Other **net** types (**wire, wand,** and **wor** or their equivalents **tri, triand,** and **trior**) declare state signals that allow multiple driving sources. The keyword indicating a **net** type determines how multiple driving source values are resolved to form a single **net** value. Shown below is a wire declaration declaring wires *w, x, y,* and *z.* This statement declares wires between gates in Fig. 4.5.

```
wire w, x, y, z;
```

The **wire** type declares three-state signals. Multiple drivers on such signals are resolved by the table in Fig. 4.9. In this **net** type, an **X** value on any driving source overrides values from all other sources. The **Z** value is the weakest and is overridden by non-**Z** values from other driving sources. Driving a **wire** with multiple **0** and **1** conflicting values resolves in the **X** value for the wire.

The **wand** and **wor** type **net**s signify wired-and and wired-or functions, respectively. Figure 4.10 shows the notations for these functions and tabulates their resolved values. The **wand** type implements a wired-and logic. For this type, a **0** value on a driving source overrides all other source values. Value **Z** is treated as null and is overridden by any other value driving a **wand net.** The **wor,** or wired-or, type performs the logical operation OR on all its driving sources. In this operation, logic value **1** on one source overrides all other source values. As in **wand,** the **Z** value is the weakest and is overridden by **0, 1,** and **X** values.

Net types **tri0** and **tri1** are similar to **wire** (**tri**) in their resolved values except when the resolved value is to become **Z** in a **wire net.** Where **Z** is generated in a **wire net, 0** is generated in **tri0** and **1** is generated in **tri1.**

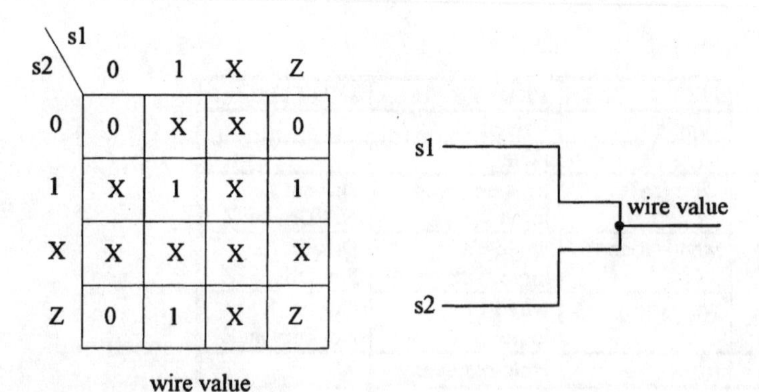

s2 \ s1	0	1	X	Z
0	0	X	X	0
1	X	1	X	1
X	X	X	X	X
Z	0	1	X	Z

Figure 4.9 **wire net** type.

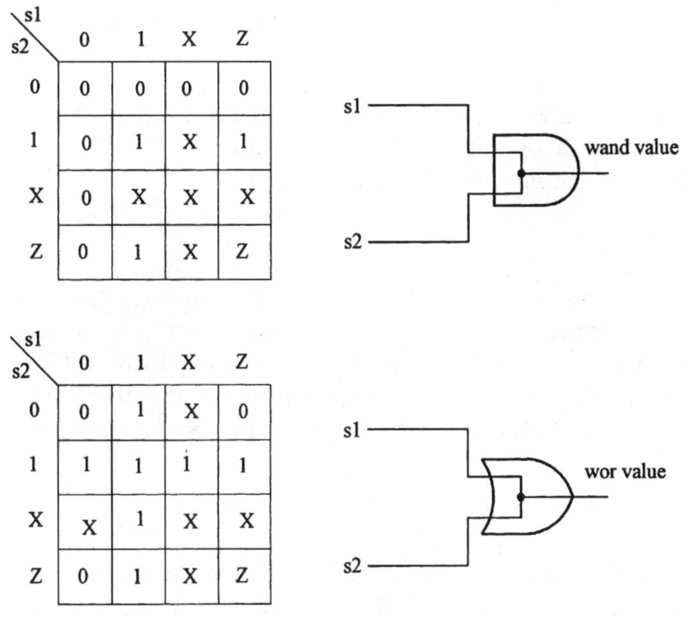

Figure 4.10 (*a*) **wand net** type, (*b*) **wor net** type.

The **trireg** type **net** behaves as a capacitive wire and holds its old value when a new resolved value is to become **Z.** As long as there is at least one driver with a **0, 1,** or **X** value, **trireg** behaves the same as **wire.** When all drivers are turned off (**Z**), a **trireg net** retains its previous value. The amount of time a **trireg net** holds a value is specified by a delay parameter in its declaration. Delay parameters will be discussed next. Chapter 5 shows examples of using **trireg** for CMOS flip-flop modeling.

In the above discussion we presented **net** declarations by **net** types. In addition to the type of the **net,** which specifies the resolution of driving-value conflicts, a **net** declaration may also include **net** delay values. Three delay values for **net** switching to **1,** to **0,** and to **Z** are specified in a set of parentheses that are followed by a # sign after the **net** type keyword. A simpler format contains a single delay value. For example, the **wire** declaration for the gate outputs of Fig. 4.5 may be done by

```
wire #3 w, x, y, z;
```

This declaration specifies a three-time-unit delay for all transitions of *w, x, y,* and *z* signals. This delay will be added to those of continuous assignments or gate outputs driving these **nets**. **Trireg net** types may also be declared with three delay parameters. Unlike the case with other **nets**, in this case the third timing parameter is not delay for the **Z** transition. Instead, this specifies the time that a declared **trireg net** holds an old value when driven by **Z.**

The initial value for all **net** types except **supply0** and **supply1** with at least one driver is **X.** A **net** with no driver assumes the **Z** value, except for **trireg,** which has the initial value **X.**

4.3.2 reg declarations

In addition to the **net** type variable declarations, Verilog also supports the **reg** data type. Unlike a **net,** which models an interconnection or a gate output, a **reg** is a variable for holding intermediate signal values or nonhardware parameters and function values. The **reg** declaration shown below declares the gate inputs of the Fig. 4.5 example:

```
reg a, b, c;
```

Because **reg** variables are not used exclusively for hardware modeling, other **reg** type declarations for more convenient forms of model parameters are provided in Verilog. These **reg** types are **integer** and **time.** An **integer** declaration declares a signed 2s-complement number, and a **time** declaration declares an unsigned **reg** variable of at least 64 bits. Verilog also allows declaration of **real** and **realtime** variables. These variables are similar in use to **integer** and **time** variables, but do not have direct bit-to-bit correspondence with **reg** type registers. The initial value of all declared **reg** variables is (**X**).

4.4 Assignments to nets and regs

The previous section discussed **net** and **reg** declarations. Because these variables represent very different entities in a hardware model, the Verilog HDL is very specific as to the way they are used and the language constructs they are used in.

Variables declared as **net** are assigned values in Verilog concurrent bodies using continuous assignment statements. On the other hand, **reg** variables are assigned values in procedural bodies. The following discusses details of these constructs.

4.4.1 Continuous assignments

In this section we will discuss simple continuous assignments, assignments with delay, strength specification, net assignments, and assignments with multiple drivers.

Simple assignments. A continuous assignment in Verilog is used only in concurrent Verilog bodies, discussed in Sec. 4.1. This assignment represents a **net** driven by a gate output or a logic function. In its simplest form, a continuous assignment begins with the assign keyword, followed by the left-hand-side net type variable, an equal sign, and a right-hand-side expression. The example shown below models gate *g2* in Fig. 4.5.

```
assign x= a & b;
```

Like gate *g2* in Fig. 4.5, this assignment becomes active only if an input of the gate or a variable on the right-hand side of the assignment changes value.

Delay specification. Continuous assignments may also include delay parameters. Using this format, a better modeling of *g2* of Fig. 4.5 is.

```
assign #12 x = a & b;
```

This assignment becomes active when *a* or *b* changes. At this time, the new value of the *a & b* expression is evaluated, and after a wait time of 12 time units, this new value is assigned to *x*.

The order in which continuous assignments appear in a Verilog concurrent body is not significant. Figure 4.11 shows four concurrent assignments corresponding to the gates of Fig. 4.5. Regardless of its position in the code, and as with the *g1, g2, g3,* and *g4* gates, each assignment waits for a right-hand-side variable to change for it to execute.

Hand simulating the Verilog code of Fig. 4.11 for the waveform of Fig. 4.7, the first change occurs on *a* at time 100 ns. This causes the assignments to *x* and *w* to execute, resulting in these **net**s changing from **0** to

```
`timescale 1ns/100ps

module fig4_5 (a, b, c, z);
input a, b, c;
output z;
wire w, x, y, z;

assign #12 x = a & b;
assign #12 w = ~a;
assign #12 y = w & c;
assign #12 z = x | y;

endmodule
```

Figure 4.11 Concurrent continuous assignments.

1 at time 112 ns. Changes on *x* and *w* cause the third and fourth assignments in Fig. 4.11 to execute. This causes a change on *y* after 12 ns and a change on *z* to **0** at the same time. When *y* changes at this time, the right-hand side of the continuous assignment with **net** *z* on its left-hand side will again be evaluated, resulting in the value **1** to being assigned to *z* 12 ns later. As a result, the waveforms shown in Fig. 4.7 will result after simulation of the Verilog code of Fig. 4.11.

As another example of the use of **nets** for modeling hardware, consider the Verilog code in Fig. 4.12. Delay values in the continuous assignments in this figure are 9 ns. This delay value models delays associated with gates *g1, g2, g3,* and *g4,* which drive wires *w, x, y,* and *z1,* respectively. On the other hand, in the **wire** declarations of Fig. 4.12, 3-ns delay values are specified for the *w, x, y,* and *z* wire segments. This models wire or line delay values for lines used for gate interconnections. The time it takes a change on the right-hand side of an assignment to affect its left-hand side is the sum of the delay associated with the left-hand-side variable and the continuous assignment delay. Therefore, all changes on wires in Fig. 4.12 occur after 12 ns, that is, the sum of the 3-ns wire delay and the 9-ns assignment delay. Simulation of the code of Fig. 4.12 results in the same waveform as simulation of that of Fig. 4.11. As in the simulation of the code of Fig. 4.11, and as expected according to the waveform in Fig. 4.7, a 12-ns hazard appears on *z* when the code of Fig. 4.12 is simulated.

Figure 4.13 presents another example illustrating delays in continuous assignments. This code is an alternative to the code of Fig. 4.11 for modeling the gate circuit in Fig. 4.5. Instead of individual continuous assignments delays of 12 ns, the assignment to output *z* in Fig. 4.13 is delayed by 36 ns, which is the worst-case path delay. Concurrency in simulation of all four continuous assignments still exists in this code, and the final value assigned to *z* is as expected. However, as shown in Fig. 4.14, because of intermediate zero-delay assignments, the expected hazard on *z* does not appear in the simulation of Fig. 4.13.

Following the events in Fig. 4.14, when *a* changes from **1** to **0** at 100 ns, *w* and *x* change to **1** and **0** in zero time—still at 100 ns time, but in the next simulation cycle. The new value of *x* causes the right-hand side of the assignment to *z* to be evaluated, and this causes a **0** to be scheduled for output *z* for 36 ns later. At the same time, the new value of *w* causes *y* to become **1** in the next simulation cycle. At this time, the right-hand side of *z* will be evaluated again, which causes a **1** to be scheduled for *z* for 36 ns later. This new scheduling for *z* overrides the previously scheduled **0** for this wire. As a result, the actual value of *z* remains **1** regardless of changes on its right-hand side.

```
`timescale 1ns/100ps

module fig4_5 (a, b, c, z);
input a, b, c;
output z;
wire #3 w, x, y, z1;

assign #9 x = a & b;
assign #9 w = ~a;
assign #9 y = w & c;
assign #9 z1 = x | y;
assign z = z1;

endmodule
```

Figure 4.12 Continuous assignments; delayed wires are used.

```
`timescale 1ns/100ps

module fig4_5 (a, b, c, z);
input a, b, c;
output z;
wire w, x, y, z;

assign x = a & b;
assign w = ~a;
assign y = w & c;
assign #36 z = x | y;

endmodule
```

Figure 4.13 Lumping delay values.

Strength specification. In addition to a logical value in the four-value system (**0**, **1**, **X**, and **Z**), Verilog allows **nets** to have strength values. Strength adds another degree of accuracy in modeling signal when the basic four values do not suffice. Strengths for **nets** are specified when assignments are done. As signals combine, new strength values are formed in the resulting signals.

Net strengths are specified by a pair of strength values bracketed by a set of parentheses, as shown below:

```
assign (strong0, strong1) x = a & b;
```

One strength value is for logic **1** and one is for logic **0,** and the order in which the strength values appear in the set of parentheses is not important. Strength value names for logic **1** end with a **1** (**supply1, strong1, pull1, weak1,**...), and those for logic **0** end with a **0** (**supply0, strong0, pull0, weak0,**...).

Time ns	Sim Cycle	a	b	c	z	w	x	y
0	+0	x	x	x	x	x	x	x
0	+1	1	1	1	x	x	x	x
0	+3	1	1	1	x	0	1	x
0	+5	1	1	1	x	0	1	0
36	+0	1	1	1	1	0	1	0
100	+2	0	1	1	1	0	1	0
100	+4	0	1	1	1	1	0	0
100	+6	0	1	1	1	1	0	1

Figure 4.14 Simulation report of code of Fig. 4.13 with zero-delay continuous assignments.

For **wire** and **tri** type **net**s, drive strength values are used, and for storage **net**s, charge strength is used. Figure 4.15 shows strengths, their values, and the **net** types they apply to.

The strength value set for **wire** and **tri** type **net**s is referred to as *drive_strength* and is specified when an assignment to **net** takes place or when net is declared. Default values for these **net**s are **strong0** and **strong1** for logic **0** and logic **1,** respectively. The strength for **trireg** type **net**s specifies how weak or strong the charge storage capability of the **net** is. Three strength values, **large, medium,** and **small,** are used for these **net** types, and the default is **medium.** Examples in Chap. 5 illustrate how different strength values are resolved on gate outputs.

Net declaration assignments. Using the **nets** declaration assignment, a **net** assignment can be done at the same time that it is being declared. An assignment made as such provides one driver for the declared **net.** More drivers can be assigned to the same **net** using continuous assignment statements. As with continuous assignments, and net declarations, strengths and timing may also be specified in a **net** declaration assignment.

The use of *net_declaration_assignment* is illustrated in Fig. 4.16. In this code, all the continuous assignments of Fig. 4.11 are replaced by a list of **net** declaration assignments providing drivers for the *w, x, y,* and *z* signals. Drive strengths and/or delay values specified in a **net** declaration assignment apply to all **net** drivers. The syntax used here consists of a list of **net** declaration assignments as part of a **net** declaration construct. The same syntax can also be used with continuous assignments. In order to be able to specify delay values for individual wires, use of separate continuous assignments, as in Fig. 4.11, or separate **net** declaration assignments is recommended.

wire (tri), wand (triand), wor (trior), tri0, tri1		trireg
Strength values	Level	Strength values
Suppply0	7	
Strong0	6	
Pull0	5	
	4	Large (0)
Weak0	3	
	2	Medium (0)
	1	Small (0)
Highz0	0	
Highz1	0	
	1	Small (1)
	2	Medium (1)
Weak1	3	
	4	Large (1)
Pull1	5	
Strong1	6	
Supply1	7	

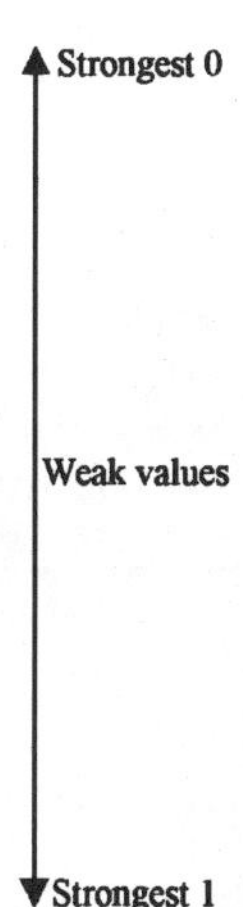

Figure 4.15 **net** types and their strengths.

```
`timescale 1ns/100ps

module fig4_5 (a, b, c, z);
input a, b, c;
output z;

wire #12 x = a & b, w = ~a, y = w & c, z = x | y;

endmodule
```

Figure 4.16 Using *net_declaration_assignment*.

Multiple drivers. A situation in hardware in which several gate outputs are connected to the same wire is modeled with continuous assignments by having multiple assignments to the same left-hand-side **net**. In this case, the **net** value is said to be driven with multiple sources simultaneously. The resulting **net** value is determined by the resolution of multiple driving values depending on the **net** types, as discussed in Sec. 4.3.1.

An example for multiple drivers using wired-or resolution is shown in Fig. 4.17. The code shown here models the circuit of Fig. 4.5 and is equivalent to the Verilog code shown in Fig. 4.11. The OR operation on the right-hand side of *z* in Fig. 4.11 is replaced by multiple assignments to the *z* **net** of **wor** type. This **net** has a delay of 12 ns, which is added to the individual driver delay values. A value assigned to *z* is first delayed by continuous assignment delay. Before this value

```
`timescale 1ns/100ps

module fig4_5 (a, b, c, z);
input a, b, c;
output z;
wire w;
wor #12 z;

assign #12 z = a & b;
assign #12 w = ~a;
assign #12 z = w & c;

endmodule
```

Figure 4.17 A **net** with multiple drivers.

Time ns	Sim Cycle	a	b	c	z	w
0	+0	X	X	X	X	X
0	+1	1	1	1	X	X
12	+0	1	1	1	X	0
24	+0	1	1	1	1	0
100	+2	0	1	1	1	0
112	+0	0	1	1	1	1
124	+0	0	1	1	0	1
136	+0	0	1	1	1	1

Figure 4.18 Simulation run result of code of Fig. 4.17.

appears on *z*, it is further delayed by the 12 ns specified in the **wor** declaration.

Figure 4.18 shows the simulation result of the Verilog code in Fig. 4.17. When *a* changes at time 100 ns, a **0** value is scheduled for *z* and appears on this output after two 12-ns delays. Therefore, *z* becomes **0** at time 124 ns. Also, *w* changes at time 112 ns, which schedules a **1** for *z* for 24 ns later. This value appears on *z* at time 136 ns. The two consecutive assignments to *z* cause a 12-ns zero glitch on this output.

Wired resolutions are useful in open-collector gate and bus structure modeling. Examples of various applications of this language feature will be presented in Chaps. 5 and 8.

4.4.2 Procedural assignments

Procedural assignments in Verilog take place in the **initial** and **always** procedural constructs, which are regarded as procedural bodies as discussed in Sec. 4.1.2. Primarily, assignments to **reg** data types take place in procedural bodies. This section discusses procedural flow control blocking assignments, nonblocking assignments, and two forms of procedural continuous assignments.

Procedural flow control. Statements in a procedural body are executed when program flow reaches them. Several flow control statements are available to control procedural flow in a procedural body. These language constructs are classified as delay control and event control and will be discussed in Chap. 9. However, because these constructs affect the way procedural assignments are done, a brief discussion is included here.

Within a module, several procedural bodies may be used. Flows into these bodies all begin at the same time at the start of simulation. An event or delay control statement in a procedural body causes program flow to be put on hold temporarily. The flow continues after an event occurs or a delay expires. Figure 4.19 shows three procedural flow control statements. In Fig. 4.19*a,* program flow stops when it reaches the *@(reset)* statement and resumes when the value of *reset* changes. In Fig. 4.19*b,* program flow resumes after the positive edge of *clk,* and in Fig. 4.19*c,* program flow resumes after being put on hold for 10 time units.

Procedural blocking assignments. A blocking assignment uses a **reg** data type on the left-hand side and an expression on the right-hand side of an equal sign. An event or delay control statement may delay execution of this statement. In addition to statements for control of program flow, procedural assignments may also contain intra-assignment delay or event control. The syntax of intra-assignment control constructs is similar to that of procedural flow control statements, discussed above, but these constructs appear on the right-hand side of an equal sign in a procedural assignment. Shown below is a procedural assignment that is delayed by 200 time units by a delay control statement and by 100 time units by an intra-assignment delay control.

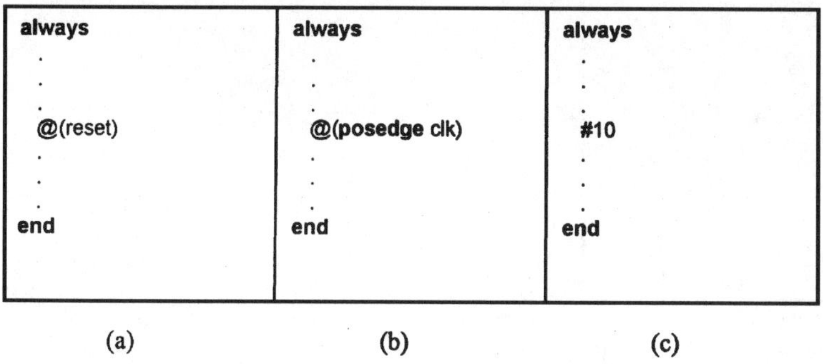

always . . . @(reset) . . . end	always . . . @(posedge clk) . . . end	always . . . #10 . . . end
(a)	(b)	(c)

Figure 4.19 Procedural flow control.

```
. . .; #200 a = #100 b;
```

In this example, after 200 time units, when program flow reaches the procedural assignment, b is evaluated, and its value is assigned to a after 100 time units. The equal sign is used here for blocking assignment. In this case, the 100-time-unit delay blocks the procedural program flow until assignment to a takes place.

Figure 4.20 shows a test bench for the circuit in Fig. 4.11. Signal a is a **reg** that is initialized to **1** at time 0 ns, set to **0** at time 100 ns, and set back to **1** at time 200 ns. The intra-assignment timing control blocks procedural flow and stops it from reaching assignment of **1** to a until a is assigned the value **0.**

Another important issue regarding procedural **reg** type assignments is the way multiple assignments from multiple procedural constructs interact. Because of the sequential flow in procedural bodies, an assignment takes place only when the program flow reaches it. If several assignments appear at the same real time in a procedural body, the last assignment overrides all others. However, if program flow in two pro-

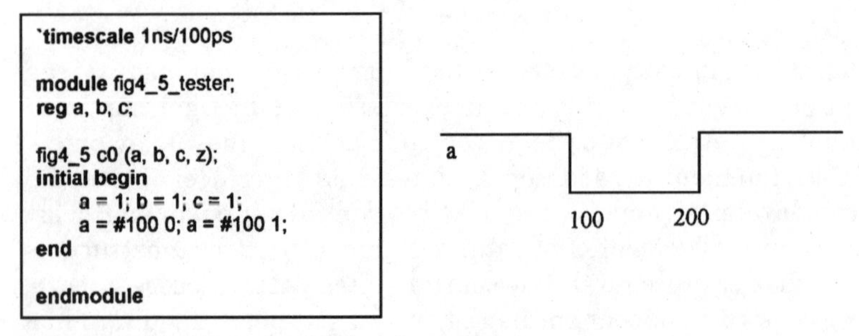

Figure 4.20 Blocking procedural assignments.

```
.
.
.
initial begin
    clk = 0;
end

always begin
    clk = ~ clk;
    #100;
end
.
.
.
```

Figure 4.21 Partial code, multiple **reg** assignments.

cedural bodies reaches assignments to the same **reg** at exactly the same time, the outcome of the value assigned to the left-hand side of the assignment will not be known.

Consider, for example, the partial code shown in Fig. 4.21 for clock generation for testing a flip-flop. The intended operation of this code may have been for *clk* to get initialized to **0,** and after that get complemented every 100 time units. However, this is not the way the code simulates.

At time **0,** flow into the **initial** and **always** blocks begins. In the **initial** block, *clk* is being set to **1,** and in the **always** block, at exactly the same time, *clk* is being complemented. Because the initial value of the **reg** type variable *clk* is **X,** the complement operation in the **always** block tries to complement **X** if *clk* is not initialized to **0** in the **initial** block. Whether *clk* is first set to **0** and then complemented, is first complemented, or is set to **X** because of the conflict is not known. This code works properly only if complementing of the *clk* is delayed until the *clk* is initialized to **0** in the **initial** block.

One way to correct this problem is to delay complementing the clock by one simulation cycle. This can be done by inserting *#0*; in the **always** block before complementing the *clk,* as shown in Fig. 4.22. This way, in the first simulation cycle, *clk* is set to **0** at time 0, and immediately after that, in the next simulation cycle, still at time 0, *clk* is set to the complement of **0.** Every time *clk* is complemented, flow in the **always** block is suspended for 100 time units, after which *clk* is complemented again for as long as the simulation run continues.

Procedural nonblocking assignments. Within a procedural block, nonblocking assignment to **reg** data types may be done. A nonblocking

```
.
.
.
initial begin
        clk = 0;
end

always begin
        #0;
        clk = ~clk;
        #100;
end
.
.
.
```

Figure 4.22 Partial code for multiple procedural assignments producing deterministic results.

assignment uses the left arrow notation <= (left angular bracket followed by the equal sign) instead of the equal sign used in blocking assignments.

A nonblocking assignment is different from a blocking assignment only in the way in which intra-assignment control constructs are treated. Unlike a blocking assignment, a nonblocking assignment does not block the program flow in a procedural construct. When flow reaches a nonblocking assignment, the right-hand side of the assignment is evaluated and will be scheduled for the left-hand-side **reg** to take place when the intra-assignment control is satisfied.

The test bench for the circuit of Fig. 4.5 presented in Fig. 4.20 is repeated in Fig. 4.23 using nonblocking assignments. The code shown in Fig. 4.23 produces the same waveform on *a* as that of Fig. 4.20, which uses blocking assignments.

The **initial** block in Fig. 4.23 is entered at time 0. At this time, *a, b,* and *c* are initialized to **1.** Then **0** is scheduled for *a,* to take place after 100 ns using a nonblocking procedural assignment. Immediately after this scheduling, and still at time 0, the procedural body flow reaches the next statement, which assigns a **1** to *a.* This assignment will be scheduled to take place 200 ns after time 0. In order to produce the same waveform as that produced by the test bench of Fig. 4.20, the assignment of a **1** to *a* uses a 200-ns intra-assignment delay. After this scheduling (still at time 0), program flow reaches the end of the **initial** block and terminates. Two values are still left pending for **reg** *a.*

Procedural continuous assignment. Using a procedural continuous assignment construct, an assignment to a **reg** type variable can be made to stop all other assignments to this variable from taking place. The procedural continuous assignment to variable *q* is done by the following procedural statement:

```
assign q= 0;
```

While *q* is not deassigned, no other assignments to *q* affect the value of this variable. Deassigning a **reg** type variable *q* is done by the following statement:

```
deassign q;
```

Figure 4.24 shows an example of the use of procedural continuous assignments. Two **always** blocks are used in this example. The first block becomes active any time *reset* changes. In this block, if *reset* is **1,** the *q* output of *flipflop* is assigned a **0.** This value is forced on *q* and cannot be overridden until *q* is deassigned. In the first **always** block, if *reset* is not active (has the value **0**), the flip-flop output *q* is deassigned. Only after *q* is deassigned can other assignments to *q* change its value.

The second **always** block in Fig. 4.24 assigns *d to q* on the positive edge of the *clk*. This assignment takes place only when reset is **0**.

While an **assign** is in effect, another **assign** to the same variable deassigns the one that is in effect and then assigns a new value to the variable.

Another form of procedural continuous assignment is provided by **force** and **release** statements. Unlike **assign** and **deassign,** which apply to **reg** type variables, **force** and **release** constructs apply to **net** and **reg** types. Forcing a value on a **net** overrides all values assigned to the **net** through continuous assignments or connected to it through gate outputs.

```
`timescale 1ns/100ps

module fig4_5_tester;
reg a, b, c;

fig4_5 c0 (a, b, c, z);
initial begin
        a = 1; b = 1; c = 1;
        a <= #100 0; a <= #200 1;
end

endmodule
```

Figure 4.23 Using nonblocking assignments.

```
`timescale 1ns/100ps

module flipflop (d, clk, reset, q);
input d, clk, reset;
output q;
reg q;
    always @(reset) begin
        if (reset == 1)
          assign q = 0;
        else
          deassign q;
    end
    always @(posedge clk) begin
          q = d;
    end
endmodule
```

Figure 4.24 Procedural continuous assignments.

```
`timescale 1ns/100ps

module flipflop_tester;
reg d, clk, reset;
    flipflop u1 (d, clk, reset, q);
    initial begin
        d = 0; #250 d = 1; #400 d = 0; #250 d = 1; #400 d = 0;
    end
    initial begin
        reset <= 0; reset <= #600 1; reset <= #1100 0;
    end
    initial begin
        clk = 0; #1500; $stop;
    end
    always begin
        #0; clk = ~ clk; #100;
    end
endmodule
```

Figure 4.25 *flipflop* test bench.

We will conclude this section by presenting a test bench for the flip-flop discussed above. The test bench and simulation run illustrating language concurrency features of Verilog are discussed here.

Figure 4.25 shows a test bench for the Verilog code in Fig. 4.24. After the module header and declarations, the *flipflop* is instantiated. Application of data to the inputs (*d, clk,* and *reset*) of the flip-flop takes place in four concurrent procedural blocks.

The first procedural block sets values to the *d* input. Because blocking procedural assignments are used in this block, the specified time values are relative. The list of values generated on *d* is shown in Fig. 4.26. The next procedural block sets values to the *reset* input of the flip-flop. Nonblocking assignments are used here, and time values specified for assigning new values to *reset* are absolute. As shown in Fig. 4.26, *reset* changes to **1** at time 600 ns and to **0** at 1100 ns.

The next two procedural blocks initialize *clk* and generate a periodic waveform on this signal. As shown in Fig. 4.26, the initial value of *clk* is set to **0** by the assignment statement in the **initial** block of Fig. 4.25. Following this assignment, after waiting a simulation cycle in the **always** block of Fig. 4.25, a **1** is assigned to *clk* by complementing its initial **0** value. This shows in the listing of Fig. 4.26 as a **1** at time 0 ns. After this initialization, the **initial** block is put on hold for 1400 ns before it terminates. Meanwhile, the **always** block repeats itself, causing *clk* to toggle every 100 ns.

The *q* output of the flip-flop is also shown in Fig. 4.26. While *reset* is active, the **assign** procedural continuous assignment forces this output to **0**, ignoring all other assignments made to this signal. Deassigning *q* when reset becomes **0** enables other assignments to *q* to affect its value.

4.5 Summary

This chapter presented important issues regarding **net** and **reg** type variables, delay values, flow control, simulation cycles, procedural blocks, and concurrency. We defined Verilog concurrent and procedural bodies and described assignment to variables with various timing configurations in these blocks. An understanding of the simulation mechanism and timing issues presented here is essential in understanding Verilog HDL. Concurrency issues and how a Verilog simulator makes us believe that execution of continuous assignments is done simultaneously were specially emphasized in this chapter. It is important to realize that timing and concurrency are the two most impor-

Time ns	Sim Cycle	d	clk	reset	q
0	+0	0	0	x	x
0	+1	0	1	x	x
0	+6	0	1	x	0
0	+7	0	1	0	0
100	+1	0	0	0	0
200	+1	0	1	0	0
250	+0	1	1	0	0
300	+1	1	0	0	0
400	+1	1	1	0	0
400	+6	1	1	0	1
500	+1	1	0	0	1
600	+1	1	1	0	1
600	+5	1	1	1	1
600	+10	1	1	1	0
650	+0	0	1	1	0
700	+1	0	0	1	0
800	+1	0	1	1	0
900	+0	1	1	1	0
900	+2	1	0	1	0
1000	+1	1	1	1	0
1100	+1	1	0	1	0
1100	+4	1	0	0	0
1200	+1	1	1	0	0
1200	+6	1	1	0	1
1300	+0	0	1	0	1
1300	+2	0	0	0	1
1400	+1	0	1	0	1
1400	+6	0	1	0	0

Figure 4.26 Simulating the *flipflop* of Fig. 4.24 using the test bench of Fig. 4.25.

tant features of hardware description languages, and that these features distinguish hardware languages from software programming languages.

Further Reading

IEEE Standard VHDL Language Reference Manual, ANSI/IEEE Std. 1076-1993, Institute of Electrical and Electronic Engineers, New York, 1994.

Navabi, Z., *VHDL: Analysis and Modeling of Digital Systems,* McGraw-Hill, New York, 1998.

Problems

4.1 Using continuous assignments, write a Verilog description for a full-adder with three inputs and two outputs. Use 5 ns delay for XOR and 3 ns for AND and OR operations. Hand simulate this code for several data inputs, showing values on all intermediate signals.

4.2 Using **net** declaration assignments, write Verilog code for the full-adder described in Prob. 4.1. Perform the hand simulation discussed in that problem.

4.3 Using procedural blocks, write Verilog code for the generation of a periodic waveform that repeats the cycle of being **0** for 200 ns, **1** for 300 ns, **0** for 100 ns, and **1** for 80 ns.

Chapter

5

Structural Specification of Hardware

To describe a system at the structural level, the components of that system are listed and the interconnections between them are specified. A term often used to describe this form of description is *netlist.* Because this level of abstraction closely corresponds to the actual hardware, it is easiest for hardware designers to understand and use. Software-oriented readers, on the other hand, should pay attention to concurrency features in the language that is introduced in this chapter.

Verilog provides language constructs for concurrent instantiation and wiring of hardware components. Hardware components may be described as user-defined primitives, built-in structures, defined modules, or modules that are defined by hierarchical instantiation of other modules. Language constructs for definition and use of modules and basic structures will be discussed.

After we introduce basic structures, we will develop a hierarchical design from simple primitive components. This design is a combinational circuit for which a simple test bench will be developed for functional testing. As a second example for further illustration of hierarchical design, a hierarchical register that is based on the same basic parts used for the combinational example will also be developed. The chapter concludes by putting the two examples together in a complete design. Overall, basic components, structural descriptions, and hierarchical design will be discussed.

Throughout this chapter, a set of notations and vocabulary is introduced for presenting and referencing Verilog HDL descriptions. This vocabulary is consistent with that used in IEEE Std. 1364-1995, which describes the Verilog HDL.

5.1 Basic Structures

To build a design from the bottom up, predefined library components or basic gate structures are assembled to form more complex structures. These predefined parts may be found in a library or used from other designs. Verilog high-level constructs allow definition of basic logic structures at any level of detail. On the other hand, built-in structures of Verilog provide the gate- and transistor-level functional and timing details needed for most digital circuit designs. Therefore, it is the choice of a logic designer whether to develop his or her own basic structures or to use already defined language built-ins.

In this section we will show several ways in which basic logic structures may be developed. First, we will introduce available Verilog built-ins and ways in which they can be used. Then we will show how other modules and user-defined primitives may be used for defining basic logic gate structures. The example that we will use is an *aoi* gate. This gate will be implemented using built-in gates, using other gate modules, and using user-defined primitives. Once a basic structure is defined, other higher-level hardware modules can be based on it. Several structures explained in this section will be used in the description of more complex functions in the rest of this chapter.

5.1.1 Using built-in structures

Verilog provides built-in structures for logic gates, pull gates, and switches. Figure 5.1 shows a complete list of Verilog built-ins. These structures are instantiated in a way similar to the way modules are instantiated within other modules. However, for more accurate timing and logic value simulation, they can be given strength and delay values along with port connections.

Figure 5.2 shows an **and** and a **nand** gate primitive that are instantiated to form an and-or-invert function within the *aoi* module. The output of this module is defined by the expression $z = a'.\ (bc)'$, which is implemented by anding the complement of a with the nand of b and c. Signal w is declared as an intermediate wire that connects the output of **nand** to the second input of **and.** The module description follows the syntax described in Fig. 3.47. The *aoi* module declaration includes two gate_instantiation constructs; the syntax details of the instantiation of **nand** are shown in Fig. 5.3.

As shown in Fig. 5.3, *gate_instantiation* begins with specification of the gate type. The gate type for **nand** as defined in the Verilog HDL IEEE standard document is *n_input_gate_type.* Other gates in this category are **and, nor, or, xnor,** and **xor.** Following the gate type name is a set of parentheses specifying the optional drive-strength construct.

LOGIC GATES			PULL GATES	SWITCHES	
n-input	n-output	Tristate		Unidirectional	Bidirectional
and	buf	bufif0	pulldown	cmos	rtran
nand	not	bufif1	pullup	nmos	rtranif0
or		notif0		pmos	rtranif1
xnor		notif1		rcmos	tran
xor				rnmos	tranif0
				rpmos	tranif1

Figure 5.1 Verilog built-in structures.

```
`timescale 1ns/100ps

module aoi (a, b, c, z);
//
input a, b, c;
output z;
wire w;
  and (strong1, strong0) #(3, 4) n1 (z, ~a, w);
  nand (strong1, strong0) #(3, 4) n2 (w, b, c);
endmodule
```

Figure 5.2 Using built-in **nand** and **and** gates.

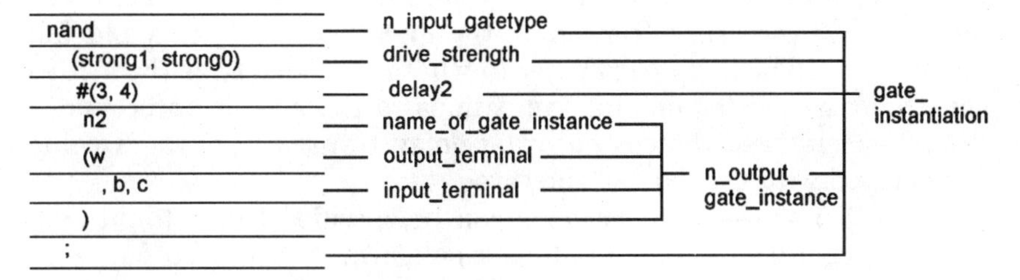

Figure 5.3 Syntax details of gate instantiation.

The next field specifies delay values in the units specified by the **`timescale** construct in Fig. 5.2. For a gate instantiation, a label or a name is optional. For the **nand** of Fig. 5.3, we have used *n2* for the *name_of_gate_instance*. Following the name, the last field of a *gate_instantiation* specifies output and input connections of the instance of a gate. Fields used with a gate instantiation and variations of these fields for other gate types will be described in the following paragraphs.

Drive strength. Built-in gate instantiations can optionally include drive strengths for the output of the gate. A pair of strengths, one for logic **0** and one for logic **1,** is specified in any order in a set of parentheses.

Allowable strengths for logic **0,** i.e., *strength0,* for the **nand** gate in Fig. 5.1 are **supply0, strong0, pull0, weak0,** and **highz0.** Similarly, allowable strengths for *strength1* for this gate are **supply1, strong1, pull1, weak1,** and **highz1.** Default strength values for basic logic gates are **strong1** and **strong0.** Strengths determine how a single set of logic values resolve when several gate outputs are wired. By use of examples in the following sections, we will demonstrate application of strength values. Figure 5.4 shows strength values for the gate types shown in Fig. 5.1.

Delays. Instantiation of a built-in gate can optionally include delay values. Figure 5.2 shows a *delay2* construct that begins with a sharp sign (#) and contains two delay values. The first delay refers to transition to logic **1,** and the second refers to transition to logic **0.** Delay values for transitions to **X** and **Z** are the smaller of the two delay values. If only one delay value is specified, it will be used for all transitions at the gate output. Obviously, not using the *delay2* construct causes the built-in gate output to make transitions in zero time.

Simulation of gates in the sections that follow will illustrate how delay values affect waveforms at the gate outputs. Figure 5.5 shows valid delay parameters for the built-in gate types shown in Fig. 5.1.

As shown in Fig. 5.5, the majority of switches use three delay values. As with the *delay2* construct, the first two specify rise and fall delay values. The third parameter in what is referred to as the *delay3* construct is used for making transitions to the **Z** logic value. As in *delay2,* when three values appear for delay, transitions to the **X** value are delayed with the least of the three.

Figure 5.5 shows that no delay can be specified for pull and tran gates. Pull gates are used for pullup or pulldown of tristate outputs, and tran gates are used as bidirectional pass gates that always conduct.

	LOGIC GATES	PULL GATES		SWITCHES
		Pullup	Pulldown	
Strength0	Supply0 Strong0 Pull0 Weak0 Highz0		Supply0 Strong0 Pull0 Weak0	No strength
Strength1	Supply1 Strong1 Pull1 Weak1 Highz1	Supply1 Strong1 Pull1 Weak1		No strength

Figure 5.4 Gate type strength values.

Delays on wires related to pull gates and tran gates can be modeled by associating delays with gates driving such gates.

Like those of other delay specifications, the gate output delay specification format allows each delay field to include minimum, typical, and maximum delay values. Shown below is instantiation of the two-input **nand** gate of Fig. 5.2 using three values for each delay parameter.

```
nand (strong1, strong0) #(2:3:4, 3:4:5) n2 (w, b, c);
```

For the rise delay, minimum, typical, and maximum values are 2, 3, and 4; and for the fall delay, they are 3, 4, and 5. During simulation, simulation switches or directives allow specification of which set of values to use. The default is usually the typical set of delay values.

Gate name. In addition to strengths and delays, gate instantiation in Fig. 5.2 also includes an optional gate name (*n2*). For a module instantiation, this label is required and is used for hierarchical naming of signals. For consistency in instantiation format, specifying a label is also allowed with instantiation of built-in gates.

Output-input connections. As shown in Fig. 5.2, the last part of a built-in gate instantiation specifies connections to the output and input terminals of a gate. The convention used in Verilog is that the connections to gate output terminals are listed first, followed by connections to the inputs. The number of outputs, inputs, and control inputs of a gate depend on the type of gate. Figure 5.6 shows terminals for the gate types of Fig. 5.1.

	LOGIC GATES		PULL GATES	SWITCHES except tran & rtran	tran rtran
	Two State	Tristate			
Rise	✓	✓	-	✓	-
Fall	✓	✓	-	✓	-
To Z	-	✓	-	✓	-

Figure 5.5 Built-in gate delay parameters.

	LOGIC GATES			PULL GATES	SWITCHES	
	n-input	n-output	Tri-state		Unidirectional	Bi-directional
Left most	Output	All output fanouts	Output	pull	Output1	Inout1
Next	All inputs	Input	Input1	-	Input1	Inout2
Right most	...		Enable control	-	Control input (two controls for cmos)	Control (no control for tran)

Figure 5.6 Order of connections to gate terminals.

Standard logic gates may be used with any number of inputs. In the output terminal connection list of **not** and **buf,** all output fanouts may be listed. The last terminal of these gates is their only input. Pull gates have only one connection. Switches, shown in the last two columns of Fig. 5.6, have a control input in addition to their input and output terminals. If it exists, connection to the control terminal appears last in the terminal connection list.

In a gate input-output connection list, signals or expressions using signals can be used. These signals must either be declared as inputs or outputs of the module that the gate is instantiated in or be explicitly declared. In the example of Fig. 5.2, *z, a, b,* and *c* are output and inputs of *aoi* that are accessible in this module, while *w,* which is used for connecting the output of **nand** to the input of **and,** is explicitly declared as **wire.**

In the preceding paragraphs, we described the use of built-ins in the definition of higher-level modules. The example we used was an *aoi* gate, and the built-in example that we used was a two-input **nand** gate. At a still lower level of abstraction, modules equivalent to some of the built-in gates may be defined using more primitive built-in structures such as transistors. Doing so enables a modeler to develop more accurate gate models. The following example demonstrates defining basic gate modules in terms of primitive built-ins and shows the level of accuracy obtained when such gates are simulated.

Figure 5.7*a* shows a two-input CMOS NAND gate. This structure uses two PMOS and two NMOS transistors. Gate labels and intermediate connection signals are shown in this figure.

Figure 5.7*b* shows the *cmos_nand* module that corresponds to the diagram of Fig. 5.7*a*. Built-in structures **nmos** and **pmos** are used with a single delay value and no strength specification. As indicated in Fig. 5.4, built-in switches do not take strength values. As shown in Fig. 5.5, switches can use up to three delay values. The single values used in the code of Fig. 5.7*b* apply to all three delay parameters. Labels *g1, g2, g4,* and *g5* are optional, and we have used them to make a better correspondence between the diagram of Fig. 5.7*a* and the code of Fig. 5.7*b*. The last field of the switch instantiations in Fig. 5.7*b* lists output, input, and control terminal connections. As indicated in Fig. 5.6, because the switches are unidirectional, the output-input connections are based on the direction of flow of data out of the switches. For N-type transistors, the *gnd* terminal is input pushing **0** into the switch, and for P-type transistors, the *vdd* terminal is considered as input pushing **1** into the switch. Control connections (transistor gate terminals) are the last of the three terminal connections.

Figure 5.8 shows simulation result of *cmos_nand* for various combinations of values on the input terminals. This list indicates that for a period of 1 ns when the NAND output switches from **1** to **0** or from **0**

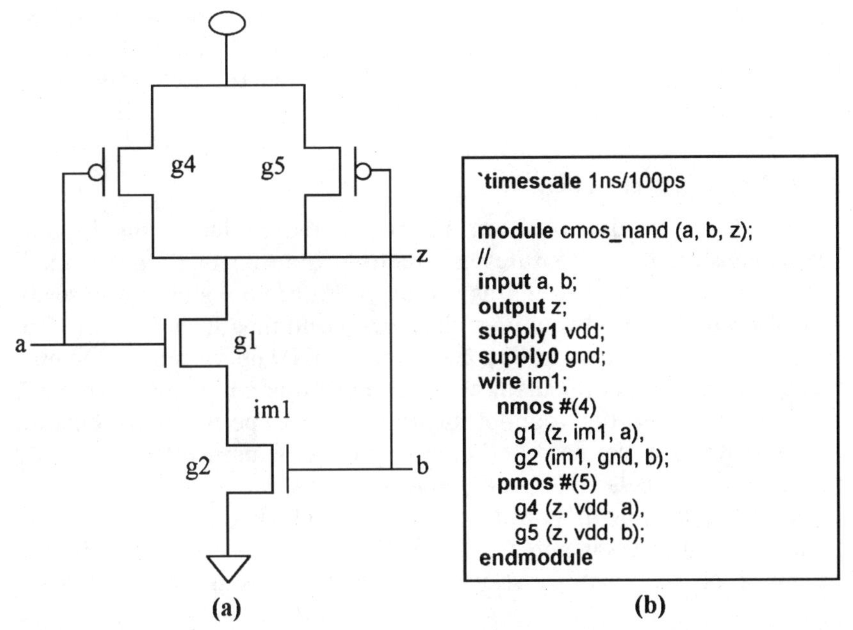

Figure 5.7 CMOS NAND gate. (*a*) Transistor circuit; (*b*) Verilog description.

ps	a	b	z
0	0	1	x
5000	0	1	1
2500000	1	1	1
2504000	1	1	x
2505000	1	1	0
3600000	1	0	0
3605000	1	0	x
3608000	1	0	1
4600000	0	0	1
5600000	0	1	1
6700000	1	1	1
6704000	1	1	x
6705000	1	1	0
7800000	1	0	0
7805000	1	0	x
7808000	1	0	1

Figure 5.8 Simulation of *cmos_nand.*

to **1,** the output becomes **X.** This temporary value is due to the use of different delay values for **nmos** and **pmos** built-in gates. When switching from **1** to **0** at 2,500,000 ps, the **nmos** switches start conducting at 2,504,000 ps while the **pmos** switches are still on. Driving *z* with both **0** and **1** values causes **X** to appear on this output.

As discussed, simulation of this structure results in details of behavior of a CMOS NAND gate that appear in built-in gate structures. These details propagate through upper-level structures that use these modules.

5.1.2 Using gate modules

In Sec. 5.1.1 we used built-in structures for implementing higher-level modules. These modules can be used in other modules in much the same way as built-in gates are used. Figure 5.9 shows a description of *aoi* of Sec. 5.1.1 that uses the *cmos_nand* module of Fig. 5.7. The first two instantiations in Fig. 5.9 form an AND operation on the output of *n2* and the complement of *a*. The output of *aoi* is taken from *n3*.

Module instantiations cannot include strength specifications, but can specify timing if the module being instantiated uses internal timing parameters. Labels that are referred to as *name_of_instance* are required in module instantiations. Note that this is different from instantiation of built-in gates, in which the gate name is optional. Syntax details of the first module instantiation in Fig. 5.9 are shown in Fig. 5.10.

```
`timescale 1ns/100ps

module aoi (a, b, c, z);
//
input a, b, c;
output z;
wire w, y;
   cmos_nand n1 (w, ~a, y);
   cmos_nand n3 (y, y, z);
   cmos_nand n2 (b, c, w);
endmodule
```

Figure 5.9 Using *cmos_nand* for implementing *aoi*.

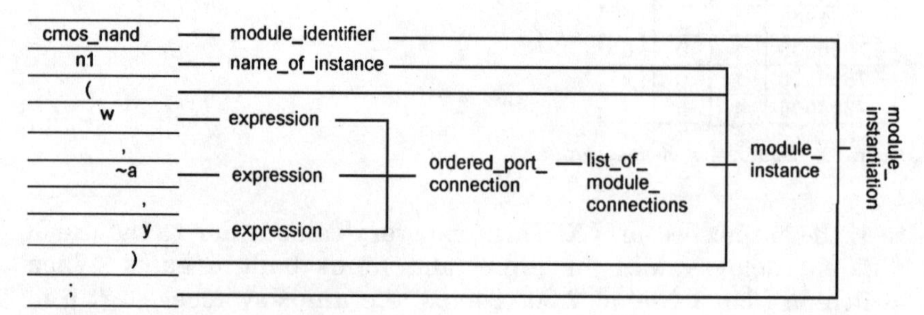

Figure 5.10 Syntax details of module instantiation.

The order of input and output port connections to an instantiated module depends on the way in which the module is defined. Since we used the last port of *cmos_nand* for its output, the three instantiations of this module in Fig. 5.9 use *y, z,* and *w* gate outputs as the last item in the list of module connections.

In the syntax shown in Fig. 5.10, *name_of_instance, ordered_port_connection,* and a set of parentheses form a *module_instance.* A *module_instance* and a *module_identifier* form the complete *module_instantiation* syntax construct.

Alternatively, instead of ordering port connections according to ports of the module being instantiated, connections to ports of a module may be named. Figure 5.11 shows another description for *aoi* in which connections to ports of the *nand_cmos* module use the names of the ports of *cmos_nand* that they are connecting to. In this case, the expression that connects to a module port, e.g., ~*a*, is enclosed by a set of parentheses preceded by a dot and the actual name of the port of the module. The term *.b(~a)* connects ~*a* to the *b* port of *cmos_nand.* This language construct is referred to as *ordered_port_connection,* and its syntax details are shown in Fig. 5.12.

```
`timescale 1ns/100ps

module aoi (a, b, c, z);
//
input a, b, c;
output z;
wire w, y;
  cmos_nand n1 (.a(w), .b(~a), .z(y));
  cmos_nand n3 (.a(y), .b(y), .z(z));
  cmos_nand n2 (.a(b), .b(c), .z(w));
endmodule
```

Figure 5.11 Using *named_port_connection.*

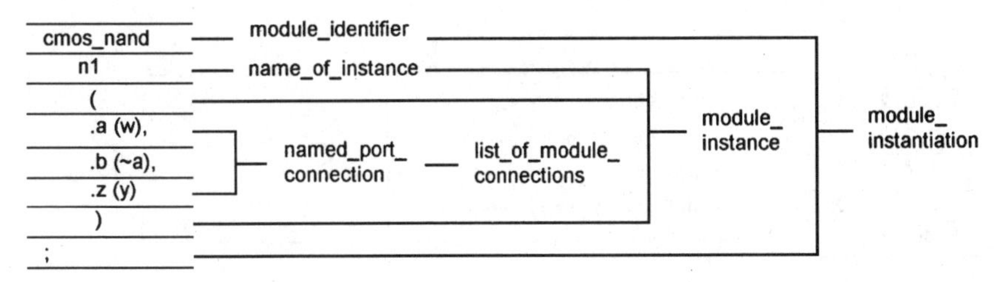

Figure 5.12 Syntax of *named_port_connection.*

5.1.3 Using user-defined primitives

In addition to the use of built-in gates and gate modules, Verilog allows the use of user-defined primitives (UDPs) for defining basic structures or library elements. UDPs offer a very simple but limited method of defining basic hardware elements.

A UDP defines a gate element with only one output in a tabular form. The output must be first in a UDP port list; only Verilog values **0, 1,** and **X** are allowed; no strength specification can be associated with a UDP; and only rise and fall delays may be used.

Figure 5.13 shows the *aoi* of Fig. 5.2 or Fig. 5.9 implemented using a UDP. The keyword **primitive** is used here instead of **module** as in the other descriptions of this chapter. The heading is followed by declaration of inputs and outputs and the UDP body. The body, shown in Fig. 5.14, can be *combinational_body* or *sequential_body.* For the *combinational_body* of Fig. 5.14, the **table** keyword is followed by four *combinational_entry* constructs.

A combinational body is similar to a truth table, listing input combinations and their corresponding output values. A *combinational_entry* lists an output value for a combination of input values. Several rows of a UDP table may be merged by using a **?** character, which takes the

```
`timescale 1ns/100ps

primitive aoi (z, a, b, c);
//
input a, b, c;
output z;
table
  0 0 ? : 1;
  0 ? 0 : 1;
  1 ? ? : 0;
  0 1 1 : 0;
endtable
endprimitive
```

Figure 5.13 Defining *aoi* using a UDP.

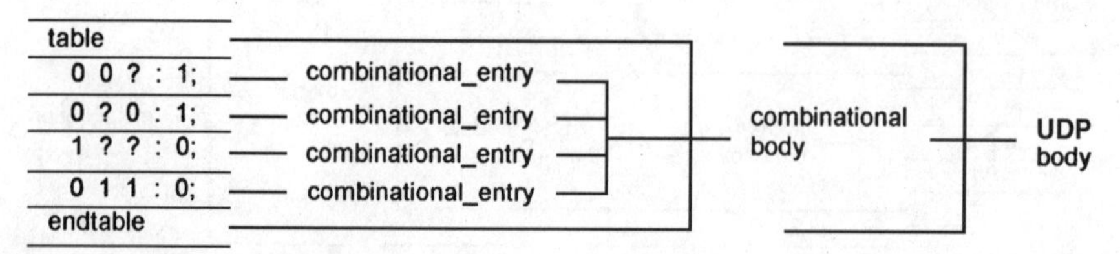

Figure 5.14 UDP syntax details.

place of a **0, 1,** or **X.** The symbol **b** substitutes for **0** or **1** only. With this interpretation, the second *combinational_entry* in Fig. 5.14 takes the place of the following expanded entries:

```
0  0  0 : 1;
0  1  0 : 1;
0  X  0 : 1;
```

A sequential UDP can be defined in a similar way, except that the present state of the output must also be listed in each entry. Figure 5.15 shows the description of a rising-edge D-type flip-flop with asynchronous preset and reset inputs.

As shown in Fig. 5.15, the output of a sequential UDP must be declared as **reg.** Each entry in a UDP table is referred to as *sequential_entry* and has the syntax shown in Fig. 5.16. Sequential entries may specify input level values or input edges. Allowed edge specifications and their equivalent notations are shown in Fig. 5.17.

This section has presented ways of defining basic structures or primitive library elements. We have also presented ways of instantiating these structures. In the sections that follow, we will discuss hierarchical formation of higher-level hardware structures using basic components. For this purpose, we will limit the use of basic components to a **not,** a two- and a three-input **nand,** and a two-input **and** built-in gates. Presentation of higher-level structures based on these components will be based on graphical symbols for individual gates. Figure 5.18 shows

```
`timescale 1ns/100ps

primitive dffpr (q, c, d, p, r);
//
input c, d, p, r;
output q;
reg q;
      table
//c  d   p     r     q     q+
  ?  ?   1     0    : ? :   1 ;
  ?  ?   0     1    : ? :   0 ;
 (01) 0  0     0    : ? :   0 ;
 (01) 1  0     0    : ? :   1 ;
 (0?) 0  0     0    : 0 :   0 ;
 (0?) 1  0     0    : 1 :   1 ;
 (?0) ?  0     0    : ? :   - ;
      endtable
endprimitive
```

Figure 5.15 UDP for a D-type flip-flop with *p* and *r* inputs.

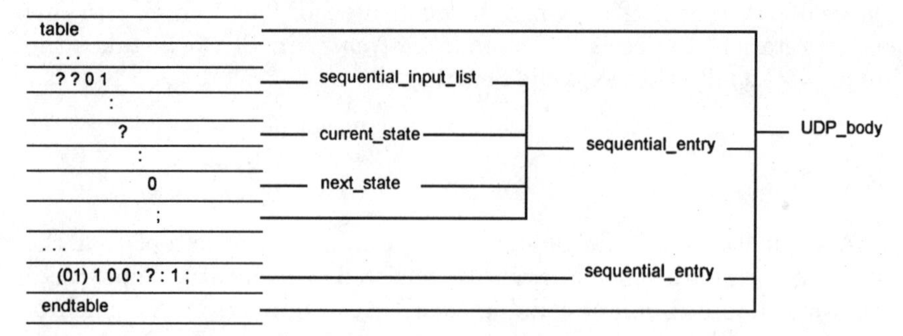

Figure 5.16 Syntax details for UDP sequential bodies.

TRANSITION	EQUIVALENT NOTATION	POTENTIAL CATEGORY
(01)	r	p
(0X)		p
(0?)		
(10)	f	n
(1X)		n
(1?)		
(??)	*	
(X0)		p
(X1)		n

Figure 5.17 Edge specification notations in sequential UDPs.

the built-in gates that we will use in our designs in this chapter, along with a logical symbol and a graphical symbol for each gate. For each gate, Fig. 5.18 shows a possible instantiation format, although specific designs that follow may use variations of the formats shown here.

Graphical gate notations show inputs with hollow boxes and outputs with solid filled boxes. Verilog also allows bidirectional lines, for which a half-solid box will be used. Inputs and outputs of the graphical notations are named. The names used here are the same as those in the instantiation of the gate and must be declared within the body that the gate is instantiated in. The notations used here indicate that internal terminal names for built-in gates are not defined. This is actually the reason that named port connection, which can be done for module instantiations, cannot be used for built-in instantiations. The last gate notation in Fig. 5.18 shows an AND gate with a barred diamond symbol near its output terminal. A diamond with an underline is often used for open-collector gate outputs where logic **1** is open. We use the symbol in Fig. 5.18*d* for logic **0** being open. Instantiation of **and** with **highz0, strong1** strength makes logic **0** to be high-impedance and logic **1** to be pulled or driven high.

The gates in Fig. 5.18 provide a sufficient set of primitives for discussing the structural description of hardware in this chapter. More elaborate timing and parameterizing designs will be discussed in later chapters.

5.2 Wiring of Primitives

Wiring of the primitive gates for generation of larger designs is demonstrated in this section. In Verilog, the operation of a design can be described in terms of its subcomponents. To completely specify this operation, we must indicate the component interconnections and use a set of available library cells. The main language constructs that support this

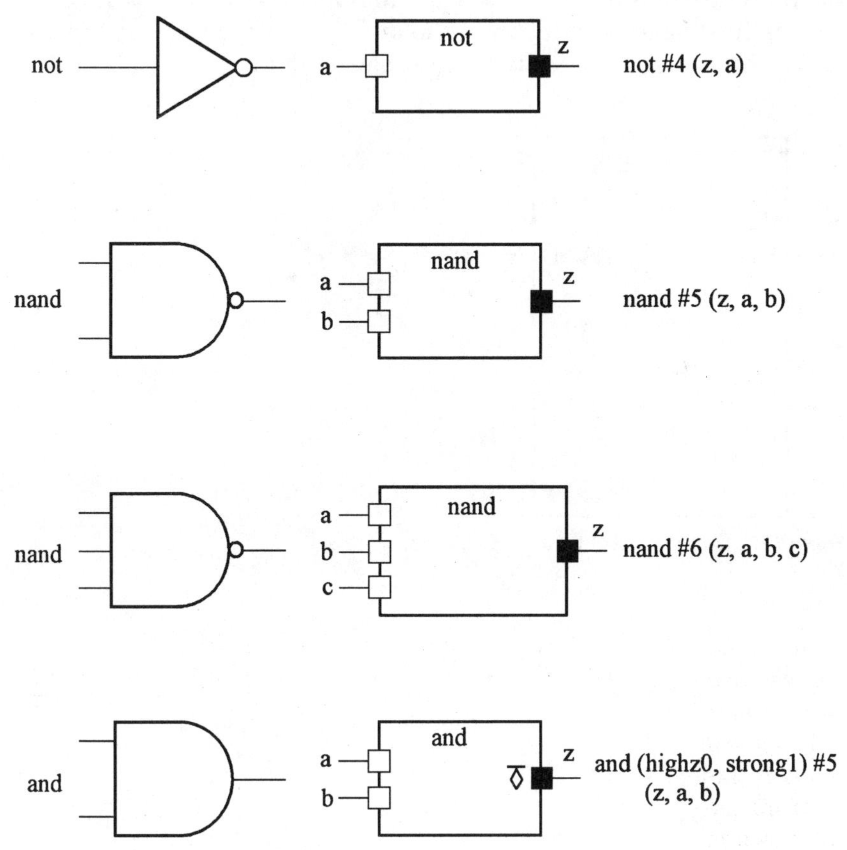

Figure 5.18 Built-in gates used in our designs.

style of hardware description are net declarations and module or gate instantiations. These constructs are discussed here. For this purpose, a single-bit comparator is designed and described in terms of the built-in gates of Fig. 5.18.

5.2.1 Logic design of comparator

A single-bit comparator circuit (*bit_comparator*) has two data inputs, three control inputs, and three compare outputs. The logical symbol for this circuit is shown in Fig. 5.19. The three control inputs provide a mechanism for generation of multibit comparators by cascading several *bit_comparators*.

The $A > B$ output is 1 if the A input is greater than the B input (AB is 10) or if A is equal to B and the $>$ input is 1. The $A = B$ output is 1 if A is equal to B and the $=$ input is 1. The $A < B$ output is the opposite of the $A > B$ output. This line becomes 1 if the A input is less than the B input (AB is 01) or if A is equal to B and the $<$ input is 1. Based on this functional description of the *bit_comparator,* Karnaugh maps for its three outputs are extracted, as shown in Fig. 5.20.

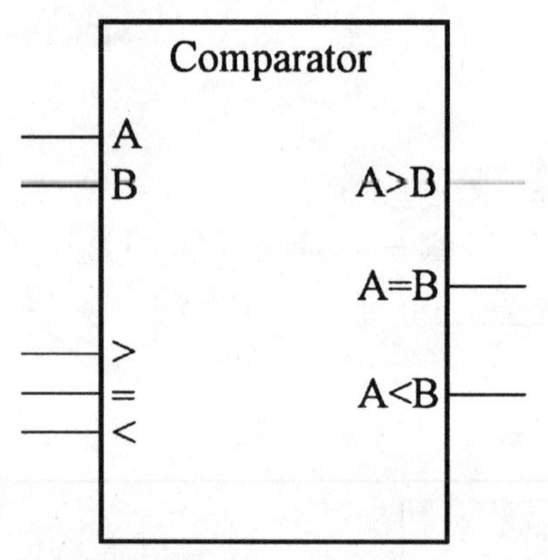

Figure 5.19 Logical symbol for a single-bit comparator.

> \ a, b	00	01	11	10
0				1
1	1		1	1

$a > b$

= \ a, b	00	01	11	10
0				
1	1		1	

$a = b$

< \ a, b	00	01	11	10
0		1		
1	1	1	1	

$a < b$

Figure 5.20 Karnaugh maps for the outputs of the single-bit comparator.

Boolean expressions for the three outputs of the *bit_comparator* resulting from applying minimization methods to the Karnaugh maps in Fig. 5.20 are shown below. To avoid confusion, these expressions use *gt, eq,* and *lt* instead of the >, =, and < symbols used in the logical symbol in Fig. 5.19. Other notational changes have been made for readability purposes.

$$\text{a\_gt\_b} = \text{a . gt} + \text{b}' \text{. gt} + \text{a . b}' \qquad (5.1a)$$

$$\text{a\_eq\_b} = \text{a . b . eq} + \text{a}' \text{. b}' \text{. eq} \qquad (5.1b)$$

$$\text{a\_lt\_b} = \text{a}' \text{. lt} + \text{b . lt} + \text{a}' \text{. b} \qquad (5.1c)$$

Using DeMorgan's theorem, Eqs. (5.1*a*) to (5.1*c*) can be transformed into Eq. (5.2*a*) to (5.2*c*), respectively. These equations have an appropriate form for all-NAND and inverter implementation, which is of course available in our library of parts (Fig. 5.18). The gate-level circuit diagram of the *bit_comparator* resulting from these equations is shown in Fig. 5.21.

$$\text{a\_gt\_b} = ((\text{a . gt})'.(\text{ b}' \text{. gt})'.(\text{ a . b}')')' \qquad (5.2a)$$

$$\text{a\_eq\_b} = ((\text{a . b . eq})'.(\text{a}' \text{. b}' \text{. eq})')' \qquad (5.2b)$$

$$\text{a\_lt\_b} = ((\text{a}' \text{. lt})'.(\text{b . lt})'.(\text{ a}' \text{. b})')' \qquad (5.2c)$$

5.2.2 Verilog description of *bit_comparator*

At this point, we have completed the design of the single-bit comparator and its definition in terms of our available primitive gates. Next, the Verilog code for this unit will be developed. A graphical notation will be developed first, from which the Verilog code will be extracted.

The Verilog description of *bit_comparator* must specify the way this unit is instantiated and the operation of the unit. Figure 5.22 shows the *bit_comparator* ports and their visibility. This notation indicates the name of this unit and the internal names of its ports. It also indicates that the names used for inputs and outputs of the bit comparator are visible only inside the component.

Figure 5.23 shows the internal structure of *bit_comparator* using the built-in gates of Fig. 5.18. This diagram is obtained by simply replacing the gates of Fig. 5.21 with the corresponding notations from Fig. 5.18. Port names that are visible inside *bit_comparator* are used for connections from *bit_comparator* ports to ports of its internal submodules. For connecting ports of submodules of *bit_comparator* to other submodule ports, wires with specific names are used. For example, wire *im3* is used for connecting the output of the two-input NAND gate in the

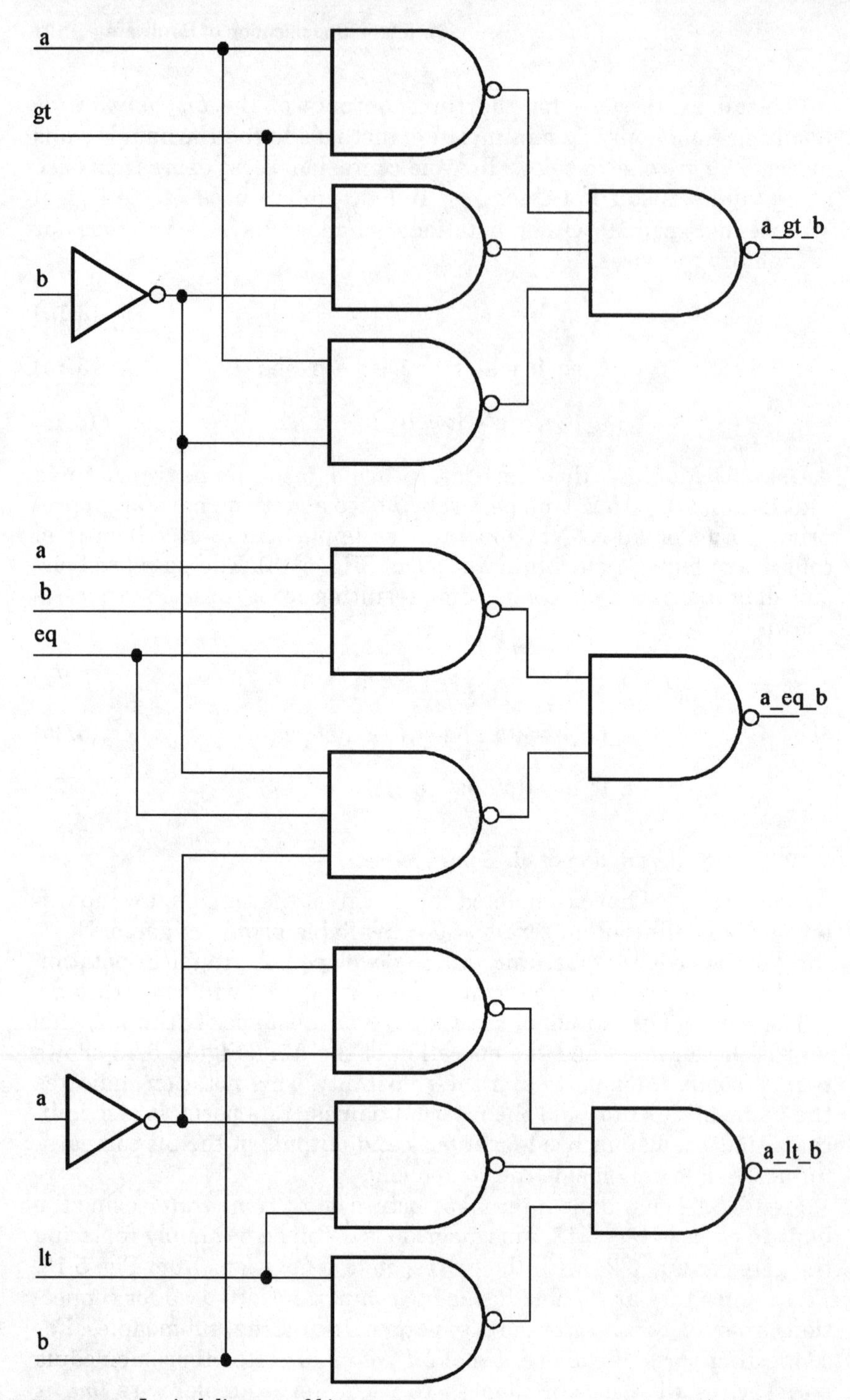

Figure 5.21 Logical diagram of *bit_comparator.*

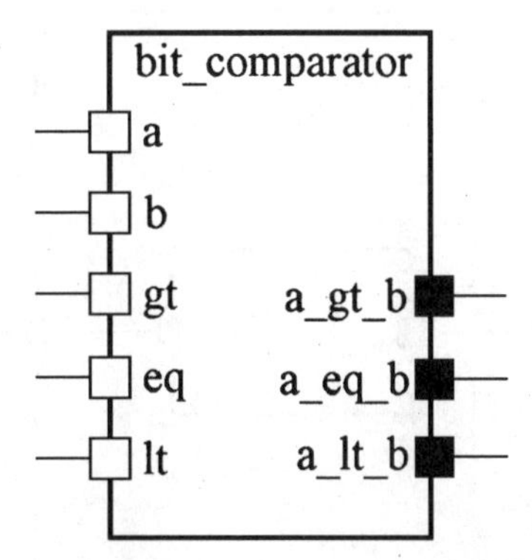

Figure 5.22 *bit_comparator* identification and ports.

upper part of Fig. 5.23 to the three-input NAND gate that drives the *a_gt_b* output of *bit_comparator.* Wires *im1* through *im10* are used for all internal wirings.

Corresponding to Fig. 5.22 and Fig. 5.23 is the Verilog code depicted in Fig. 5.24. The first line of this code defines a time scale of 1 ns with 100-ps resolution. Starting with **module,** the first nine lines specify the name of the module and its ports. This part of the code corresponds to the notation shown in Fig. 5.22, which provides the information that is needed for using the *bit_comparator* in upper-level designs. For example, to connect a wire to *a_gt_b* port of *bit_comparator* in an instance of this unit, the syntax shown below may be used:

```
bit_comparator label (..., .a_gt_b (connection to a_gt_b), ...)
```

Starting with line 10, the Verilog code of Fig. 5.24 specifies the internal structure of *bit_comparator.* Line 10 declares all intermediate wires used for interconnection of built-in gates. Wires *im1* through *im10* declared here are used in the Verilog code in a way similar to the way these wires are used in the diagram of Fig. 5.23. For example, *im3* connects the output of *g2* **nand** gate on line 17 to the second input of *g5* **nand** gate on line 21. Using the name *im3* in both places shows this connection.

Figure 5.25 shows syntax details of a *module_declaration* consisting of ports, declarations, and instantiations. At the top level, a *module_declaration* consists of a *module_identifier, list_of_ports,* and *module_item* bracketed by **module** and **endmodule** keywords. The *list_of_ports* construct consists of actual module ports bracketed by a set of parentheses. Declarations that follow that are part of

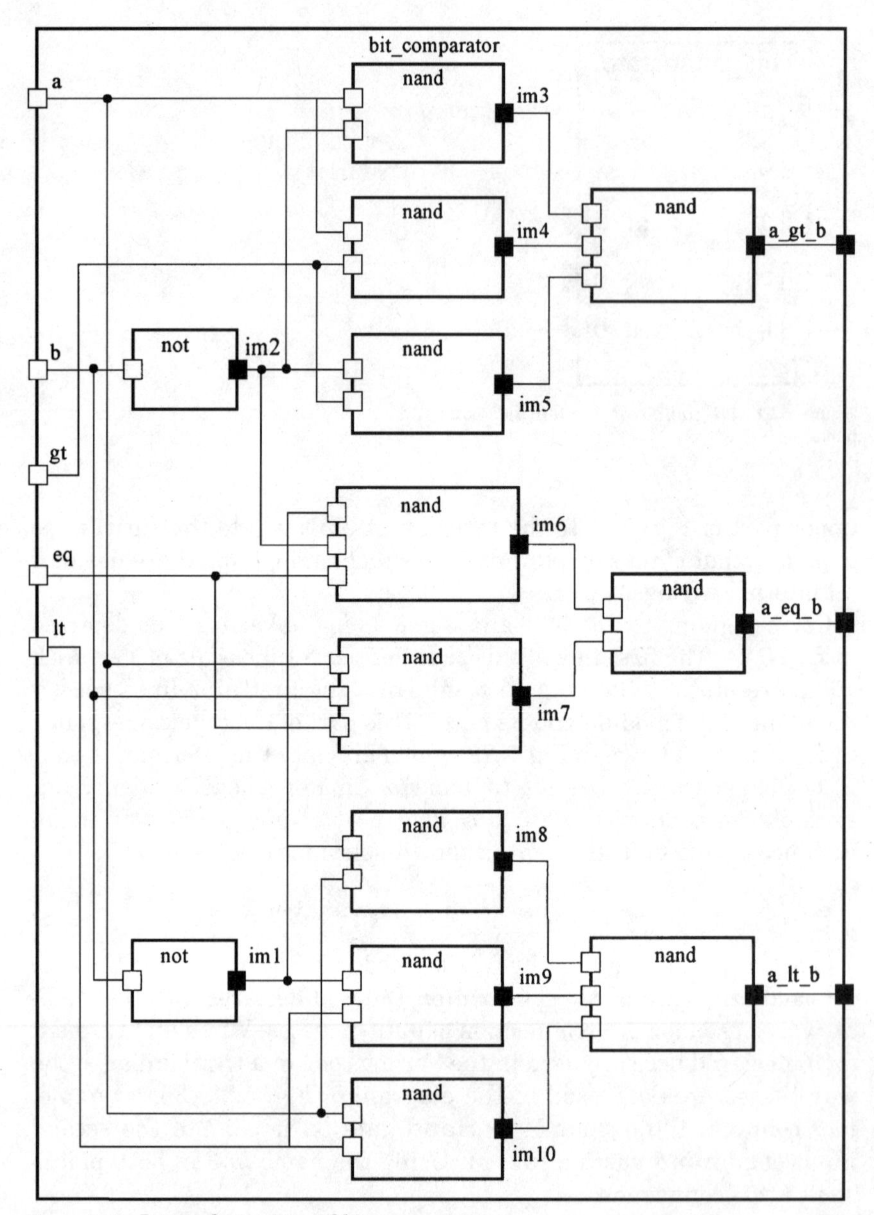

Figure 5.23 Internal structure of *bit_comparator.*

a *module_item* determine which ports of the *list_of_ports* are inputs and which are outputs of a module.

The *module_item* construct in Fig. 5.25 consists of *module_item_declaration* and *gate_instantiation* constructs. The declaration declares intermediate signals in addition to module inputs and outputs. The

wire type is used for all **net** declarations. Instantiation of submodules in Fig. 5.24 is achieved by *gate_instantiation* syntax constructs, the details of which have been illustrated in Sec. 5.1.

The code for the bit comparator of Fig. 5.24 is a complete simulation model that can be compiled and simulated for verification of the correctness of our design. Verilog simulators provide mechanisms for applying test inputs and observing output waveforms. Figure 5.26 shows the result of a simulation run of the *bit_comparator* module.

```
`timescale 1ns/100ps

module bit_comparator (
a, b,                 // data inputs
gt, eq, lt,           // previous greater than, equal, less than
a_gt_b,               // greater
a_eq_b,               // equal
a_lt_b);              // less than
//
input a, b, gt, eq, lt;
output a_gt_b, a_eq_b, a_lt_b;
wire im1, im2, im3, im4, im5, im6, im7, im8, im9, im10;

  // a_gt_b output
  not #(4)
    g0 (im1, a),
    g1 (im2, b);
  nand #(5)
    g2 (im3, a, im2),
    g3 (im4, a, gt),
    g4 (im5, im2, gt);
  nand #(6)
    g5 (a_gt_b, im3, im4, im5);

  // a_eq_b output
  nand #(6)
    g6 (im6, im1, im2, eq),
    g7 (im7, a, b, eq);
  nand #(5)
    g8 (a_eq_b, im6, im7);

  // a_lt_b output
  nand #(5)
    g9 (im8, im1, b),
    g10 (im9, im1, lt),
    g11 (im10, b, lt);
  nand #(6)
    g12 (a_lt_b, im8, im9, im10);
endmodule
```

Figure 5.24 Verilog code for *bit_comparator.*

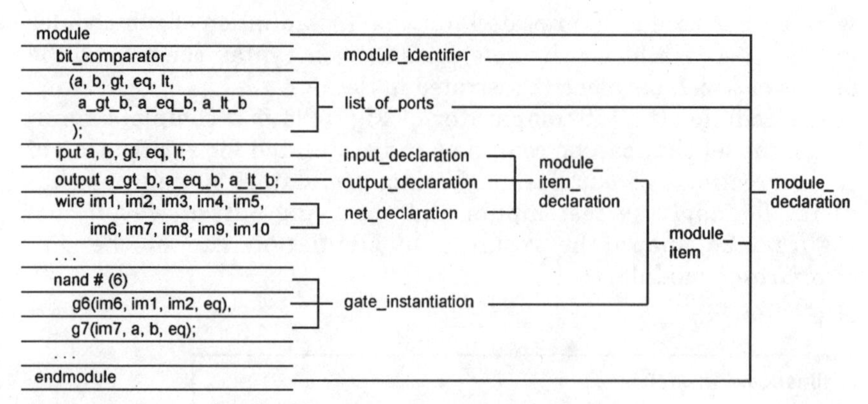

Figure 5.25 Syntax details of *bit_comparator* module declaration.

This simulation run tests the *bit_comparator* for *lt, eq,* and *gt* values of 010 and all four combinations of *a* and *b*. A complete test of this circuit would require application of 32 test vectors for all combinations of the five inputs of the circuit. Through the use of Verilog test benches, exhaustive testing or an adaptive application of a test can be done. Developing test benches in Verilog will be discussed later in this chapter.

Before we show how our *bit_comparator* may be used in larger designs, we will develop Verilog code for several other implementations of this unit. By these implementations, we will illustrate ways in which built-in gates may be used for modeling wired logic. The output of the **and** gate in Fig. 5.18 becomes **strong1** for logic **1** and high impedance for logic **0.** A possible hardware that behaves this way is shown in Fig. 5.27.

The *z* output of the structure of Fig. 5.27 will be driven by V_{dd} (logic **1**) only when both *a* and *b* are **1.** Otherwise, the *z* output will assume value **Z.** Wiring the outputs of several of these structures with a pull-down resistor implements a wired-or logic. Figure 5.28 shows hardware for $a.b + c.d$.

Figure 5.29 shows Verilog code for a *bit_comparator* using the built-in **and** gate of Fig. 5.18 and wired-or logic. Each **and** gate in the *bit_comparator* uses (**highz0, strong1**) strengths, which causes logic **0** to appear as **highz** (**Z** value). Circuit outputs *a_gt_b, a_eq_b,* and *a_lt_b* are declared and used in the same way. We will use the *a_gt_b* output as an example to illustrate how the logic expressions of Eqs. (5.1*a*), (5.1*b*), and (5.1*c*) are generated on these outputs.

The first three **and** gate instances in Fig. 5.29, i.e., gates *g2, g3,* and *g4,* connect the *a_gt_b* output to the outputs of the three **and** gates. Since *a_gt_b* is a wired-or **net,** if any of the outputs of *g2, g3,* or *g4* becomes **1,** the *a_gt_b* output will become **1.** On the other hand, if none are **1,** i.e., if all are at logic **0** and appear as value **Z,** then the *a_gt_b* output wants to

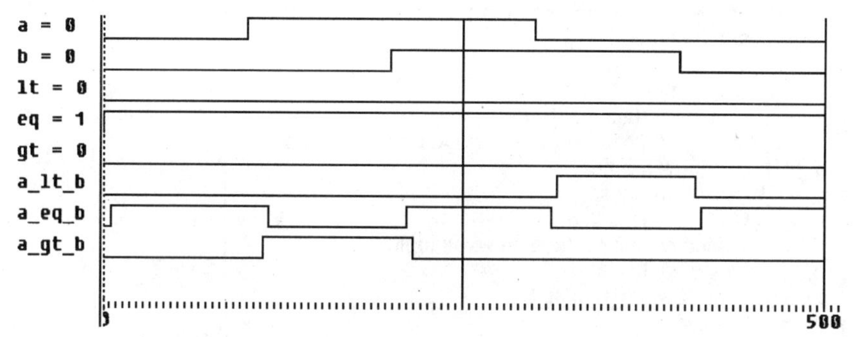

Figure 5.26 *bit_comparator* simulation run.

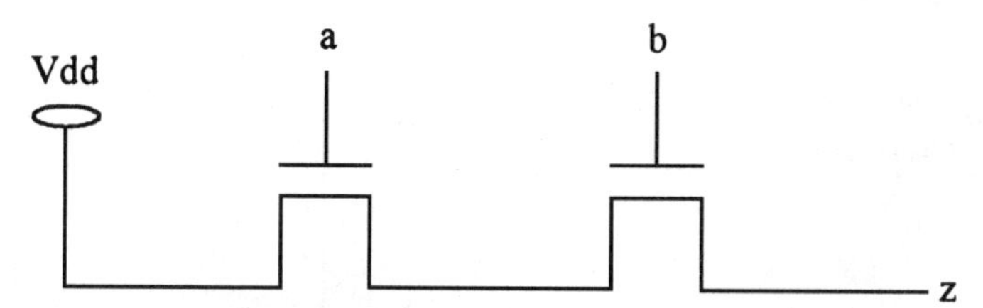

Figure 5.27 Anding *a* and *b*, logic **0** appears as **Z.**

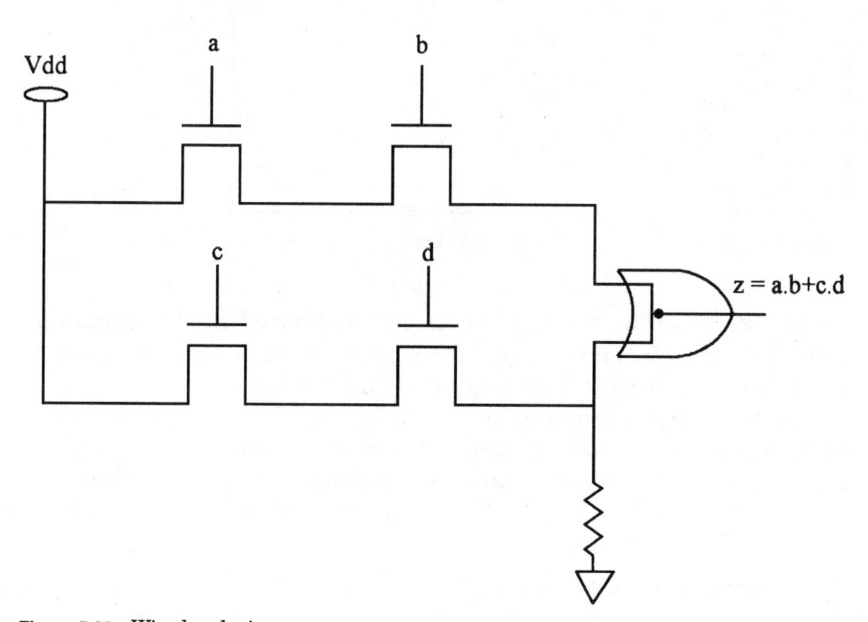

Figure 5.28 Wired-or logic.

```
`timescale 1ns/100ps

module bit_comparator (
a, b,              // data inputs
gt, eq, lt,        // previous greater than, equal, less than
a_gt_b,            // greater
a_eq_b,            // equal
a_lt_b);           // less than
// Cannot specify delay values on wired logic.
input a, b, gt, eq, lt;
output a_gt_b, a_eq_b, a_lt_b;
wire im1, im2;
wor #6 a_gt_b, a_lt_b;
wor #5 a_eq_b;
  pulldown (a_gt_b), (a_eq_b), (a_lt_b);

  // a_gt_b output
  not #(4)
    g0 (im1, a),
    g1 (im2, b);
  and (highz0, strong1) #(5)
    g2 (a_gt_b, a, im2),
    g3 (a_gt_b, a, gt),
    g4 (a_gt_b, im2, gt);

  // a_eq_b output
  and (highz0, strong1) #(6)
    g6 (a_eq_b, im1, im2, eq),
    g7 (a_eq_b, a, b, eq);

  // a_lt_b output
  and (highz0, strong1) #(5)
    g9 (a_lt_b, im1, b),
    g10 (a_lt_b, im1, lt),
    g11 (a_lt_b, b, lt);
endmodule
```

Figure 5.29 *bit_comparator* using wired-or logic.

become **Z.** However, the **pulldown** gate connected to this output by instantiating **pulldown** in Fig. 5.29 causes high-impedance **Z** values on these lines to be pulled down to ground and become **0.**

Simulation of the Verilog code in Figs. 5.24 and 5.29 produces the same exact results. Note that to make the timing of *bit_comparator* of Fig. 5.29 the same as that of the original code, delay parameters have been used with wired-or declarations. Pullup gates cannot have delay parameters.

5.3 Wiring Iterative Networks

In addition to language constructs for definition and instantiation of modules, Verilog includes higher-level constructs that can be used for definition of repetitive hardware at the structural level. Such constructs are discussed in this section. The example used is a 4-bit comparator, which is referred to as a *nibble_comparator.* This circuit uses the *bit_comparator* circuit.

5.3.1 Design of a 4-bit comparator

A 4-bit comparator with two 4-bit data inputs, three control inputs, and three compare outputs is shown in Fig. 5.30. The functionality of this circuit is similar to that of the *bit_comparator.* For the *nibble_comparator,* the discussion in Sec. 5.2.1 applies to 4-bit positive numbers instead of single bits of data. The $A > B$ output is **1** when data on the A input, treated as a 4-bit positive number, are greater than the 4-bit positive number on B, or when data on A and B are equal and the $>$ input is **1.** This arrangement makes it possible to wire together several *bit_comparators*, *nibble_comparators*, or both for building comparators of any size.

The *nibble_comparator* can easily be built by cascading four *bit_comparators* as shown in Fig. 5.31. In this circuit, the compare outputs of each of the *bit_comparators*—for example, $A > B$, $A = B$, and $A < B$ of bit 1—are connected to similarly named control inputs of a more significant bit—for example, $>$, $=$, and $<$ of bit 2. The control inputs of the least significant bit—that is, $>$, $=$, and $<$ of bit 0—are considered to be the control inputs of the *nibble_comparator,* and the compare outputs of the most significant bit—that is, $A > B$, $A = B$, and $A < B$ of bit 3—are considered to be the compare outputs of the 4-bit comparator circuit. With this arrangement, when two positive numbers are compared, more significant comparator bits generate appropriate outputs if they can be determined by their corresponding data bits. A comparator bit uses the outputs of a less significant bit

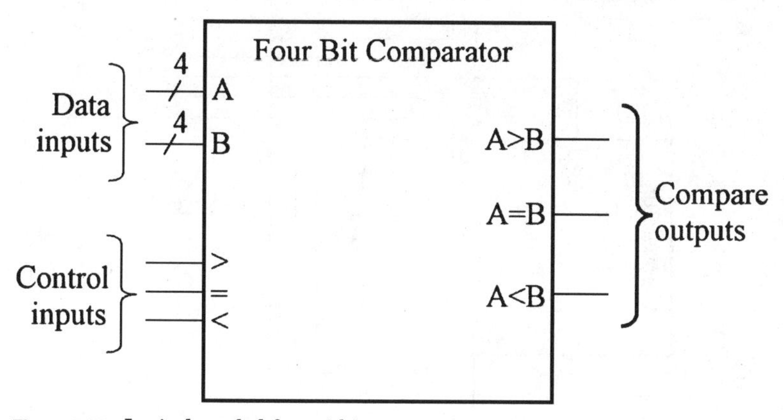

Figure 5.30 Logical symbol for a 4-bit comparator.

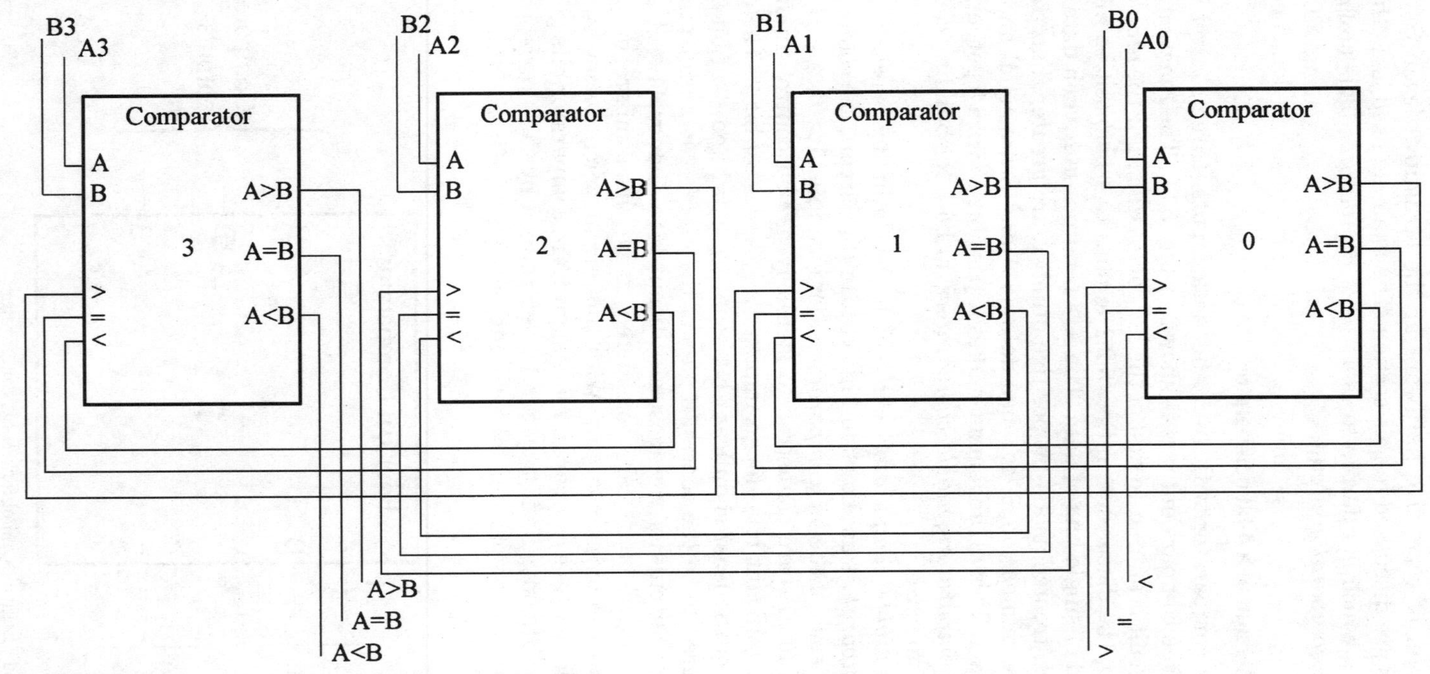

Figure 5.31 A 4-bit comparator using four single-bit comparators.

only if its own data bits are equal. With this scheme, the compare outputs of the *nibble_comparator* are generated faster because they do not have to depend on all *bit_comparators* if the result can be determined by a few most significant bits. For example, when comparing 0100 and 0011, the result is determined by bits 2 and 3, and there is no need for propagation of data through all four *bit_comparators*.

5.3.2 Verilog description of a 4-bit comparator

Figure 5.32 shows a graphical notation for our *nibble_comparator* illustrating its name and its input-output terminals. The 4-bit inputs of the comparator use a 3:0 range, enclosed by square brackets. This notation is used in Verilog for vector and array declarations, bit-select (vector indexing), and part-select (vector slicing). These terminals are internally known to the *nibble_comparator* and may be used for wiring submodules of this 4-bit comparator. Figure 5.33 shows details of this wiring in terms of four single-bit comparators.

Details of the structure of *nibble_comparator* shown in Fig. 5.33 are derived from the schematic diagram in Fig. 5.31. All signals in this diagram have names assigned to them. For those signals that are not primary ports of the *nibble_comparator*, intermediate names *im(0)* through *im(8)* are used. This figure indicates that four copies of the *bit_comparator* module, shown in Fig. 5.24 or Fig. 5.29, are used for structural-level implementation of the *nibble_comparator*. The Verilog description for this circuit directly corresponds to the diagram in Fig. 5.33 and is shown in Fig. 5.34.

Module declarations in Fig. 5.34 include input-output declarations and a declaration for a 9-bit intermediate vector. The inputs are declared as 4-bit vectors, with their leftmost bit numbered 3 and their rightmost bit

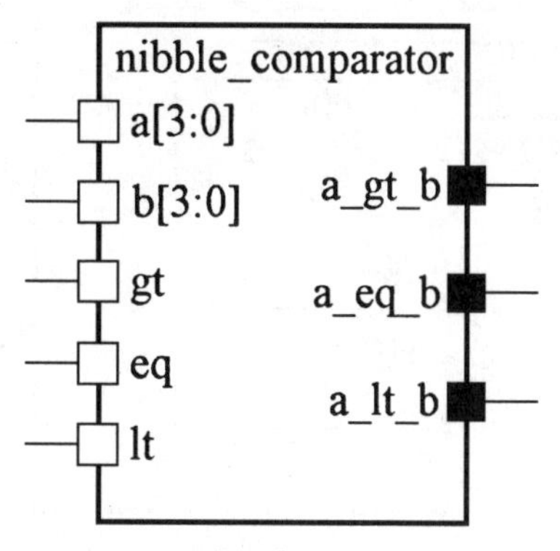

Figure 5.32 Graphical notation for *nibble_comparator*.

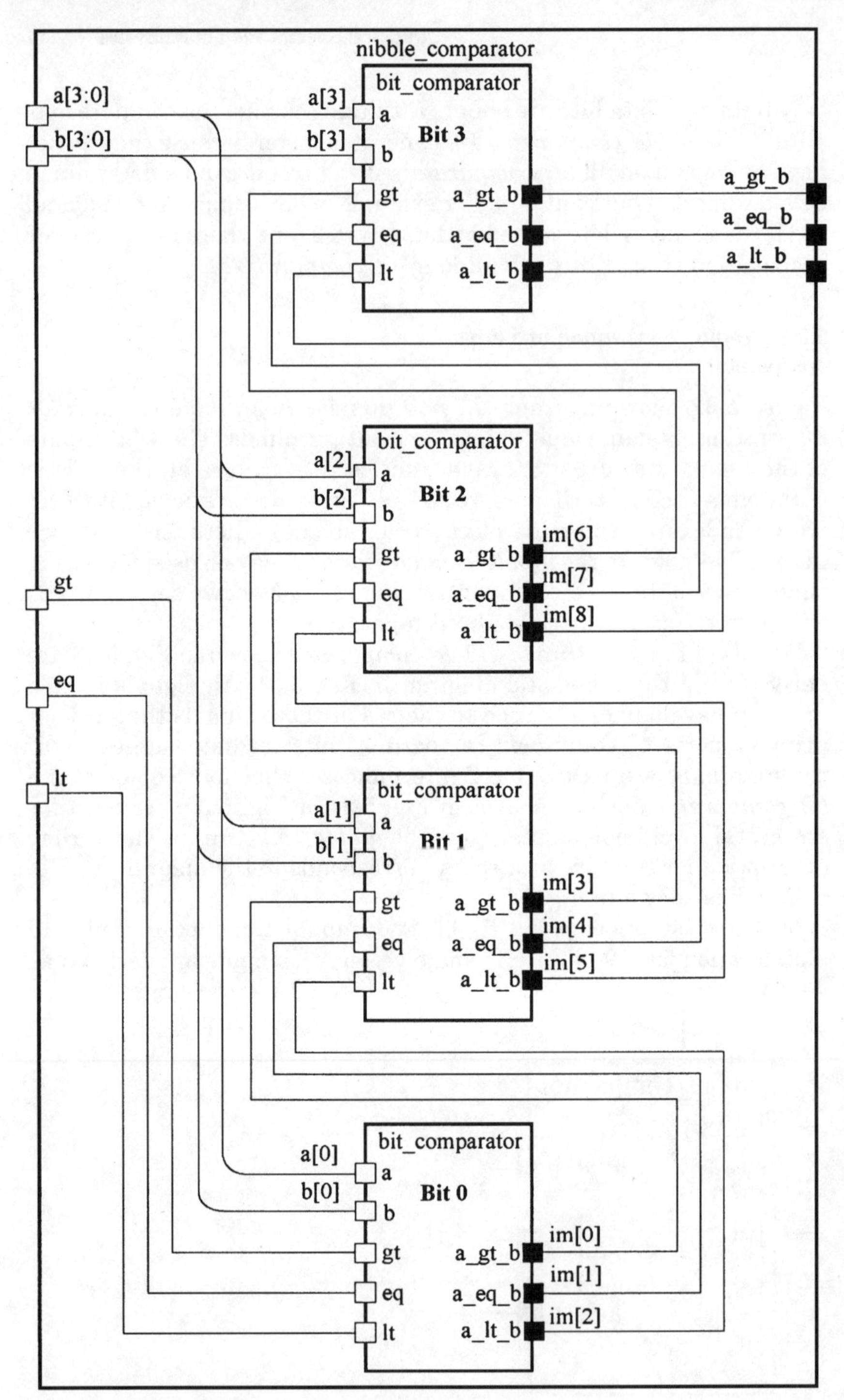

Figure 5.33 Wiring details of *nibble_comparator*.

```
module nibble_comparator (
a, b,             // data inputs
gt, eq, lt,       // previous greater than, equal, less than
a_gt_b,           // greater
a_eq_b,           // equal
a_lt_b);          // less than
//
input [3:0] a, b;
input gt, eq, lt;
output a_gt_b, a_eq_b, a_lt_b;
wire [0:8] im;

bit_comparator cc[3:0] ( a, b,
  {im[6],im[3],im[0],gt}, {im[7],im[4],im[1],eq}, {im[8],im[5],im[2],lt},
  {a_gt_b,im[6],im[3],im[0]}, {a_eq_b,im[7],im[4],im[1]}, {a_lt_b,im[8],im[5],im[2]});

endmodule
```

Figure 5.34 Verilog code for the iterative structure of *nibble_comparator.*

being 0. Declaring *im* as a vector enables use of iterative constructs for instantiating multiple copies of the *bit_comparator* module.

The module item in Fig. 5.34 includes a single *module_instantiation* that causes instantiation of all four necessary *bit_comparators*. To understand the use of this construct in this format, we will look at *bit_comparators* in the diagram of Fig. 5.33 as an array of instances of this module that are named *cc[3], cc[2], cc[1],* and *cc[0].* This is reflected in the Verilog code as

```
bit_comparator cc[3:0] ( ... )
```

The *a* and *b* inputs of this array of modules consist of 4-bit *a* and *b* vectors that are primary inputs of the *nibble_comparator.* Therefore, in the instantiation of *bit_comparator* in Fig. 5.34, inputs *a[3]* through *a[0]* are applied to the first inputs of the *cc[3]* through *cc[0]* instances of *bit_comparator,* and *b[3]* through *b[0]* to their second inputs. The third set of inputs to the array of *bit_comparators* in Fig. 5.33 (the *gt* inputs) consists of *im[6], im[3], im[0],* and *gt* signals put together in this order to connect to instances *cc[3]* through *cc[0]* of *bit_comparator.* This converting of signals is done by the concatenation operator in the Verilog code of Fig. 5.34. Concatenation, shown in bold in this figure, forms a 4-bit vector that connects to the implied vector formed by *gt* inputs of the four instances of the *bit_comparator* module. Other connections to vectorized ports of instances of *bit_comparator* are done in the same way.

In Verilog, the format shown here is often referred to as an array of instances or vectorized instantiation. Although this format is described in the 1364 IEEE standard, not many simulation tools support it. With simulators that do not handle arrays of instances, all necessary copies of a module must be explicitly instantiated. Figure 5.35 shows an alternative description for the *nibble_comparator* circuit. Vectorized instantiation of Fig. 5.34 expands to explicit instantiations of Fig. 5.35.

5.4 Modeling a Test Bench

Testing the *nibble_comparator* involves generating a test bench description and using it to provide stimuli to the input ports of the 4-bit comparator. A test bench must contain the circuit under test and should have sources for providing data to its inputs. Containment of the *nibble_comparator* as well as application of waveforms to its inputs can be modeled in Verilog. Development of a test bench for the comparator circuit requires the use of language constructs that are generally not considered to be at the structural level. In order to stay within the scope of this chapter, we develop a simple test bench that requires the use of only simple assignments and module instantiations.

5.4.1 Verilog description of a simple test bench

The test bench for the *nibble_comparator* is shown in Fig. 5.36. It provides waveforms for the *a* and *b* inputs and connects the *gt, eq,* and *lt* control inputs to *gnd, vdd,* and *gnd,* respectively. This programs the

```
module nibble_comparator (
a, b,              // data inputs
gt, eq, lt,        // previous greater than, equal, less than
a_gt_b,           // greater
a_eq_b,            // equal
a_lt_b);           // less than
//
input [3:0] a, b;
input gt, eq, lt;
output a_gt_b, a_eq_b, a_lt_b;
wire [0:8] im;
   bit_comparator c0 (a[0], b[0], gt, eq, lt, im[0], im[1], im[2]);
   bit_comparator c1 (a[1], b[1], im[0], im[1], im[2], im[3], im[4], im[5]);
   bit_comparator c2 (a[2], b[2], im[3], im[4], im[5], im[6], im[7], im[8]);
   bit_comparator c3 (a[3], b[3], im[6], im[7], im[8], a_gt_b, a_eq_b, a_lt_b);
endmodule
```

Figure 5.35 Explicit instantiation of four *bit_comparators*.

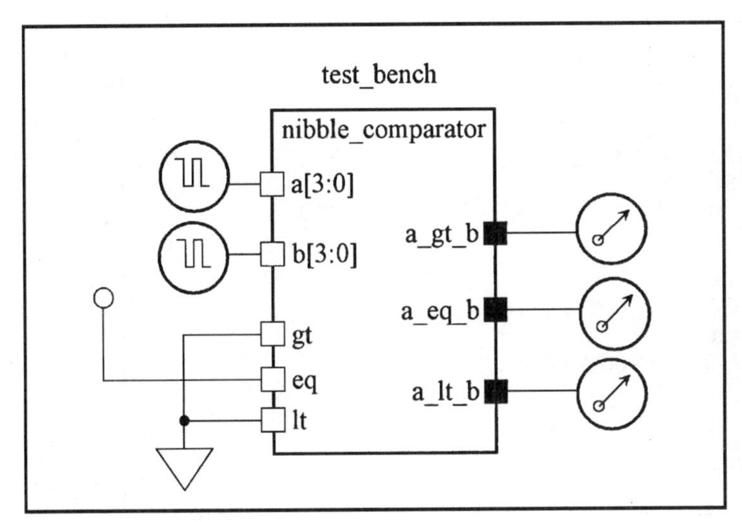

Figure 5.36 Graphic symbol for test bench for *nibble_comparator.*

comparator such that the *a_gt_b* (or *a_lt_b*) output becomes **1** only when *a* is greater than *b* (or *a* is less than *b*). Had the *gt* control input been connected to *vdd,* its corresponding output would become **1** when *a* is greater than or equal to *b*.

The Verilog description of the test bench is shown in Fig. 5.37. The test bench specifies the time scale as 10 ns with 1-ns resolution. The *test_nibble_comparator* module declares internal 4-bit signals *a* and *b* as **reg** so that they can be assigned values using procedural statements in a Verilog procedural body. Control inputs of *nibble_comparator, gt,* and *lt* are declared as **supply0,** which models the ground connections of these wires. The *eq* input wire is declared as **supply1,** modeling its connection to the supply voltage. The comparator outputs that are driven by instantiated *nibble_comparator* are declared as simple **net**s using **wire** type declaration. The initial procedural block in Fig. 5.37 assigns values to *a* and *b* vectors every 500 ns. Initially, at time 0, *a* and *b* are assigned **0.** The flow of activities in the initial block waits 500 ns by the *#50* construct, after which 4-bit decimal 15 values (*4'd 15*) are assigned to both *a* and *b*. The line in Fig. 5.37 that follows this line causes activity flow to wait another 500 ns. This is then followed by another 500-ns wait and assignment of 1110 and 1111 to *a* and *b*. Such activities continue until activity flow reaches the last statement in the **initial** block (**$finish**), which makes the simulator exit. Using system task **$stop** instead of **$finish** would suspend the simulation instead of exiting from it. If an **initial** block does not contain either of these two system tasks, it terminates when it reaches its last statement, and no more activities will be originated by this block.

```
`timescale 10ns/1ns

module test_nibble_comparator;
reg [3:0] a, b;
supply0 gt, lt; supply1 eq;
wire a_gt_b, a_eq_b, a_lt_b;
  nibble_comparator u1 (a, b, gt, eq, lt, a_gt_b, a_eq_b, a_lt_b);
  initial
  begin
          a = 4'd0; b = 4'd0;
          #50 a = 4'd15; b = 4'd14;
          #50 ;
          #50 a = 4'd14; b = 4'd15;
          #50 ;
          #50 b = 4'd12;
          #50 ;
          #50 a = 4'd10;
          #50 a = 4'd00; b = 4'd15;
          #50 a = 4'd15;
          #50 a = 4'd00;
          #50 b = 4'd00;
          #50 a = 4'd15;
          #50 $finish;
  end
endmodule
```

Figure 5.37 *nibble_comparator* test bench.

The test bench used here is a simple one, and we will continue using this style until more convenient data application methods are described in the later chapters. Chapter 7 describes several procedures that simplify development of test benches. Reading data from external files is discussed in Chaps. 7 and 9.

Before analysis of simulation results, a point that is worth mentioning is that the names used for local signals are *a* and *b,* which are the same as those of the *nibble_comparator* input ports. This is done only for clarity; any other name could be used for the local signals of the test bench. The port connection list determines the connection between *a* of the *nibble_comparator* and the local *a* of the test bench. Where options exists, descriptions should generally be written for better readability.

5.4.2 Simulation

The result of the simulation run in which the *test_nibble_comparator* module is the top unit is shown in Fig. 5.38. Simulation begins at 0 ns and ends at 6015 ns. This table shows values of signals and the times at which each value appears on a signal. Times at which signals change values are clearly shown. Using this list format instead of the waveform of Fig. 5.26 is advantageous when timing details are of concern.

Initial values on the outputs of the circuit are **X** or unknown. At time 48, the **0000** values on *a* and *b* cause the *a_eq_b* output to become **1**. The longest delay in the circuit occurs when a value from the least significant *bit_comparator* propagates to circuit outputs. Such a situation occurs at time 1500 ns for the less-than case. For this case, it takes 48 ns for the circuit to reach steady state. As seen from the simulation run at times 4500 and 5500 ns, the worst-case delay for the *a_eq_b* output is also 48 ns. The fastest that this circuit can produce results is when its two operands are different only in their most significant bits. This occurs at times 5000 and 6000 ns, where the circuit produces appropriate outputs and reaches steady state only 15 ns later. Between the worst case of 48 ns and the best case of 15 ns, other delay

ns	a	b	gt	lt	eq	a_gt_b	a_eq_b	a_lt_b
0	0000	0000	Z	Z	Z	X	X	X
0	0000	0000	0	0	1	X	X	X
44	0000	0000	0	0	1	0	X	0
48	0000	0000	0	0	1	0	1	0
500	1111	1110	0	0	1	0	1	0
544	1111	1110	0	0	1	1	1	0
548	1111	1110	0	0	1	1	0	0
1500	1110	1111	0	0	1	1	0	0
1544	1110	1111	0	0	1	0	0	0
1548	1110	1111	0	0	1	0	0	1
2500	1110	1100	0	0	1	0	0	1
2533	1110	1100	0	0	1	0	0	0
2537	1110	1100	0	0	1	1	0	0
3500	1010	1100	0	0	1	1	0	0
3522	1010	1100	0	0	1	0	0	0
3526	1010	1100	0	0	1	0	0	1
4000	0000	1111	0	0	1	0	0	1
4500	1111	1111	0	0	1	0	0	1
4544	1111	1111	0	0	1	0	1	1
4548	1111	1111	0	0	1	0	1	0
5000	0000	1111	0	0	1	0	1	0
5011	0000	1111	0	0	1	0	0	0
5015	0000	1111	0	0	1	0	0	1
5500	0000	0000	0	0	1	0	0	1
5544	0000	0000	0	0	1	0	0	0
5548	0000	0000	0	0	1	0	1	0
6000	1111	0000	0	0	1	0	1	0
6011	1111	0000	0	0	1	1	1	0
6015	1111	0000	0	0	1	1	0	0

Figure 5.38 Simulation report for simulating the test bench in Figure 5.27. All events are observed.

values occur when bits 1 or 2 of the two operands are different. The event on the *b* input at 2500 ns, for example, causes the two operands to differ only in bit 1. The propagation of values, therefore, must occur through bits 1, 2, and 3 of the comparator, causing a 37-ns delay before the circuit reaches steady state at 2537 ns time. At 3500 ns, the event on the *a* input causes bit 2 of the two operands to be different. The result of this comparison becomes available at 3526 ns after propagating through bits 2 and 3.

The circuit we are simulating has several levels of hierarchical nesting. The *test_nibble_comparator* module contains the *nibble_comparator,* which contains four instances of the *bit_comparator,* each of which contains several instances of **not** and **nand** gates. Events that occur on the ports of the outermost modules pass through the intermediate modules and reach the innermost gates. Evaluation of these gate outputs may result in generation of events on the output terminals of the modules that they are enclosed in. Such events will cause other modules to evaluate their outputs or pass the events down to subcomponents within them. For an event in the *test_nibble_comparator,* module invocation of built-in gates continues until the circuit reaches steady state. Outputs of a module are evaluated only if an event occurs on at least one of its inputs. Consider, for example, the situation at 3500 ns, when bit 2 of *a* changes. This event causes events on input ports of *bit_comparator* number 2 (see Fig. 5.33), which travel downward to eventually cause events on inputs of the gates (see Fig. 5.23) of this comparator bit. After these gates are evaluated, they cause events upward to reach the outputs of comparator bit 2 (Fig. 5.33). Since the outputs of this bit are connected to the input ports of bit 3, a similar downward and then upward propagation of events occurs inside this comparator bit. In the simulation of a Verilog description, such events occur, and at each stage only the affected statements are evaluated.

5.5 Sequential Example

At the end of this chapter, we will develop a hierarchical design consisting of combinational and sequential parts. In order to have enough ammunition for this example, and also to present other gate-level designs, this section will develop latches and flip-flops based on Verilog built-in gates. Gate-level and switch-level designs will be presented. A static design based on the gates of Fig. 5.18 will be used in designs in the next section.

5.5.1 Gate-level latch

Consider the clocked, level-sensitive, set-reset latch of Fig. 5.39. The structural-level Verilog description of this circuit, using **nand** primitive gates, is depicted in Fig. 5.40. This description uses four instances of a **nand** gate and generates the output of the latch on the output.

5.5.2 Switch-level circuit

Using transmission gates and capacitances, other even less abstract implementations of latches and flip-flops are possible in Verilog. Consider for example, the dynamic latch in Fig. 5.41. The Verilog code shown in Fig. 5.42 corresponds to this structure. For the two inverters shown and for complementing the clock *c*, gates *g1, g2,* and *g3* are used. Outputs of both **cmos** gates (*g4* and *g5*) drive the *im1* intermediate node. The *g4* gate conducts when *c* is **1** and pushes the value of *d* into *im1,* and *g5* conducts when *c* is **0,** which causes *im1* to be driving itself

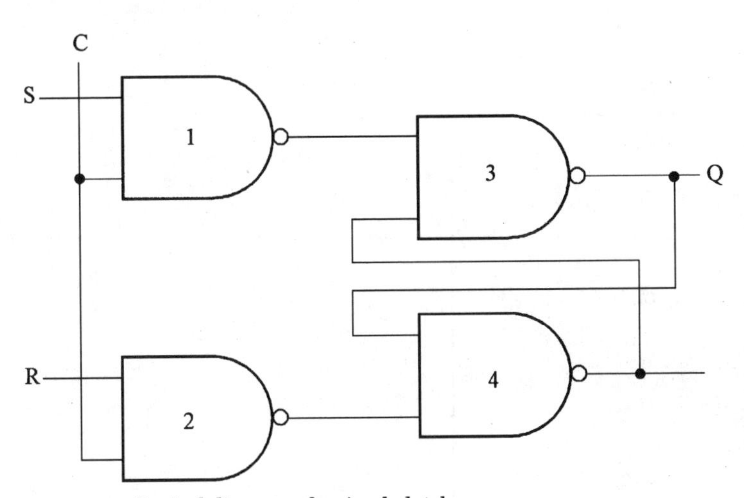

Figure 5.39 Logical diagram of a simple latch.

```
`timescale 1ns/100ps

module sr_latch (s, r, c, q);
//
input s, r, c;
output q;
wire im1, im2, im3;
  nand #(5)
    g1 (im1, s, c),
    g2 (im2, r, c),
    g3 (q, im1, im3),
    g4 (im3, q, im2);
endmodule
```

Figure 5.40 Verilog code for a set-reset latch.

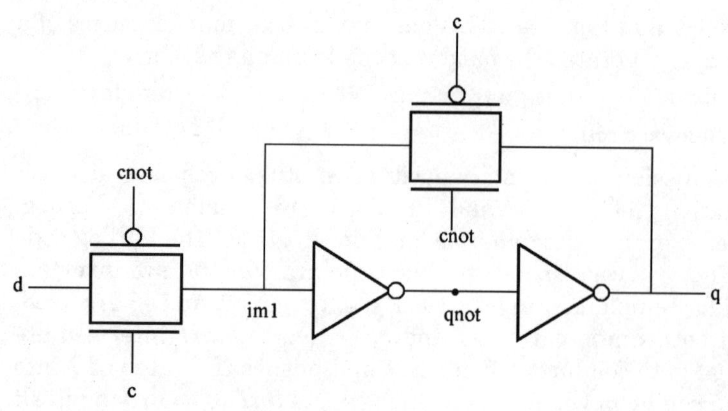

Figure 5.41 Dynamic D-type latch.

```
`timescale 1ns/100ps

module dd_latch (d, c, q);
//
input d, c;
output q;
wire im1, cnot, qnot;
  not #(5)
    g1 (cnot, c),
    g2 (qnot, im1),
    g3 (q, qnot);
  cmos #(3)
    g4 (im1, d, c, cnot),
    g5 (im1, q, cnot, c);
endmodule
```

Figure 5.42 Verilog code for dynamic D-type latch.

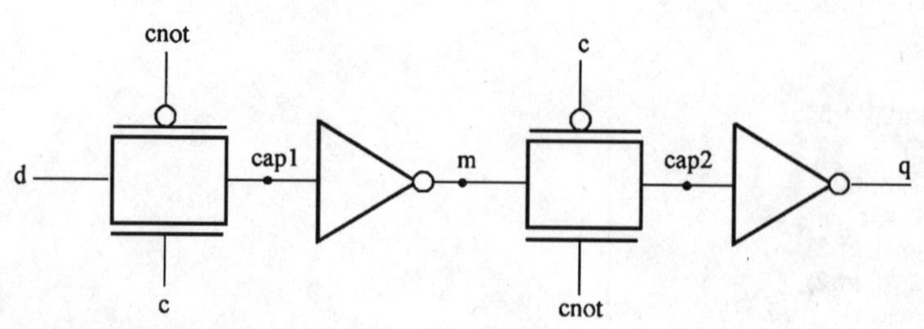

Figure 5.43 Dynamic master-slave D flip-flop.

after passing through two inverters and through output *q*. When *c* is **0,** output *q* holds its previous value.

Another switch-level design, this one for a D-type master-slave flip-flop, is shown in Fig. 5.43. As with the design of Fig. 5.41, for the flip-flop of Fig. 5.43, **cmos** and **not** gates are used. However, unlike the design of Fig. 5.41, this circuit does not have a refresh loop. Therefore, for the value of *d* to be saved in the flip-flop, gate capacitances as inputs of inverters must be properly modeled in a corresponding Verilog code.

Figure 5.44 shows the Verilog code for a CMOS D-type master-slave flip-flop. Gate capacitances at the inputs of the *g2* and *g3* **not** gates are named *cap1* and *cap2*. These wires are declared as **trireg** with a 96-ns charge hold time and 0 rise and fall times. When *g4* conducts, the value of *d* will be saved in *cap1* and the complement of it will appear on *m*. At this time, *g5* is driving its output with high impedance, which causes **0** or **1** values at this output to remain intact. When *c* becomes **0,** *g5* conducts, causing *cap1* to be disconnected from the changes on input *d,* and loads *cap2* with the value saved at node *m*. At this time, the complement of *m* constitutes the *q* output of the flip-flop.

5.6 Top-Down Wiring

The descriptions of the previous sections were intended to illustrate features of Verilog for describing hardware modules at the structural level. We presented several designs in a bottom-up fashion. Gates were put together to generate a bit comparator, and bit comparators were wired to create a nibble comparator. The same methodology was used in creating latches.

```
`timescale 1ns/100ps

module dd_master_slave (d, c, q);
//
input d, c;
output q;
wire m, cnot;
trireg #(0, 0, 96) cap1, cap2;
  not #(5)
    g1 (cnot, c),
    g2 (m, cap1),
    g3 (q, cap2);
  cmos #(3)
    g4 (cap1, d, c, cnot),
    g5 (cap2, m, cnot, c);
endmodule
```

Figure 5.44 Verilog code for CMOS D flip-flop.

As another example for structural description of hardware, this section presents on implementation of a design in a top-down fashion, as discussed in Chap. 3. Low-level components in this implementation will be those developed in the previous section of this chapter, namely a latch and a bit comparator.

5.6.1 Sequential comparator

The example we will use is a sequential comparator comparing consecutive sets of data on an 8-bit input bus. The *gt_compare* (or *eq_compare*) output becomes **1** when the new data on the bus are greater than (or equal to) the old data. Data on the input bus are synchronized with a clock signal.

The top level of this circuit using an 8-bit latch and an 8-bit comparator is shown in Fig. 5.45. As shown in this figure, an 8-bit latch saves old data on the *i* bus and a comparator compares saved old data and new data on the bus before it latches into the latch. Two outputs of the *byte_comparator* drive the outputs of the *old_new_comparator* circuit.

Figure 5.46 shows the Verilog module code of this circuit. The wiring shown in Fig. 5.45 is done in Verilog using module instantiations in *old_new_comparator* in Fig. 5.46. A net declaration in this module declares *con1* for intermediate wires.

Instantiation of *byte_comparator* in Fig. 5.46 contains a connection with no wire name for the last port of *byte_comparator*. Describing a port connection as such leaves that port open. As shown in Fig. 5.45, the less-than output of *byte_comparator* is not being utilized and is left

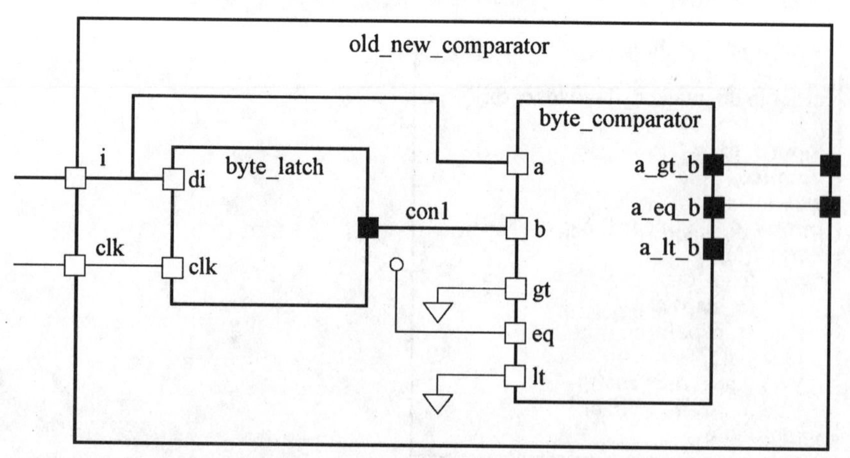

Figure 5.45 Components of the sequential comparator (*old_new_comparator*).

open. Alternatively, the *named_port_connection* construct (Fig. 5.12) may be used to name only those ports to which connections are made.

5.6.2 Byte latch

Figure 5.47 shows the Verilog description of the 8-bit latch used in the design of *old_new_comparator*. Eight instantiations of *d_latch* are shown in this description. If this code is written for an IEEE-compliant Verilog simulator that supports arrays of instances, the explicit instantiations in Fig. 5.47 could be replaced by

```
d_latch c[7:0] (di, c, q0);
```

This coding alternative is more compact and can easily be modified for other size vectors or latches. The *d_latch* in the design of the

```
module old_new_comparator (i, clk, gt_compare, eq_compare);
//
input [7:0] i;
input clk;
output gt_compare, eq_compare;
wire [7:0] con1;
supply1 vdd;
supply0 gnd;
  byte_latch l (i, clk, con1);
  byte_comparator c (con1, i, gnd, vdd, gnd, gt_compare, eq_compare, );
endmodule
```

Figure 5.46 *old_new_comparator* Verilog description.

```
`timescale 1ns/100ps

module byte_latch (di, c, qo);
input [7:0] di;
input c;
output [7:0] qo;
  d_latch c7 (di[7], c, qo[7]);
  d_latch c6 (di[6], c, qo[6]);
  d_latch c5 (di[5], c, qo[5]);
  d_latch c4 (di[4], c, qo[4]);
  d_latch c3 (di[3], c, qo[3]);
  d_latch c2 (di[2], c, qo[2]);
  d_latch c1 (di[1], c, qo[1]);
  d_latch c0 (di[0], c, qo[0]);
endmodule
```

Figure 5.47 Verilog description of *byte_latch*.

byte_latch is made of an inverter (Fig. 5.18) and a clocked *sr_latch* (Fig. 5.40), as shown in Fig. 5.48.

5.6.3 Byte comparator

The Verilog description for the *byte_comparator* used in the wiring of Fig. 5.45 is shown in Fig. 5.49. This code uses eight instantiations of *bit_comparator* of Fig. 5.29 and is similar to the code for *nibble_comparator* in Fig. 5.35. As in the *byte_latch,* an array of instances could be used to replace the explicit instantiations in Fig. 5.49.

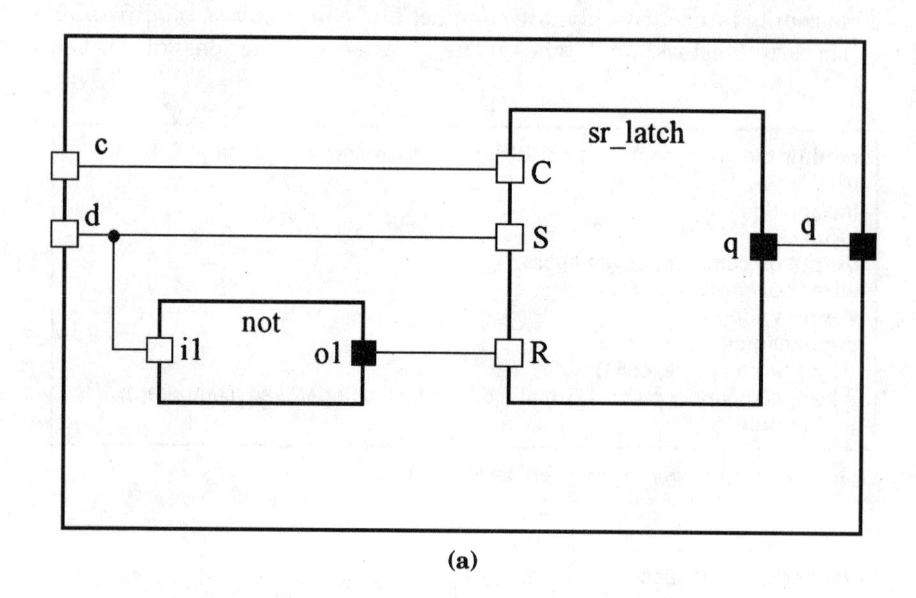

(a)

```
`timescale 1ns/100ps

module d_latch (d, c, q);
input d, c;
output q;
wire dbar;
  sr_latch c1 (d, dbar, c, q);
  not #4 c2 (dbar, d);
endmodule
```

(b)

Figure 5.48 Design of *d_latch*. (*a*) Graphical wiring; (*b*) Verilog description.

```
module byte_comparator (
a, b,            // data inputs
gt, eq, lt,      // previous greater than, equal, less than
a_gt_b,          // greater
a_eq_b,          // equal
a_lt_b);         // less than
//
input [7:0] a, b;
input gt, eq, lt;
output a_gt_b, a_eq_b, a_lt_b;
wire [0:20] im;
  bit_comparator c0 (a[0], b[0], gt, eq, lt, im[0], im[1], im[2]);
  bit_comparator c1 (a[1], b[1], im[00], im[01], im[02], im[03], im[04], im[05]);
  bit_comparator c2 (a[2], b[2], im[03], im[04], im[05], im[06], im[07], im[08]);
  bit_comparator c3 (a[3], b[3], im[06], im[07], im[08], im[09], im[10], im[11]);
  bit_comparator c4 (a[4], b[4], im[09], im[10], im[11], im[12], im[13], im[14]);
  bit_comparator c5 (a[5], b[5], im[12], im[13], im[14], im[15], im[16], im[17]);
  bit_comparator c6 (a[6], b[6], im[15], im[16], im[17], im[18], im[19], im[20]);
  bit_comparator c7 (a[7], b[7], im[18], im[19], im[20], a_gt_b, a_eq_b, a_lt_b);
endmodule
```

Figure 5.49 *byte_comparator* Verilog description.

5.7 Summary

A structural description for a design consists of a wiring specification of its subcomponents. In this chapter, the definition and usage of components in larger designs was illustrated. We also presented several methods of defining basic primitive structures and using predefined built-in structures. Arrays of instances were introduced as a convenient way to describe repetitive hardware structures. The last part of this chapter presented a top-down design using basic gates and components presented in the earlier sections. Using simple gates, the reader should now be able to design larger digital circuits with many levels of module nesting.

Further Reading

IEEE Standard Hardware Description Languages Based on the Verilog Hardware Description Language, IEEE Std. 1346-1995, Institute of Electrical and Electronic Engineers, New York, 1996.

Palnitkar, Samir, *Verilog HDL,* Sunsoft Press, Prentice-Hall, Upper Saddle River, N.J., 1996.

Thomas, Donald E., and Philip R. Moorby, *The Verilog Hardware Description Language,* (3d ed.), Kluwer Academic Publishers, Norwell, Mass., 1996.

Problems

5.1 Write a test bench for the *bit_comparator* of Sec. 5.2. Find the worst-case delay for this circuit. Why is this delay not equal to one-fourth of the worst-case delay of the *nibble_comparator?* Analyze the timings of both circuits and answer this question.

5.2 Write a description of a full-adder using the built-in gates presented in this chapter. What is the worst-case delay for this circuit?

5.3 Write a Verilog description for a package of four NAND gates, using the **nand** built-in gate for each of the gates.

5.4 Use the *four_nand2* package of Prob. 5.3 to describe a clocked SR latch. The solution to this problem depends on code developed in Prob. 5.3.

5.5 Treating *four_nand2* of Prob. 5.3 as a packaged IC, use this package to develop a clocked D-type flip-flop. Use Verilog module of this flip-flop and four_nand2 packages to develop Verilog code for a 2-bit binary up-counter. Generate a test-bench to test this counter.

5.6 Use only two *four_nand2* packages of Prob. 5.3 to describe a BCD prime number detector. The output is to be **1** when the input BCD number is a prime number. The number 1 is a prime number. The solution to this problem depends on code developed in Prob. 5.3.

5.7 Using only XOR gates, write a Verilog description for an 8-bit even/odd parity checker. The circuit has an 8-bit input vector and two outputs. The *odd* output is to become **1** when the number of 1s on the input is odd. The *even* output is the opposite of the *odd* output. Use an array of instances.

5.8 Describe an 8-bit adder using the full-adder description in Prob. 5.2. Take advantage of array of instances if your simulator allows it. Write a test bench for this adder, and find the worst-case delay. The solution to this problem depends on code developed in Prob. 5.2.

5.9 If the inputs to the adder in Prob. 5.8 are 2s-complement numbers, an overflow may occur when two positive or two negative numbers are added. Use this adder in a design of an 8-bit adder with an overflow indication output. Do not modify the description of the original adder; rather, use it in a top-level design that instantiates the adder of Prob. 5.8 as well as gates for the overflow detection hardware. The solution to this problem depends on code developed in Probs. 5.2 and 5.8.

5.10 Using built-in switches and gates, modify the master-slave D flip-flop of Sec. 5.5.2 to add a refresh path when the clock is **0.**

5.11 Write a Verilog description for a master-slave JK flip-flop. Use static built-in **nand** gates.

5.12 Use the *JK* flip-flop in Prob. 5.11 to design a 1-bit modular binary counter. Based on this module, build a 4-bit binary ripple up-counter. Design this counter in such a way that it can easily be cascaded for building larger counters. The solution to this problem depends on code developed in Prob. 5.11.

5.13 A 1-bit module of an n-bit synchronous binary counter with serial enable input can be formed using a T flip-flop and an AND gate. The AND gate ANDs the outputs of the previous stage of the counter (Qi-1) with the output of its AND gate (Eni-1). (*a*) Using the flip-flop of Prob. 5.11, write a Verilog description for a 1-bit module of such a counter. (*b*) Write an 8-bit synchronous counter description using the counter module of part *a*.

Chapter

6

Design Organization and Parametrization

In a digital system design environment, functional design of a digital system is often done independently of the physical characteristics or the technology in which the design is being implemented. Hardware designers who implement their design in an FPGA and designers who use pretested layouts of CMOS cells share many top-level design stages. There is still more sharing when hardware designers implement their designs in chips that differ only in speed or use various series in the same logic family. Component characteristics differ based on technology, power consumption, speed, temperature range, and packaging. Often, in a parts library, there is only one model for all logically equivalent modulator components. Designers using these generic modules customize module descriptions to fit their specific design or implementation environments. For such purposes, parametrizing of modules is helpful.

Verilog provides language constructs for parametrizing and customizing designs. These constructs enable the designer to generate a functional design independent of the specific technology and to customize this generic design at a later stage. By use of parameters, Verilog also allows a design to be parametrized so that the specific timing, number of bits, or current and loading are determined when the design is configured.

Another language issue that can influence the organization of a design is the use of subprograms. Like any high-level language, Verilog allows the definition and usage of *functions* and *tasks*. In addition to having important hardware implications, these language constructs greatly improve the readability and organization of a hardware description.

This chapter discusses subprograms, design parametrization, and parameter specification. We will explain postdesign specifications of physical characteristics, like timing, and describe various methods that can be used to configure a design for a specific set of parameters.

In most examples we will use timing parameters to illustrate various parameter specification techniques. However, most of these techniques can be applied to other parameters as well.

6.1 Definition and Usage of Subprograms

In many programming languages, subprograms are used to simplify coding, modularity, and readability of descriptions. Verilog uses subprograms for these applications as well as for those that are more specific to hardware descriptions. Regardless of the application, behavioral softwarelike constructs are allowed in subprograms. Verilog allows the use of *functions* and *tasks.* Functions return values and cannot alter the values of their arguments. A task, on the other hand, is used as a statement and can alter the values of its arguments.

In addition to constructs for definition and usage of user-defined subprograms, Verilog provides several utility system tasks and functions. These system utilities can only be used the way they are defined in the language, and are used for data generation, monitoring, file handing, and timing control.

6.1.1 A functional single-bit comparator

The *bit_comparator* in Chap. 5 was designed at the gate level to specify the interconnection of all its gates using module instantiations. A simpler description can be developed by using the boolean equations [Eqs. (6.1*a*) to (6.1*c*) of the three outputs of this circuit.

$$a\_gt\_b = a \,.\, gt + b' \,.\, gt + a \,.\, b' \tag{6.1a}$$

$$a\_eq\_b = a \,.\, b \,.\, eq + a' \,.\, b' \,.\, eq \tag{6.1b}$$

$$a\_lt\_b = b \,.\, lt + a' \,.\, lt + b \,.\, a' \tag{6.1c}$$

As is evident from the equations of *a_gt_b* and *a_lt_b,* rearranged and repeated here for reference (and also from the schematic diagrams in Fig. 5.21), one expression used with different signal names can be made to express both of these outputs. The *a_eq_b* output, however, requires a separate expression. Figure 6.1 shows a Verilog description that is based on these equations and one that takes advantage of the similarities between the *a_gt_b* and *a_lt_b* outputs.

For the *bit_comparator* in Fig. 5.22, the module description in Fig. 6.1 is an alternative to the gate-level description in Fig. 5.24. The module item declaration part of this *bit_comparator* contains two

```
`timescale 1ns/100ps

module bit_comparator (a, b, gt, eq, lt, a_gt_b, a_eq_b, a_lt_b);

  function fgl;
  input w, x, gl;
    begin
      fgl = w & gl | ~x & gl | w & ~x;
    end
  endfunction

  function feq;
  input w, x, eq;
    begin
      feq = w & x & eq | ~w & ~x & eq ;
    end
  endfunction

  input a, b, gt, eq, lt;
  output a_gt_b, a_eq_b, a_lt_b;
  assign #12 a_gt_b = fgl (a, b, gt);
  assign #12 a_eq_b = feq (a, b, eq);
  assign #12 a_lt_b = fgl (b, a, lt);
endmodule
```

Figure 6.1 A functional *bit_comparator* using the same function for two outputs.

function declarations. The first function returns a 1-bit value which is a function of *w, x,* and *gl* (*g*reater or *l*ess). The *a_gt_b* and *a_lt_b* outputs use this function. The other function in the declaration part of the module of Fig. 6.1 returns a 1-bit value for the expression for the *a_eq_b* output. In the module item part, three function calls are used for the three outputs of the *bit_comparator* on the right-hand sides of continuous assignment statements. When a function is called, it returns its calculated value in zero time. Because of this, and in order for this code to better represent the actual circuit, assignment of function values to appropriate outputs is delayed by 12 ns. This delay value is a rough estimate of the worst-case propagation delay of each *bit_comparator* and is based on the 48-ns worst-case delay of the *nibble_comparator* discussed in Chap. 5.

The syntax details of the *fgl* function are shown in Fig. 6.2. The construct for definition of a function is *function_declaration.* This construct is bracketed by **function** and **endfunction** keywords. *Function_identifier* follows the **function** keyword and is then followed by declarations of function. A function cannot have output declarations. Following the declaration item, the statement construct constitutes the main body of a function. In our *fgl* example of Fig. 6.2, a single *blocking_assignment* forms the *statement* of the function.

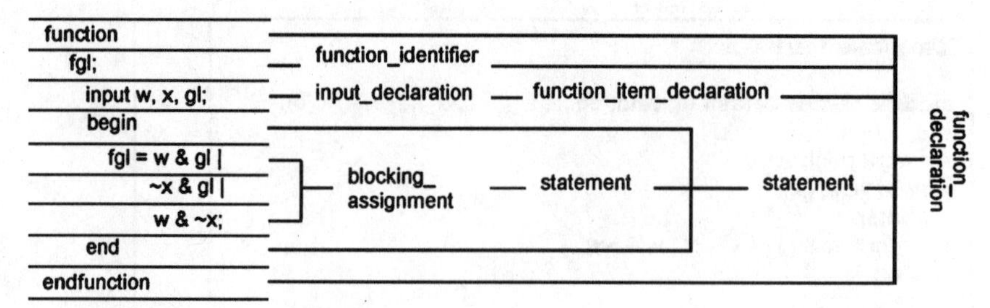

Figure 6.2 Syntax details of a function declaration.

6.1.2 Using tasks in a test bench

The main purpose of the test bench developed for the 4-bit comparator in Chap. 5 was to apply data to the 4-bit input of the *nibble_comparator.* We will discuss several system- and user-defined tasks that can significantly simplify this process and will use the *nibble_comparator* of Chap. 5 to demonstrate this topic.

Figure 6.3 shows a test bench for our *nibble_comparator* example that uses the **$readmemh** system task. As before, the declarations of this module include variables for connecting to ports of the *nibble_comparator* that is being tested. In addition, two tables of 13 nibbles (4-bit words), *atable* and *btable,* are declared as **reg** in the declaration item part of the *test_nibble_comparator* module. These tables will be loaded with values to be applied to the *a* and *b* ports of the *nibble_comparator.* The *module_item* in Fig. 6.3 consists of a module instantiation and an *initial_construct.* The statements in the initial block read *avalues.dat* and *bvalues.dat* external files, assign their contents to *atable* and *btable,* respectively, and assign 4-bit entries of these tables to the *a* and *b* ports of *nibble_comparator* every 500 ns.

Figure 6.4 shows the contents of *avalues.dat* and *bvalues.dat* external files. Entries in these files are in hex format and are read by enabling the **$readmemh** system task. When the **$readmemh** system task is enabled with ("avalues.dat", atable) arguments, the entire contents of *avalues.dat* is read and loaded into the *atable.* The general form of this task is

```
$readmemh (file, memory, begin_address, end_address)
```

When this task is used with all parameters, file contents will be loaded into memory starting from *begin_address* and ending in *end_address.* If *end_address* is not specified, the last memory location is assumed. If both beginning and ending addresses are not specified, location 1 and the last location of memory are assumed for the beginning and ending addresses. A variation of this system task is **$readmemb,** which reads binary data

```
`timescale 10ns/1ns

module test_nibble_comparator;
reg [3:0] a, b;
supply0 gt, lt; supply1 eq;
wire a_gt_b, a_eq_b, a_lt_b;
reg [4:1] atable [13:1], btable [13:1];
integer i;
  nibble_comparator u1 (a, b, gt, eq, lt, a_gt_b, a_eq_b, a_lt_b);
  initial
  begin
        $readmemh ("avalues.dat", atable);
        $readmemh ("bvalues.dat", btable);
        for (i=1; i<=13; i=i+1)
        begin
                a = atable [i];
                b = btable [i];
                #50;
        end
  end
endmodule
```

Figure 6.3 Using the **$readmemh** system task in the *nibble_comparator* test bench.

avalues.dat	bvalues.dat
0	0
F	E
F	E
E	F
E	F
E	C
E	C
A	C
0	F
F	F
0	F
0	0
F	0

Figure 6.4 File contents read by **$readmemh** task invocations.

instead of the hex data read by **$readmemh.** Appendix A shows a complete list of Verilog system tasks and a brief description for each. In the text we will discuss only system tasks that we use in our examples.

To demonstrate the definition and usage of user-defined tasks, another test bench, shown in Fig. 6.5, is developed for the *nibble_comparator.* As in the test bench in Fig. 6.3, an initial block in the code of Fig. 6.5

applies data to the *a* and *b* ports of *nibble_comparator*. This is achieved by enabling the *generate_data* user task. Definition of this task is shown in the second half of Fig. 6.5.

The *generate_data* task calls the **$random** system function to obtain 8-bit random data. These data are divided into two 4-bit parts that are assigned to the outputs of the *generate_data* task. For more random generation of data, if the two halves happen to be the same, one output (*target1*) will be complemented and assigned to the other output (*target2*).

Figure 6.6 shows syntax details of a user-defined task. As in function declaration, the body of a task consists of a statement construct. As shown in Fig. 6.2, a statement uses **begin** and **end** keywords for bracketing procedural statements that are enclosed in it.

Enabling of tasks can take place only within procedural bodies of Verilog. A task is enabled when the flow of program reaches it. The statement part of a task or a function is itself considered procedural. Tasks can enable other tasks or functions, and functions can enable only other functions. The **$random** function is a system function that is used in our example user-defined task.

```
`timescale 10ns/1ns

module test_nibble_comparator;
reg [3:0] a, b;
supply0 gt, lt; supply1 eq;
wire a_gt_b, a_eq_b, a_lt_b;
integer i;

  nibble_comparator u1 (a, b, gt, eq, lt, a_gt_b, a_eq_b, a_lt_b);
  initial
        for (i=1; i<=13; i=i+1)
        begin
                generate_data (a, b);
                #50;
        end
  //
  task generate_data;
  output [3:0] target1, target2;
        begin   : rangen
        reg [7:0] values;
                values = $random;
                target1 = values [7:4];
                target2 = values [3:0];
                if (target1 == target2)
                  target2 = ~ target1;
        end
  endtask
endmodule
```

Figure 6.5 Using user-defined tasks.

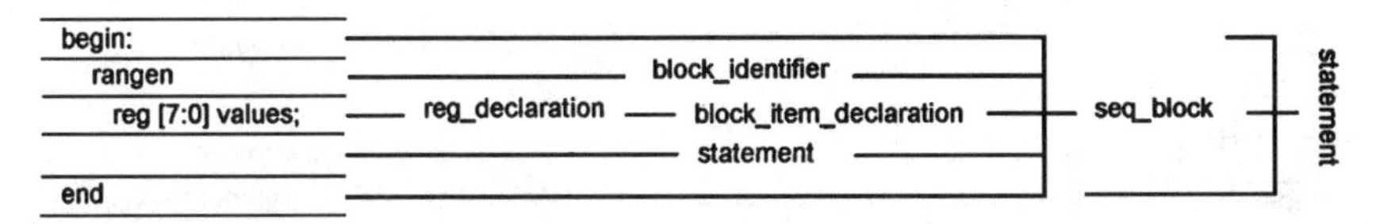

Figure 6.6 Statement describing operation of the *generate_data* task.

6.1.3 Language aspects of tasks and functions

As seen in the previous section, the part of a user task or function that performs the operation of the subprogram is referred to as *statement* in the syntax definition of Verilog. In order to better understand language constructs that are allowed for the definition of subprograms, the details of the syntax of *statement* will be discussed here.

Figure 6.6 zooms in on the statement used for definition of the *generate_data* task in Fig. 6.5. This statement is a sequential block (the syntax name is *seq_block*) that begins and ends with **begin** and **end** keywords and is identified as *rangen*. Declaration of *values,* which is done by the *block_item_declaration,* is visible only to the statements in this block. The *rangen* block consists of a *statement* that is further zoomed in on and described in Fig. 6.7. The *statement* in Fig. 6.7 consists of three *blocking_assignments* and one *conditional_statement,* each of which is considered to be a *statement.* The first *statement* in this figure uses a function call on the right-hand side of a blocking assignment.

The conditional statement in Fig. 6.7 begins with an **if** keyword and again consists of a statement in its body. Another statement that is useful in describing behavioral procedures is the loop statement. The initial block in Fig. 6.5 uses this statement.

The recursive use of *statement* in Figs. 6.6 and 6.7 indicates that the operation of tasks may be described by nesting several statements. Statements may be conditional, loop, case, and other high-level behavioral constructs of Verilog.

6.1.4 Use by multiple modules

Verilog allows tasks, functions, and even portions of code to be shared among various modules. This utility is most useful for use of subprograms by several descriptions.

The language construct for this purpose is the **`include** compiler directive. When **`include** *"a_file.ext"* is used, the contents of the file enclosed in double quotes (*a_file.ext*) will be inserted into the Verilog code that contains the **`include** directive at the location at which this

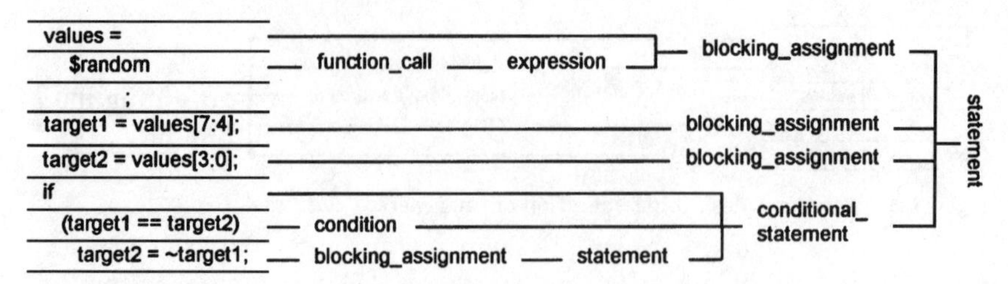

Figure 6.7 Statement within *rangen* block of Fig. 6.6.

directive appears. During compilation of a module, the code of the file is compiled as if it were placed in the module. Several modules may use the same **`include** statement to insert the same file. Full paths may be used, but if only the filename to include appears between double quotes, a compilation default directory will be assumed.

Figure 6.8 shows the *nibble_comparator* test bench of Fig. 6.5 rewritten to include the *generate_data* task at compile time instead of explicitly including it in the code. The **`include** directive refers to the *gen_data.v* file, which must be present when the nibble comparator of Fig. 6.8 is being compiled. The *gen_data.v* file is shown in Fig. 6.9.

In addition to saving disk space and editing efforts, the **`include** directive is useful for sharing libraries of parts and adding flexibility to designs by using various libraries.

6.2 Design Parametrization

Component models can be parametrized for better utilization and to make them usable in different design environments. The specific behavior of these models is dependent on the parameters that are determined by the design entities that use them. Verilog built-in gates allow timing parameters to be passed to them when they are being instantiated. The number of parameters and the way they affect the behavior of a built-in gate are determined by the language and cannot be altered. However, Verilog allows specification of parameters with any design module. The number of parameters, the way they affect the behavior of a module, and their default values are determined by the way the module is written. Module parameters can specify timing, physical size, resistance, capacitance, power consumption, or any other electrical or nonelectrical characteristic of a module. For the purpose of illustration, we will use only timing parameters in this section. How module parameters are defined and alternative ways of associating values to them will be described here.

We will begin our discussion of parameters with those of built-in gates, which are limited to the way they are assigned values. This will be followed by a discussion of the more general topic of defining module parameters and alternative methods for specifying them.

Figure 6.10 shows gates used in the designs of the previous chapter. These gates were used in Chap. 5 with single delay parameters that were assigned values when they were instantiated. As shown in Fig. 6.10, these built-in gates can have up to two parameters, for the values of which other parameters or expressions may be used. *tplh* (time of propagation from low to high) and *tphl* (time of propagation from high to low) will be used for rise and fall delay parameters. Figure 6.10 also shows a symbolic notation with *tplh* and *tphl* parameters for each built-in gate. These notations will be used to show how values are passed to these parameters.

```
`timescale 10ns/1ns

module test_nibble_comparator;
reg [3:0] a, b;
supply0 gt, lt; supply1 eq;
wire a_gt_b, a_eq_b, a_lt_b;
integer i;
  `include "gen_data.v"
  nibble_comparator u1 (a, b, gt, eq, lt, a_gt_b, a_eq_b, a_lt_b);
  initial
    for (i=1; i<=13; i=i+1)
    begin
        generate_data (a, b);
        #50;
    end
endmodule
```

Figure 6.8 Using **`include** to insert a user task.

```
task generate_data;
    output [3:0] target1, target2;
    begin    : rangen
        reg [7:0] values;
        values = $random;
        target1 = values [7:4];
        target2 = values [3:0];
        if (target1 == target2)
            target2 = ~ target1;
    end
endtask
```

Figure 6.9 *gen_data.v* file that is included in Fig. 6.8.

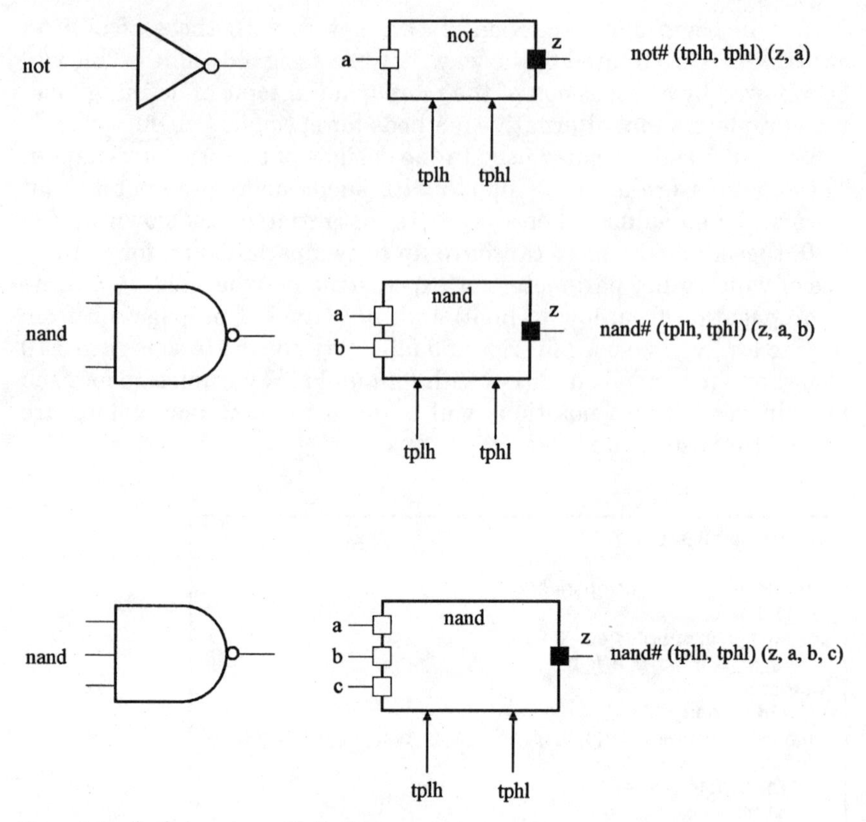

Figure 6.10 Built-in gates with timing parameters.

6.2.1 Using fixed values

The simplest and least flexible way of specifying the parameters of a built-in gate or a module is to use constants. The example in Fig. 6.11 shows specification of the rise and fall parameters of the gates of *bit_comparator.* Parameters specified as such are associated with gate or module parameters according to the way they are ordered. For two-delay value gates, the first parameter is always the rise and the second one the fall delay.

A three-value format may also be used for each parameter, representing the minimum, typical, and maximum values a parameter may take. Figure 6.12 shows another description for the *bit_comparator* in which minimum, typical, and maximum delay values are specified for the **not** and **nand** built-in gates. Specified delay values are associated with gate types and apply to all gate instances that follow them. For example, *g9, g10,* and *g11* are all of type **nand** and use 5:6:7 and 3:4:5 for rise and fall delays.

In simulating these gates, typical values will be used as default values, unless **min** or **max** simulation switches are set. Simulation of the *bit_comparator* of Fig. 6.12 using the default setting yields the exact same results as simulation of the module of Fig. 6.11.

6.2.2 Module parameters and parameter passing

Unlike built-in gate parameters, which are fixed in the way they affect the functionality of a gate, modules can be given parameters and interpreted in accordance with the way a modeler codes them into his or her module. Figure 6.13 shows a description of *bit_comparator* that declares six parameters in its declaration part, using a parameter declaration construct. Each parameter has a constant assigned to it as its

```
`timescale 1ns/100ps

module bit_comparator (
a, b,            // data inputs
gt, eq, lt,      // previous greater than, equal, less than
a_gt_b,          // greater
a_eq_b,          // equal
a_lt_b);         // less than
//
input a, b, gt, eq, lt;
output a_gt_b, a_eq_b, a_lt_b;
wire im1, im2, im3, im4, im5, im6, im7, im8, im9, im10;

  // a_gt_b output
  not #(5, 3)
    g0 (im1, a),
    g1 (im2, b);
  nand #(6, 4)
    g2 (im3, a, im2),
    g3 (im4, a, gt),
    g4 (im5, im2, gt);
  nand #(7, 5)
    g5 (a_gt_b, im3, im4, im5);

  // a_eq_b output
  nand #(7, 5)
    g6 (im6, im1, im2, eq),
    g7 (im7, a, b, eq);
  nand #(6, 4)
    g8 (a_eq_b, im6, im7);

  // a_lt_b output
  nand #(6, 4)
    g9 (im8, im1, b),
    g10 (im9, im1, lt),
    g11 (im10, b, lt);
  nand #(7, 5)
    g12 (a_lt_b, im8, im9, im10);
endmodule
```

Figure 6.11 Using fixed delay values for built-in gates.

```
`timescale 1ns/100ps

module bit_comparator (
a, b,            // data inputs
gt, eq, lt,      // previous greater than, equal, less than
a_gt_b,          // greater
a_eq_b,          // equal
a_lt_b);         // less than
//
input a, b, gt, eq, lt;
output a_gt_b, a_eq_b, a_lt_b;
wire im1, im2, im3, im4, im5, im6, im7, im8, im9, im10;

  // a_gt_b output
  not #(4:5:6, 2:3:4)
    g0 (im1, a),
    g1 (im2, b);
  nand #(5:6:7, 3:4:5)
    g2 (im3, a, im2),
    g3 (im4, a, gt),
    g4 (im5, im2, gt);
  nand #(6:7:8, 4:5:6)
    g5 (a_gt_b, im3, im4, im5);

  // a_eq_b output
  nand #(6:7:8, 4:5:6)
    g6 (im6, im1, im2, eq),
    g7 (im7, a, b, eq);
  nand #(5:6:7, 3:4:5)
    g8 (a_eq_b, im6, im7);

  // a_lt_b output
  nand #(5:6:7, 3:4:5)
    g9 (im8, im1, b),
    g10 (im9, im1, lt),
    g11 (im10, b, lt);
  nand #(6:7:8, 4:5:6)
    g12 (a_lt_b, im8, im9, im10);
endmodule
```

Figure 6.12 Using *min:typ:max* parameters.

default value. Once parameters are declared, it is up to the modeler to decide how these parameters influence the operation of a module. In our example in Fig. 6.13, module parameters are passed to built-in gates for their rise and fall delay values. If a parameter is to be interpreted as a delay value, the **`timescale** directive determines its unit and precision.

Figure 6.14 shows syntax details of the *parameter_declaration* language construct. As shown, *parameter_declaration* can declare several parameters, and each parameter is required to have a constant expression assigned to it. If not overwritten, this constant expression will be the value of the parameter.

In addition to showing specification of parameters for modules, the example of Fig. 6.13 also shows how expressions, and not only constants, may be used for built-in gate parameters. Here expressions are simply parameters that belong to the *bit_comparator* module. In effect, this example shows how parameters may be passed from upper-level descriptions to their lower-level components. Passing six timing parameters from *bit_comparator* to **not** and **nand** gates is graphically represented in Fig. 6.15. This figure shows that the *tplh* and *tphl*

```
`timescale 1ns/100ps
// timed and parameterized

module bit_comparator (
        a, b,           // data inputs
        gt, eq, lt,     // previous greater than, equal, less than
        a_gt_b,         // greater
        a_eq_b,         // equal
        a_lt_b);        // less than
        //
input a, b, gt, eq, lt;
output a_gt_b, a_eq_b, a_lt_b;
parameter       tplh1 = 5, tphl1 = 3,
                tplh2 = 6, tphl2 = 4,
                tplh3 = 7, tphl3 = 5;
wire im1, im2, im3, im4, im5, im6, im7, im8, im9, im10;
  // a_gt_b output
  not #(tplh1, tphl1)
    g0 (im1, a),
    g1 (im2, b);
  nand #(tplh2, tphl2)
    g2 (im3, a, im2),
    g3 (im4, a, gt),
    g4 (im5, im2, gt);
  nand #(tplh3, tphl3)
    g5 (a_gt_b, im3, im4, im5);

  // a_eq_b output
  nand #(tplh3, tphl3)
    g6 (im6, im1, im2, eq),
    g7 (im7, a, b, eq);
  nand #(tplh2, tphl2)
    g8 (a_eq_b, im6, im7);

  // a_lt_b output
  nand #(tplh2, tphl2)
    g9 (im8, im1, b),
    g10 (im9, im1, lt),
    g11 (im10, b, lt);
  nand #(tplh3, tphl3)
    g12 (a_lt_b, im8, im9, im10);
endmodule
```

Figure 6.13 Modules with parameters and passing parameters.

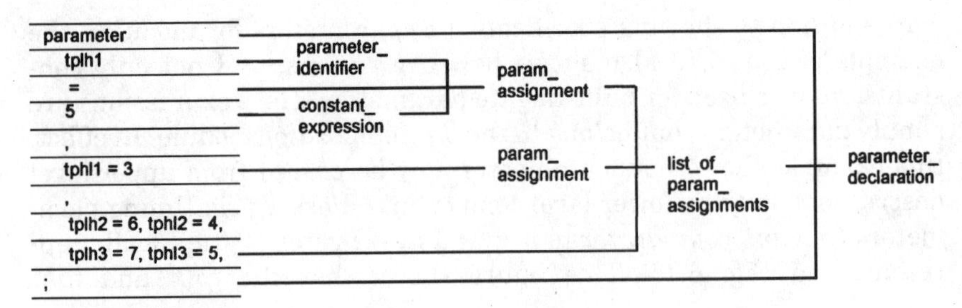

Figure 6.14 Syntax details of *parameter_declaration.*

parameters of *bit_comparator* not only can be used internally but are also visible from outside of *bit_comparator.* This topic will be discussed in the following section.

6.2.3 Passing parameters to modules

Like all *bit_comparator* descriptions we have discussed so far, the *bit_comparator* in Fig. 6.13 can be instantiated using its identifier, name of instance, and list of module interconnections. Therefore either of the *nibble_comparators* in Figs. 5.34 and Fig. 5.35 can be used to instantiate the *bit_comparator* of Fig. 6.13. If this is done, internal values for the *tplh* and *tphl* parameters of *bit_comparator* will remain as declared by the parameter declaration in Fig. 6.13. On the other hand, using the *parameter_value_assignment* construct with module instantiation overrides internal module parameters of the module being instantiated.

Shown in Fig. 6.16 is a *nibble_comparator* that instantiates the *bit_comparator* of Fig. 6.13. Each instantiation begins with a module identifier and is followed by a *parameter_value_assignment* construct. This construct begins with a sharp sign (#) and encloses six constants, which form expressions for values of the six parameters of *bit_comparator.* The values (6, 4, 7, 5, 8, 6) will override the values of the first six parameters declared within the *bit_comparator* module of Fig. 6.13. Assignment of parameter values with module parameters must be done in the order in which the parameters are declared. An instantiation with fewer parameter values than those of the instantiated module will override only those that appear first in the declaration. Other module parameters will use their default values.

Parameter value assignments are less flexible than port connections, which can have the option of using an ordered list (similar to parameters) or a named list. Figure 6.17 shows a *nibble_comparator* instantiating *bit_comparator*s using *named_port_connections.* For

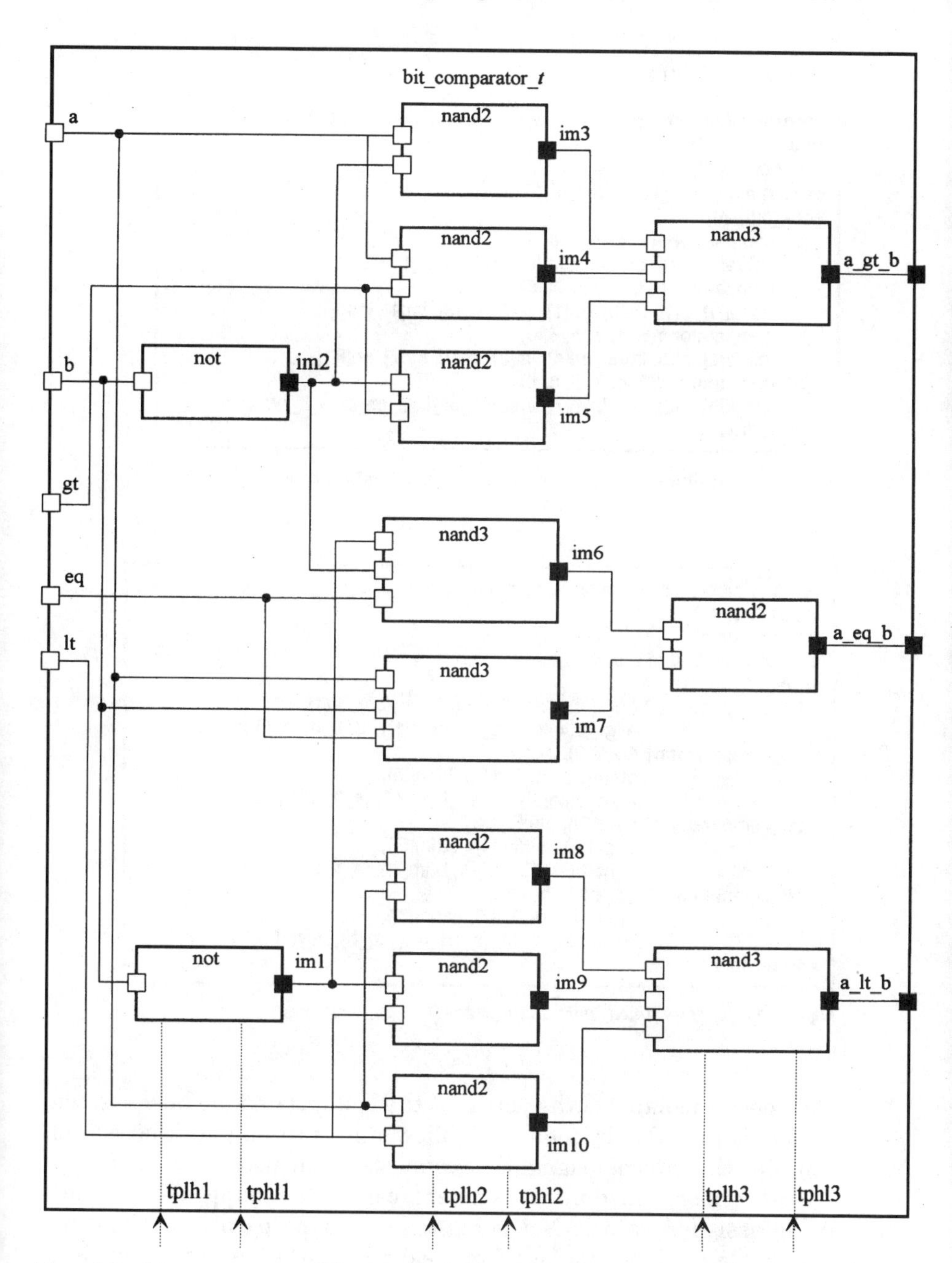

Figure 6.15 *bit_comparator_t* of Fig. 6.13. Dotted lines with arrows indicate parameters.

```
`timescale 10ns/1ns

module nibble_comparator (a, b, gt, eq, lt, a_gt_b, a_eq_b, a_lt_b);
input [3:0] a, b;
input gt, eq, lt;
output a_gt_b, a_eq_b, a_lt_b;
wire [0:8] im;
  bit_comparator #(6, 4, 7, 5, 8, 6)
      c0 (a[0], b[0], gt, eq, lt, im[0], im[1], im[2]);
  bit_comparator #(6, 4, 7, 5, 8, 6)
      c1 (a[1], b[1], im[0], im[1], im[2], im[3], im[4], im[5]);
  bit_comparator #(6, 4, 7, 5, 8, 6)
      c2 (a[2], b[2], im[3], im[4], im[5], im[6], im[7], im[8]);
  bit_comparator #(6, 4, 7, 5, 8, 6)
      c3 (a[3], b[3], im[6], im[7], im[8], a_gt_b, a_eq_b, a_lt_b);
endmodule
```

Figure 6.16 Assigning parameter values at instantiation time.

```
module nibble_comparator (a, b, gt, eq, lt, a_gt_b, a_eq_b, a_lt_b);
input [3:0] a, b;
input gt, eq, lt;
output a_gt_b, a_eq_b, a_lt_b;
wire [0:8] im;
  bit_comparator c0 (.a(a[0]), .b(b[0]), .gt(gt), .eq(eq), .lt(lt),
                     .a_gt_b(im[0]), .a_eq_b(im[1]), .a_lt_b(im[2]) );
  bit_comparator c1 (.a(a[1]), .b(b[1]),
                     .gt(im[0]), .eq(im[1]), .lt(im[2]),
                     .a_gt_b(im[3]), .a_eq_b(im[4]), .a_lt_b(im[5]) ),
  bit_comparator c2 (.a(a[2]), .b(b[2]),
                     .gt(im[3]), .eq(im[4]), .lt(im[5]),
                     .a_gt_b(im[6]), .a_eq_b(im[7]), .a_lt_b(im[8]) );
  bit_comparator c3 (.a(a[3]), .b(b[3]),
                     .gt(im[6]), .eq(im[7]), .lt(im[8]),
                     .a_gt_b(a_gt_b), .a_eq_b(a_eq_b), .a_lt_b(a_lt_b) );
endmodule
```

Figure 6.17 Using *named_port_connections*.

each port, the name of the port that the instantiated module is to connect to is preceded by a dot and followed by a set of parentheses that contains the internal signal that connects to the port.

Furthermore, module parameters cannot be skipped over when being assigned values. Not being able to name parameters and not being able to skip some of them means that when writing a module, one must order parameters intelligently. Declaration of parameters that are more apt to change from design to design must precede the declaration of other parameters.

6.3 Hierarchical Parameter Specification

The previous section showed module parameter specifications and assignments of values to these parameters at the time a module is instantiated. We also showed how parameters can be passed to upper-level modules to be specified along with instantiation of higher-level modules. The example of Fig. 6.16 showed passing parameters through two levels of hierarchy and assignment of values to parameters when they are invoked.

For large designs with many levels of hierarchy and many low level gate and transistor timing and other physical parameters, keeping records of all parameters by passing them through parameters of many levels of hierarchy becomes too cumbersome. Besides, different designs may require alteration of different sets of parameters, and pulling all parameters to upper levels so that a design can assign values to a small subset of them is not a very efficient modeling technique.

Verilog provides a mechanism for specifying parameters of low-level modules from higher-level modules without having to go through explicit parameters of all modules in between. To make this possible, hierarchical naming is used, which applies to parameters as well as **net** and **reg** type design variables.

From higher-level modules, a parameter or a variable nested in several levels of hierarchy may be referenced by its name preceded by the instance names of all modules loading to the parameter, separated by dots. Names formed as such are referred to as hierarchical names. As an example, consider parameter *p* in a module instantiated with *i* instance name. As shown in Fig. 6.18, module *i* is instantiated by another module, which itself has an instance name *ii* in a higher-level module. Parameter *p* must be referred to as *i.p* in instance *ii* and as *ii.i.p* in the higher-level module.

6.3.1 Simple hierarchy

Figure 6.19 shows a description for *nibble_comparator* that instantiates four copies of the *bit_comparator* of Fig. 6.13. Unlike in Fig. 6.16, instantiation of *bit_comparators* does not include parameter value assignments. Instead, the *parameter_override* construct, which begins with the **defparam** keyword, is used to assign values to a subset of the parameters of the *bit_comparator* of Fig. 6.13. Each parameter assignment uses hierarchical names for *tplh1* parameters in four instances of *bit_comparator*. Each name is preceded by the instance name of the module within which the parameter is being assigned a value. Assignments specified in **defparam** override default module parameters. In our example, the 20-ns value for *tplh1* of *bit_comparator* over-

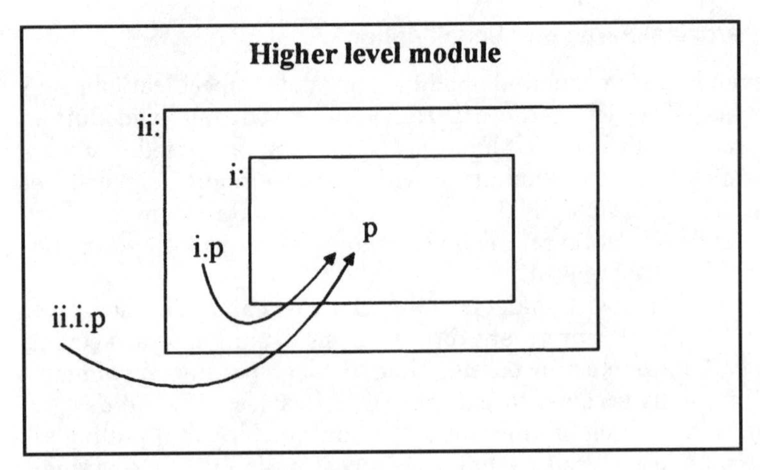

Figure 6.18 Hierarchical names.

rides the 5-ns default value in all four instances. Via the *tplh1* parameter in *bit_comparator,* the 20-ns value is passed to the **not** built-in gate (instances *g0* and *g1*) rise internal parameter.

Instead of fixed values being assigned to hierarchical parameters, a parameter defined in a module may be used for values used in the **defparam** statement. Figure 6.20 shows *tplh2* parameters for the four instances of *bit_comparator* receiving *delval* to override their default values. In the modules instantiating *passed_nibble_comparator* of Fig. 6.20, *delval* may be specified with parameter value assignment as part of module instantiation, or it may be specified using its hierarchical name. Furthermore, values for *tplh2* from modules instantiating *passed_nibble_comparator* may be passed through the *delval* parameter of this module, or they may be specified hierarchically using appropriate paths leading to *tplh2* of *bit_comparator*s.

6.3.2 Parameters of nested modules

Initially, a designer may use gates based on their functionality rather than on timings or specific technology. These designs may be used in upper-level designs, which may cause them to be buried under several levels of design hierarchy. Use of hierarchical names can be extended beyond referencing parameters of immediate modules, as was done in the examples of Sec. 6.3.1; these names can be used to set parameter values for an entire design or a test bench.

We will demonstrate the use of hierarchical names in writing modules that are primarily used for specifying lower-level parameters. For this purpose, we will use the *test_nibble_comparator* test bench in Fig. 6.3 instantiating the *nibble_comparator* of Fig. 6.17 (using *u1* instance name), which instantiates four copies of *bit_comparator* (Fig. 6.13) using *c0, c1, c2,* and *c3* instance names. The intent is to change some of the parameters of *bit_comparator*s without having to make any changes that would require recompilation of any of the upper-level modules. For this purpose, a top-level module must be developed to configure the *test_nibble_comparator* for specific parameter values of nested *bit_comparator* instances.

Figure 6.21 graphically shows what is to be done. A top-level Verilog module, *configured_test_nibble_comparator,* wraps around *test_nibble_comparator* to assign values to *bit_comparator* parameters and to make ports of *nibble_comparator* visible in this top-level module. Figure 6.22 shows the Verilog code for the configured test bench. The test bench is instantiated using the *ut* label. Because the *test_nibble_comparator* module is a top-level module, an empty list of port connections is used for it. In its declaration part, the description of Fig. 6.22 declares *a, b, a_gt_b, a_lt_b,* and *a_eq_b* wires and assigns signals of the same names with the *ut* component to these local wires of the *configured_test_nibble_comparator* module. For this purpose, hierarchical names are used for signals that are visible within the module instantiated with the *ut* name.

```
module nibble_comparator (a, b, gt, eq, lt, a_gt_b, a_eq_b, a_lt_b);
input [3:0] a, b;
input gt, eq, lt;
output a_gt_b, a_eq_b, a_lt_b;
wire [0:8] im;
defparam        c0.tplh1 = 20,  c1.tplh1 = 20,   c2.tplh1 = 20,   c3.tplh1 = 20;
  bit_comparator c0 (a[0], b[0], gt, eq, lt, im[0], im[1], im[2]);
  bit_comparator c1 (a[1], b[1], im[0], im[1], im[2], im[3], im[4], im[5]);
  bit_comparator c2 (a[2], b[2], im[3], im[4], im[5], im[6], im[7], im[8]);
  bit_comparator c3 (a[3], b[3], im[6], im[7], im[8], a_gt_b, a_eq_b, a_lt_b);
endmodule
```

Figure 6.19 Using **defparam** for hierarchical parameter specification.

```
module passed_nibble_comparator (a, b, gt, eq, lt, a_gt_b, a_eq_b, a_lt_b);
input [3:0] a, b;
input gt, eq, lt;
output a_gt_b, a_eq_b, a_lt_b;
wire [0:8] im;
parameter delval = 1;
        defparam
        c0.tplh2 = delval, c1.tplh2 = delval, c2.tplh2 = delval, c3.tplh2 = delval;

        bit_comparator c0 (a[0], b[0], gt, eq, lt, im[0], im[1], im[2]);
        bit_comparator c1 (a[1], b[1], im[0], im[1], im[2], im[3], im[4], im[5]);
        bit_comparator c2 (a[2], b[2], im[3], im[4], im[5], im[6], im[7], im[8]);
        bit_comparator c3 (a[3], b[3], im[6], im[7], im[8], a_gt_b, a_eq_b, a_lt_b);
endmodule
```

Figure 6.20 Parameter used in **defparam.**

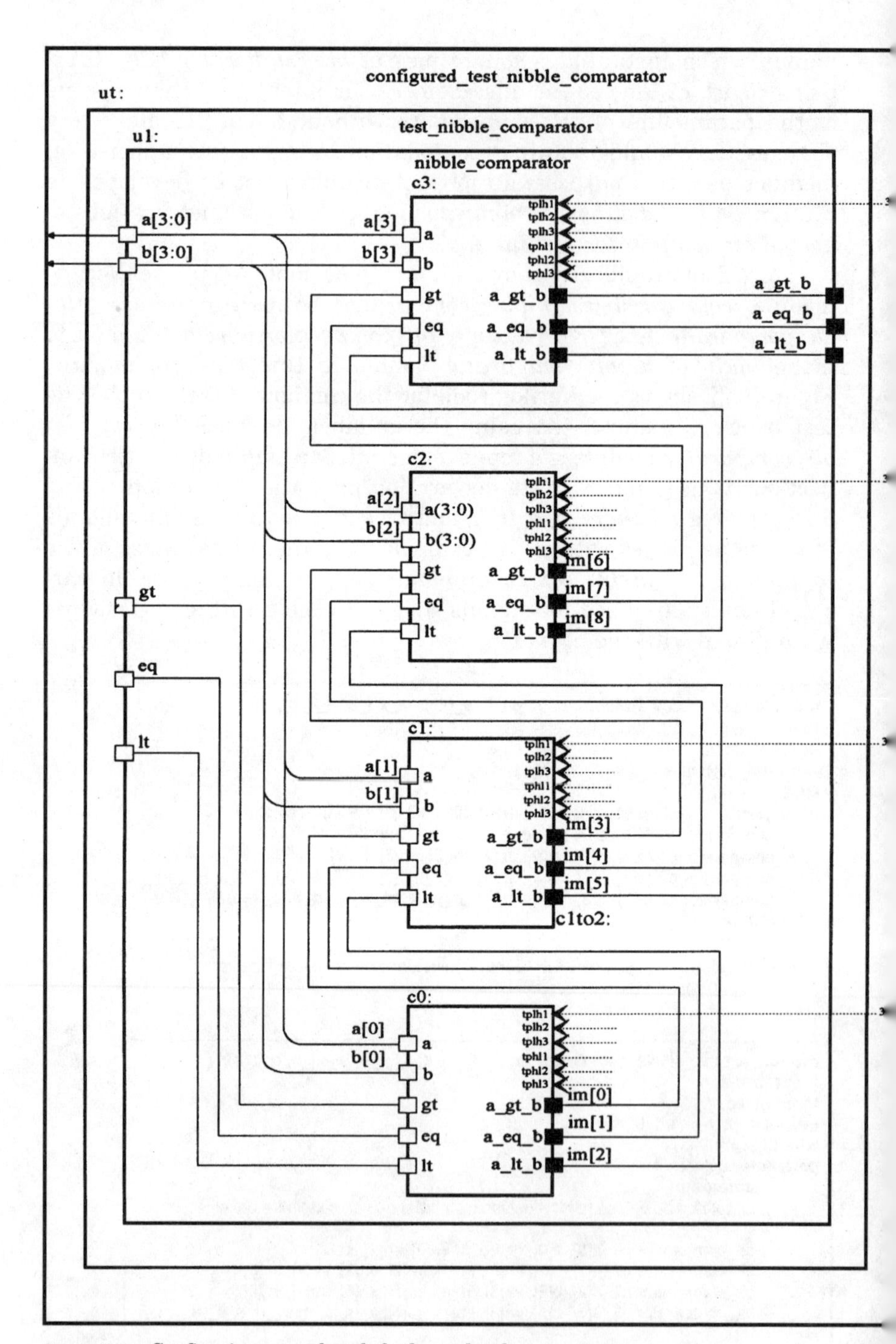

Figure 6.21 Configuring a test bench for lower-level component parameters.

```
module configured_test_nibble_comparator;
wire [3:0] a = ut.a, b = ut.b;
wire a_gt_b = ut.a_gt_b, a_eq_b = ut.a_eq_b, a_lt_b = ut.a_lt_b;
        test_nibble_comparator ut ();
        defparam
                ut.u1.c0.tplh1 = 30,
                ut.u1.c1.tplh1 = 30,
                ut.u1.c2.tplh1 = 30,
                ut.u1.c3.tplh1 = 30;
endmodule
```

Figure 6.22 Configuring a test bench using hierarchical names.

Following the *ut* instantiation in Fig. 6.22, a **defparam** statement uses the hierarchical names of the parameters of *bit_comparator* to assign new values to them. The *ut.u1.c0.tplh1*=30 parameter assignment passes through the *ut* test bench (Fig. 6.3), the *u1* 4-bit comparator (Fig. 6.17), and the *c0* bit comparator (Fig. 6.13) to reach the *tplh1* parameter. This parameter assignment overrides the *tplh1* default value of 5 ns and is passed to built-in **not** gates for their rise delay parameters.

6.3.3 An *n*-bit register example

Thus far, we have presented parameter declarations and the use of hierarchical names for assigning values to these parameters. The comparator example illustrated many of the language features for upper-level parameter assignments and use of hierarchy. A more hierarchical example will be discussed here.

Built-in **not** and **nand** gates with single delay parameters will be used here in developing a configurable 4-bit register. We will show how gate parameters are set and how a module configured for specific assignment to parameters of its underlying gates may itself be used in upper-level descriptions.

Figure 6.23 shows the gate-level description of *sr_latch*. This latch uses four instances of the built-in **nand** gate and uses a *delval* parameter that has a default of 0 for rise and fall delay values of the **nand** gates.

Obviously, cross-coupling NAND gates with zero delay values will not provide the function of an *sr_latch*. Therefore, *delval* must be overwritten in any design that uses this module.

A D-type latch consists of an inverter and an SR latch. Figure 6.24 shows the Verilog code for such a component. The **not** gate uses the *delval* parameter for its rise and fall delays. Note that this parameter is declared in *d_latch* and is different from the parameter with the same name in the *sr_latch* of Fig. 6.23. The primary input *d* of *d_latch*

```
`timescale 1ns/100ps

module sr_latch (s, r, c, q);
//
input s, r, c;
output q;
parameter delval = 0;
wire im1, im2, im3;
  nand #(delval)
    g1 (im1, s, c),
    g2 (im2, r, c),
    g3 (q, im1, im3),
    g4 (im3, q, im2);
endmodule
```

Figure 6.23 *sr_latch* using dummy parameter values.

is connected to the *s* input of the *sr_latch,* and its complement *dbar* is connected to the *r* input.

Our next level in this design process is the formation of a 4-bit register by wiring four D-type latches. This is done in the *nibble_latch* module in Fig. 6.25. Note that this figure does not specify parameter values for its lower-level gates; it is if instantiated, 0 values will be assumed for all such parameters.

In order to use the *nibble_latch* of Fig. 6.25, it has to be configured for the actual delay values of the **not** and **nand** gates that it uses in its underlying structures. Figure 6.26 graphically represents what has to be done to set the delay parameters of these gates. As shown, a new module, *configured_nibble_latch,* instantiates *nibble_latch,* passes its inputs to it, and uses outputs of the instantiated module as its own. In addition, 6 ns delay is passed from *configured_nibble_latch* through *nibble_latch* to the **nand** gate rise and fall parameters, and 3 ns is passed to **not** delays. The 3-ns value passes through *nibble_latch,* then through *d_latch,* and then reaches the **not** gates. The 6-ns value also passes through *nibble_latch* and *d_latch* instantiations, but it also has to pass through *sr_latch* before it reaches the **nand** gates.

Figure 6.27 shows the Verilog code that corresponds to the graphical representation in Fig. 6.26. So that values will reach appropriate parameters, hierarchical names of parameters as seen from *configured_nibble_latch* are used. For a value of 6 to reach *delval* of the **nand** gate in bit 2 of *sr_latch,* the *cn.c2.c1.delval* hierarchical name is used. The instance name *cn* is that of *nibble_latch* in *configured_nibble_latch, c2* is the instance name of bit 2 of *d_latch* in *nibble_latch,* and *c1* is the instance name of *sr_latch* in *d_latch.* Because **not** gates are immediately within *d_latch* instances, a shorter path name is used for their parameters.

The last step in this demonstration is showing that our *configured_nibble_latch* actually works and also showing how it should be used in higher-level designs. For this purpose, the test bench of Fig. 6.28 is developed. The *u1* instance in this figure refers to the configured latch of Fig. 6.26 with gate delay parameter values that override the default 0 values of Figs. 6.23 and 6.24.

An **initial** block in this test bench initializes the clock input and assigns several test values to the *di* input of the 4-bit latch. An **always** block toggles the circuit clock every 100 ns (10·10 ns).

6.3.4 Iterative parity checker

We will close this section with another example for hierarchical parameter specification. In this example, an array of instances will be used.

The design consists of a chain of **xor** gates for odd and even parity clocking; as shown in Fig. 6.29, the final output of the circuit provides the *odd* parity and is inverted for generating the *even* parity output.

The Verilog code for this circuit is shown in Fig. 6.30. All **xor** gates are instantiated using an array of instances *x[6:0]*. Outputs and inputs for these seven gates are formed by concatenation of the signals shown in Fig. 6.29 from right to left. The expression *{odd, im[5:0]}* wires *odd*

```
`timescale 1ns/100ps

module d_latch (d, c, q);
input d, c;
output q;
parameter delval = 0;
wire dbar;
  sr_latch c1 (d, dbar, c, q);
  not #delval c2 (dbar, d);
endmodule
```

Figure 6.24 *d_latch* using *sr_latch* and a **not** gate with dummy parameter values.

```
module nibble_latch (di, c, qo);
input [3:0] di;
input c;
output [3:0] qo;
  d_latch c3 (di[3], c, qo[3]);
  d_latch c2 (di[2], c, qo[2]);
  d_latch c1 (di[1], c, qo[1]);
  d_latch c0 (di[0], c, qo[0]);
endmodule
```

Figure 6.25 *nibble_latch* Verilog code using *d_latch*.

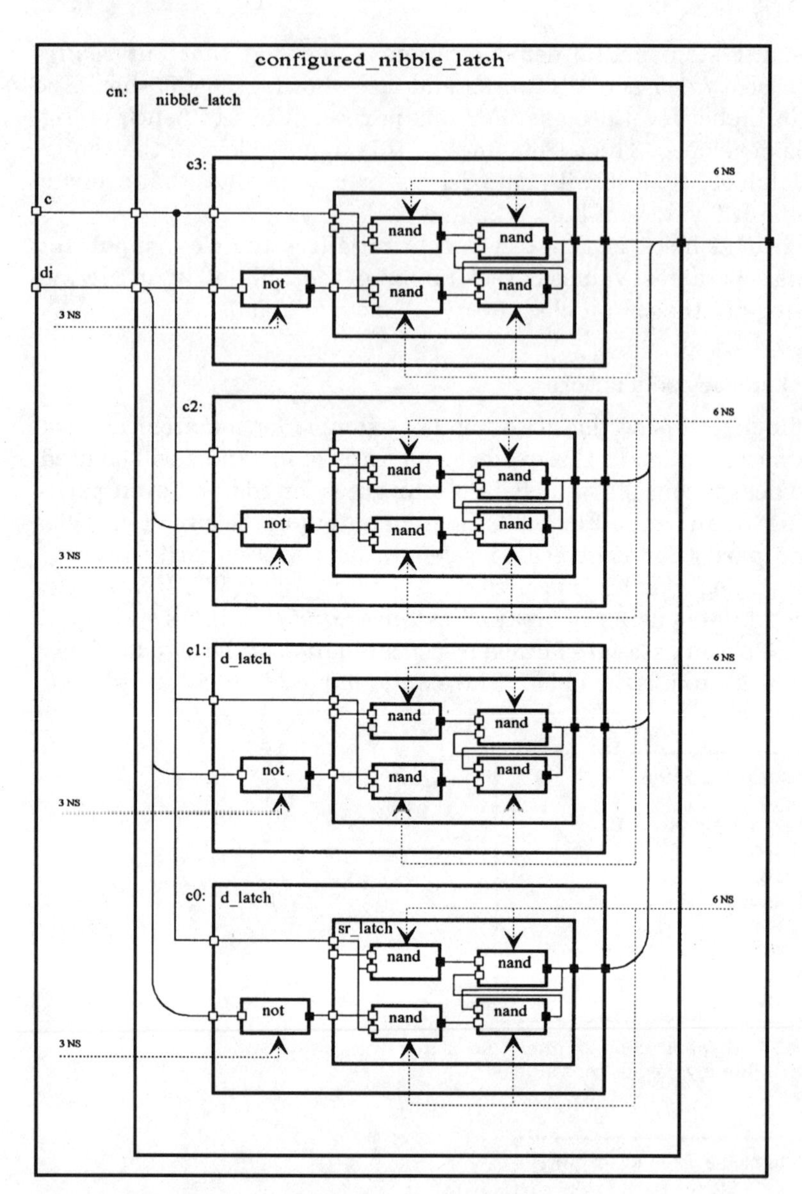

Figure 6.26 Configuring the *nibble_latch* module.

to the *x[6]* instance of **xor** (the rightmost gate in Fig. 6.29) and wires *im[0]* to the output of the leftmost **xor** gate. An instance of a **not** gate in Fig. 6.30 complements the *odd* output to generate the *even* output. The test bench shown in Fig. 6.31 instantiates the parity circuit with *u1* instance name and assigns 5 ns and 2 ns to the *delval1* and *delval2* parameters of this circuit.

6.4 Path Delay Specification

The discussion in the previous sections of this chapter concentrated on parameter declaration and specification. We used only gate delay values to demonstrate the use of parameters and ways in which they can be assigned values. The discussion, however, applies to parameters other than delay and timing.

Applying these techniques to timing, we have demonstrated how delay values that are distributed among various gates and modules of a circuit may be defined and assigned values. Delay values defined as such are referred to as distributed delays in Verilog.

```
module configured_nibble_latch (di, c, qo);
input [3:0] di;
input c;
output [3:0] qo;
        nibble_latch cn (di, c, qo);
        defparam
                cn.c3.c1.delval = 6,
                cn.c3.delval = 3,
                cn.c2.c1.delval = 6,
                cn.c2.delval = 3,
                cn.c1.c1.delval = 6,
                cn.c1.delval = 3,
                cn.c0.c1.delval = 6,
                cn.c0.delval = 3;
endmodule
```

Figure 6.27 Configuring *nibble_latch* for specific underlying gate delay values.

```
`timescale 10ns/1ns

module test_nibble_latch;
reg [3:0] di;
reg clk;
wire [3:0] qo;
configured_nibble_latch u1 (di, clk, qo);
  initial
  begin
    clk = 1'b0;
    #15 di = 8'd0; #20 di = 8'd66; #25 di = 8'd89; #20 di = 8'd76;
    #10 $stop;
  end
  always #10 clk = ~ clk;
endmodule
```

Figure 6.28 Hierarchically instantiating *configured_nibble_latch*.

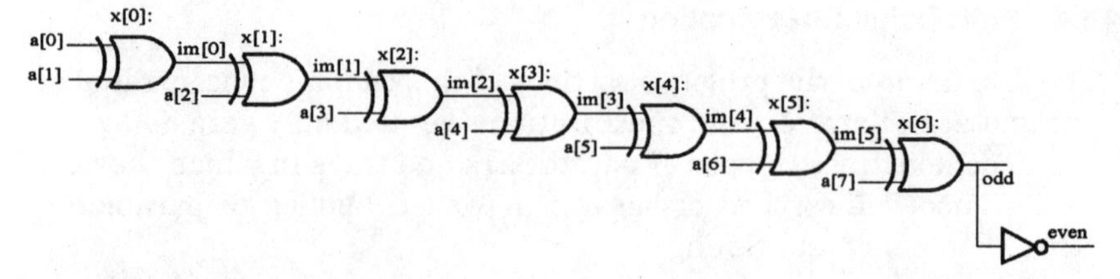

Figure 6.29 Parity generator/checker circuit.

```
`timescale 1ns/100ps

module parity (a, odd, even);
input [7:0] a;
output odd, even;
parameter delval1 = 0.0, delval2 = 0.0;
wire [5:0] im;
        xor #(delval1) x[6:0] {odd, im[5:0]}, {im[5:0], a[0]}, a[7:1];
        not #(delval2) (even, odd);
endmodule
```

Figure 6.30 Iterative parity circuit.

```
`timescale 10ns/1ns

module test_parity;
reg [7:0] xi;
wire odd, even;
parity u1 (xi, odd, even);
defparam u1.delval1 = 5, u1.delval2 = 2;
  initial
  begin
    #15 xi = 8'h0; #20 xi = 8'hAB;
    #25 xi = 8'h47; #20 xi = 8'hB6;
    #10 $stop;
  end
endmodule
```

Figure 6.31 Testing the iterative parity circuit.

The Verilog HDL also allows definition and specification of path delays. Language constructs and tasks for this purpose allow timing check and specification of delays based not on individual delays of gates and submodules of a module, but on pin-to-pin timing characteristics of these higher-level modules.

6.4.1 Full path specification

For demonstrating path delay specifications, we will use the comparator and latch examples of the previous sections. The functional *bit_comparator* shown in Fig. 6.32 uses 0 delay values for all input-to-output transitions and does not use any timing parameters to allow these values to change. Higher-level comparator structures will be based on this module.

The basic structure for specifying delays of input(s) to output(s) of a module (pin-to-pin delays) is the **specify** block. Figure 6.33 shows a *nibble_comparator* module based on the *bit_comparator* of Fig. 6.32. In addition to instantiating four 1-bit comparators, the module in this figure uses a **specify** block to set pin-to-pin delays of the *a* and *b* inputs to the *a_gt_b, a_eq_b,* and *a_lt_b* outputs of the circuit to 9.1 ns, 11.1 ns, and 9.1 ns, respectively. As a result of this specification, transitions on *a_gt_b* and *a_lt_b* outputs caused by changes on any of the bits of the *a* or *b* vector will be delayed by 9.1 ns, and transitions on the *a_eq_b* output will be delayed by 11.1 ns. This specification does not specify any delay values for transitions on circuit outputs caused by events on the *gt, eq,* or *lt* circuit inputs. As a result, zero delay values will be assumed.

The syntax details of the **specify** block of Fig. 6.33 are shown in Fig. 6.34. This block comprises three *simple_path_declarations,* one for each of the outputs of the circuit. Each such construct specifies pin-to-pin delays for sources on the left of *> to destinations on the right.

```
`timescale 1ns/100ps

module bit_comparator (a, b, gt, eq, lt, a_gt_b, a_eq_b, a_lt_b);

  function fgl;
  input w, x, gl;
    begin
      fgl = w & gl | ~x & gl | w & ~x;
    end
  endfunction

  function feq;
  input w, x, eq;
    begin
      feq = w & x & eq | ~w & ~x & eq ;
    end
  endfunction

  input a, b, gt, eq, lt;
  output a_gt_b, a_eq_b, a_lt_b;
  assign a_gt_b = fgl (a, b, gt);
  assign a_eq_b = feq (a, b, eq);
  assign a_lt_b = fgl (b, a, lt);
endmodule
```

Figure 6.32 *bit_comparator* with no delays and no parameters.

```
`timescale 1ns/100ps

module nibble_comparator (a, b, gt, eq, lt, a_gt_b, a_eq_b, a_lt_b);
input [3:0] a, b;
input gt, eq, lt;
output a_gt_b, a_eq_b, a_lt_b;
wire [0:8] im;
  specify
      (a, b *> a_eq_b) = 11.1;
      (a, b *> a_lt_b) = 9.1;
      (a, b *> a_gt_b) = 9.1;
  endspecify
  bit_comparator c0 (a[0], b[0], gt, eq, lt, im[0], im[1], im[2]);
  bit_comparator c1 (a[1], b[1], im[0], im[1], im[2], im[3], im[4], im[5]);
  bit_comparator c2 (a[2], b[2], im[3], im[4], im[5], im[6], im[7], im[8]);
  bit_comparator c3 (a[3], b[3], im[6], im[7], im[8], a_gt_b, a_eq_b, a_lt_b);
endmodule
```

Figure 6.33 Using the basic form of the **specify** block.

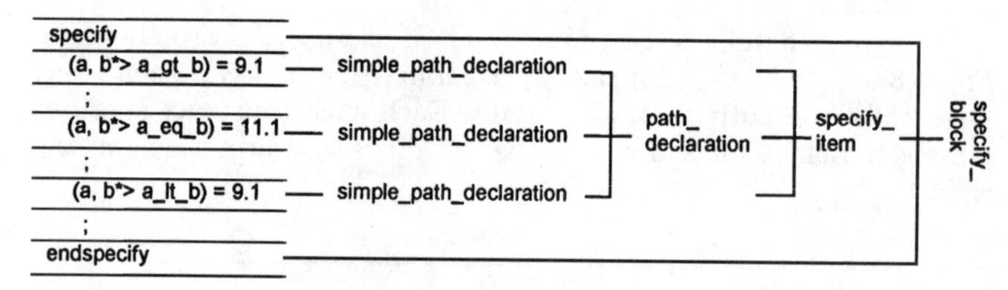

Figure 6.34 Top-level syntax details of a **specify** block.

Syntax details of a *simple_path_declaration* are shown in Fig. 6.35. As shown, a list of path inputs on the left of *> and a list of path outputs on the right of *> are allowed. This construct, which is called a *full_path_description,* specifies the delay value for all bits on the left-hand side of *> to all bits on the right-hand side of this symbol. The number of bits on the right and left do not have to be the same, nor do the number of items in the right-hand-side list and the number in the left-hand-side list have to match.

Figure 6.35 also shows a *path_delay_value* construct enclosed in the *simple_path_declaration.* Instead of a constant, expressions for rise, fall, and *z* transition delay values may also be used. The constant value of 9.1 in Fig. 6.35 applies to all these transitions. Furthermore, a *path_delay_value* may use min:typ:max format for the delay values.

Full path description uses the *> symbol and is a general form of path specification. It can be used for specifying transitions to and from any number of scalars and vectors. Another form of path description will be discussed next.

6.4.2 Parallel path specification

A more restrictive form of path specification uses the parallel connection symbol, =>. This form allows only one source on the left and one destination with an equal number of bits on the right-hand side of the => symbol.

Figure 6.36 shows a **specify** block using parallel and full path specifications. The parallel paths specify delays from the *gt, eq,* and *lt* inputs to their corresponding *a_gt_b, a_eq_b,* and *a_lt_b* outputs. Each of these constructs specifies delays from one input to one output of the circuit. If vectors are used on the two sides of =>, the specified delay applies to like bits of vectors on the right and left. The **specify** block in Fig. 6.36 also specifies full path delay for all bits of the *a* and *b* inputs to the *a_gt_b, a_eq_b,* and *a_lt_b* outputs. Figure 6.37 shows syntax details of a parallel path specification. Note that single items are used on the right and left of => instead of the lists used with *> in Fig. 6.35.

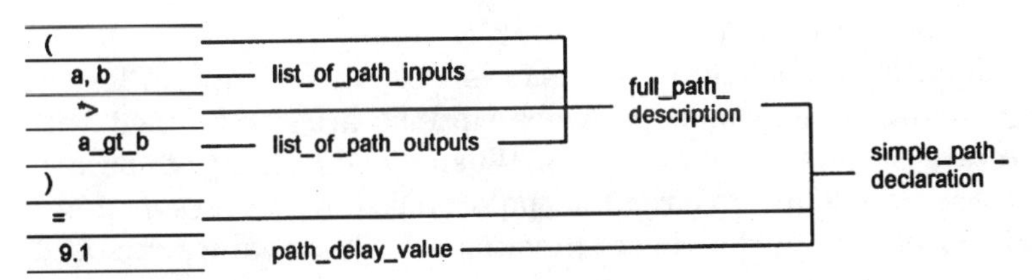

Figure 6.35 Syntax details of *simple_path_declaration.*

```
`timescale 1ns/1ps

module nibble_comparator (a, b, gt, eq, lt, a_gt_b, a_eq_b, a_lt_b);
input [3:0] a, b;
input gt, eq, lt;
output a_gt_b, a_eq_b, a_lt_b;
wire [0:8] im;
     specify
          specparam data2out = 6.666;
          specparam control2out = 8.888;
          (gt => a_gt_b) = control2out;
          (eq => a_eq_b) = control2out;
          (lt => a_lt_b) = control2out;
          (a, b *> a_gt_b, a_eq_b, a_lt_b) = data2out;
     endspecify
  bit_comparator c0 (a[0], b[0], gt, eq, lt, im[0], im[1], im[2]);
  bit_comparator c1 (a[1], b[1], im[0], im[1], im[2], im[3], im[4], im[5]);
  bit_comparator c2 (a[2], b[2], im[3], im[4], im[5], im[6], im[7], im[8]);
  bit_comparator c3 (a[3], b[3], im[6], im[7], im[8], a_gt_b, a_eq_b, a_lt_b);
endmodule
```

Figure 6.36 Parallel and full path delay specifications in a *nibble_comparator.*

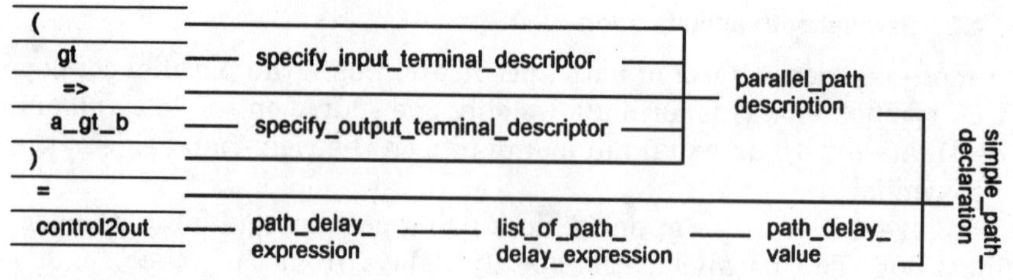

Figure 6.37 Parallel path description.

Instead of using numerical values, the **specify** block of our example uses specify parameters. These parameters are declared using the **specparam** statement and can be used only in the block in which they are declared. The simulation run shown in Fig. 6.38 shows 6666-ps output delays when *a* or *b* inputs change, and 8888-ps delays when *gt, lt,* or *eq* inputs change.

6.4.3 State dependency

Verilog allows more accurate modeling of timing using state-dependent path declaration. Figure 6.39 shows another description for the *nibble_comparator* in which path descriptions are conditioned with relative values of *a* and *b* data inputs. When *a* and *b* change to become equal, transition delays to circuit outputs caused by changes on *a* or *b* are 6666 ps; otherwise, for these transitions, 7777-ps values are used.

6.4.4 Distributed and path delays

The examples we have used thus far have all been based on the *bit_comparator* in Fig. 6.32, which uses zero delay values for all its outputs. Had this not been the case, and if the *nibble_comparator* path delays were specified in addition to the *bit_comparator* nonzero distributed delays, the maximum of the two effective values would be used for *nibble_comparator* outputs.

As an example, consider the Verilog description of the *bit_comparator* of Fig. 6.1. For all distributed delay values for individual outputs of this circuit, 12 ns is used. Simulation of the *nibble_comparator* of Fig. 6.36 using four instances of the module of Fig. 6.1, and using the same test data used in the simulation report of Fig. 6.38 would result in the listing shown in Fig. 6.40.

The 12-ns distributed delay and its multiples, for multiple comparator bit propagations (i.e., 24 ns, 36 ns, 48 ns), are larger than every path delay value specified in the *nibble_comparator* of Fig. 6.36. Therefore, all outputs are delayed with delay values from the *bit_comparator* of Fig. 6.1, and outputs change only at times that are multiples of 12 ns.

TIME	SIGNALS							
(ps)	a	b	gt	lt	eq	a_gt_b	a_eq_b	a_lt_b
0	0000	0000	0	0	1	x	x	x
6666	0000	0000	0	0	1	0	1	0
500000	1111	1110	0	0	1	0	1	0
506666	1111	1110	0	0	1	1	0	0
1000000	1111	1110	0	0	1	1	0	0
1500000	1110	1111	0	0	1	1	0	0
1506666	1110	1111	0	0	1	0	0	1
2000000	1110	1111	0	0	1	0	0	1
2500000	1110	1100	0	0	1	0	0	1
2506666	1110	1100	0	0	1	1	0	0
3000000	1110	1100	0	0	1	1	0	0
3500000	1010	1100	0	0	1	1	0	0
3506666	1010	1100	0	0	1	0	0	1
4000000	0000	1111	0	0	1	0	0	1
4500000	1111	1111	0	0	1	0	0	1
4506666	1111	1111	0	0	1	0	1	0
4700000	1111	1111	1	0	0	0	1	0
4708888	1111	1111	1	0	0	1	0	0
4800000	1111	1111	0	1	1	1	0	0
4808888	1111	1111	0	1	1	0	1	1
5000000	0000	1111	0	1	1	0	1	1
5006666	0000	1111	0	1	1	0	0	1
5500000	0000	0000	0	1	1	0	0	1
5506666	0000	0000	0	1	1	0	1	1
6000000	1111	0000	0	1	1	0	1	1
6006666	1111	0000	0	1	1	1	0	0
6500000	1111	0000	0	1	1	1	0	0

Figure 6.38 Simulation run for *nibble_comparator* with **specify** block.

```
`timescale 1ns/1ps

module nibble_comparator (a, b, gt, eq, lt, a_gt_b, a_eq_b, a_lt_b);
input [3:0] a, b;
input gt, eq, lt;
output a_gt_b, a_eq_b, a_lt_b;
wire [0:8] im;
        specify
                specparam data2out_eq = 6.666;
                specparam data2out_ne = 7.777;
                specparam control2out = 8.888;
                (gt => a_gt_b) = control2out;
                (eq => a_eq_b) = control2out;
                (lt => a_lt_b) = control2out;
                if (a == b) (a, b *> a_gt_b, a_eq_b, a_lt_b) = data2out_eq;
                if (a > b) (a, b *> a_gt_b, a_eq_b, a_lt_b) = data2out_ne;
                if (a < b) (a, b *> a_gt_b, a_eq_b, a_lt_b) = data2out_ne;
        endspecify
  bit_comparator c0 (a[0], b[0], gt, eq, lt, im[0], im[1], im[2]);
  bit_comparator c1 (a[1], b[1], im[0], im[1], im[2], im[3], im[4], im[5]);
  bit_comparator c2 (a[2], b[2], im[3], im[4], im[5], im[6], im[7], im[8]);
  bit_comparator c3 (a[3], b[3], im[6], im[7], im[8], a_gt_b, a_eq_b, a_lt_b);
endmodule
```

Figure 6.39 Path descriptions that depend on a and b values.

TIME (ps)	SIGNALS							
	a	b	gt	lt	eq	a_gt_b	a_eq_b	a_lt_b
0	0000	0000	0	0	1	x	x	x
48000	0000	0000	0	0	1	0	1	0
500000	1111	1110	0	0	1	0	1	0
548000	1111	1110	0	0	1	1	0	0
1000000	1111	1110	0	0	1	1	0	0
1500000	1110	1111	0	0	1	1	0	0
1548000	1110	1111	0	0	1	0	0	1
2000000	1110	1111	0	0	1	0	0	1
2500000	1110	1100	0	0	1	0	0	1
2536000	1110	1100	0	0	1	1	0	0
3000000	1110	1100	0	0	1	1	0	0
3500000	1010	1100	0	0	1	1	0	0
3524000	1010	1100	0	0	1	0	0	1
4000000	0000	1111	0	0	1	0	0	1
4500000	1111	1111	0	0	1	0	0	1
4548000	1111	1111	0	0	1	0	1	0
4700000	1111	1111	1	0	0	0	1	0
4748000	1111	1111	1	0	0	1	0	0
4800000	1111	1111	0	1	1	1	0	0
4848000	1111	1111	0	1	1	0	1	1
5000000	0000	1111	0	1	1	0	1	1
5012000	0000	1111	0	1	1	0	0	1
5500000	0000	0000	0	1	1	0	0	1
5548000	0000	0000	0	1	1	0	1	1
6000000	1111	0000	0	1	1	0	1	1
6012000	1111	0000	0	1	1	1	0	0
6500000	1111	0000	0	1	1	1	0	0

Figure 6.40 Maximum of distribution and path delays is used.

6.4.5 Vector path delay specification

The discussion above presented path delay specifications for scalar inputs and outputs and for vector inputs and scalar outputs. The discussion here presents an example with a vector input and a vector output.

Based on the *sr_latch* of Fig. 6.23 and the *d_latch* of Fig. 6.24, the *byte_latch* description shown in Fig. 6.41 is developed. Eight concurrent module instances wire eight *d_latch*es together as an 8-bit latch structure. For this structure, the **specify** block shown in Fig. 6.41 specifies rise and fall delay values (*d2q_r* and *d2q_f*) for changes occurring on bits of *qo.*

When an event occurs on bit *1* of *di,* bit *j* of *qo* is affected after 4444 ps rise and 3333 ps fall delay. Because this structure is a transparent latch, while the clock input is active, data input changes could change the outputs of the latch. The **specify** block in the *byte_latch* module also defines delays on outputs when the *c* input changes. For the single bit of *c* to 8 bits of *qo,* full path delay construct is used.

6.5 Summary

This chapter provides tools for better hardware descriptions and design organization. The chapter began with the definition of subpro-

```
`timescale 1ns/1ps

module byte_latch (di, c, qo);
input [7:0] di;
input c;
output [7:0] qo;
wire dbar;
        specify
                specparam d2q_r = 4.444;
                specparam d2q_f = 3.333;
                specparam c2q_r = 2.222;
                specparam c2q_f = 1.111;
                (di => qo) = (d2q_r, d2q_f);
                (c *> qo) = (c2q_r, c2q_f);
        endspecify
  d_latch c7 (di[7], c, qo[7]);
  d_latch c6 (di[6], c, qo[6]);
  d_latch c5 (di[5], c, qo[5]);
  d_latch c4 (di[4], c, qo[4]);
  d_latch c3 (di[3], c, qo[3]);
  d_latch c2 (di[2], c, qo[2]);
  d_latch c1 (di[1], c, qo[1]);
  d_latch c0 (di[0], c, qo[0]);
endmodule
```

Figure 6.41 *byte_latch* with path delay specifications.

grams, and it emphasized the use of functions and tasks for simplifying descriptions. Two main issues were discussed: (1) using functions to describe boolean expressions and (2) using tasks to write better test bench models. User-defined tasks and system tasks were used in several test benches in this chapter. Next, design parametrization and specifying parameters of designs were discussed in great detail. Several alternatives for this purpose were presented, including one that allows parameters of a built-in gate or low-level module to be specified from higher-level modules. Hierarchical naming was discussed in this regard. Discussions of parameters also included methods for overriding existing module parameters. Use of language constructs for specifying path delays in clocked and combinational circuits were discussed and compared with distributed module delay specifications. For all demonstrations, complete Verilog examples were presented. In many cases, test benches were also shown for better understanding of models and expected waveforms. Throughout the chapter, when new syntax constructs were presented, their details according to the formal language grammar were also presented.

Further Reading

Hill, F. J., and G. R. Peterson, *Digital Systems: Hardware Organization and Design,* 3d ed., John Wiley, New York, 1987.

IEEE Standard Hardware Description Language Based on the Verilog Hardware Description Language, IEEE Std. 1364-1995, Institute of Electrical and Electronic Engineers, New York, 1996.

Palnitkar, S., *Verilog HDL: A Guide to Digital Design and Synthesis,* Prentice-Hall, Upper Saddle River, N.J., 1996.

Smith, Douglas J., *A Practical Guide for Designing, Synthesizing and Simulating ASICs and FPGAs Using VHDL or Verilog,* Doone Publications, June 1996.

Thomas, D. E., and P. R. Moorby, *The Verilog Hardware Description Language,* 3d ed., Kluwer Academic Publishers, Norwell, Mass., 1996.

Wakerly, J. F., *Digital Design Principles and Practices,* 2d ed., Prentice-Hall, Englewood Cliffs, N.J., 1993.

Problems

6.1. Write a function for the carry output of a full-adder.

6.2 Write a function for the sum output of a full-adder.

6.3 Using boolean expressions for the outputs, write a function, *inc_bits,* that returns the 4-bit increment of its 4-bit input vector. Do not use the Verilog add operation.

6.4 Using the carry and sum functions of Probs. 6.1 and 6.2, write a functional description of a full-adder. Use a module with *a, b,* and *ci* inputs and *s* and *co* outputs. In this module, include the necessary functions. Use 21-ns and 18-ns delays for the sum and carry outputs, respectively. The solution to this problem depends on the code developed in Probs. 6.1 and 6.2.

6.5 Modify the *generate_data* task so that it generates 4-bit numbers from 0 to 15 on one target and numbers in the opposite direction on the other target.

6.6 Modify the task of Prob. 6.5 such that the starting numbers for the two targets will be included in its argument list.

6.7 Develop a task that places random bits on its target. Name this task *generate_bits.*

6.8 Use the *generate_bits* tasks of Prob. 6.7 in a test bench for the full-adder in Prob. 6.4.

6.9 Write a description for a fall-adder using two-input NAND and XOR gates, using rise and fall delay parameters. Declare parameters with a default value of 0 for all four delay parameters.

6.10 Configure the fall-adder in Prob. 6.9 in the top-level *configured_full_adder* module to set delay values to 4, 6, 5, and 7 ns for rise and fall of **nand** and **xor** gates.

6.11 Develop a test bench for the full-adder of Prob. 6.10. Take advantage of the task in Prob. 6.7 to apply random data to all three inputs.

6.12 Use an array of instances to describe an 8-bit adder using the full-adder in Prob. 6.9. Configure this adder for the delay values given in Prob. 6.10.

6.13 Use the full-adder in Prob. 6.10 for design of a 4-bit adder. Use **defparam** to override values given in Prob. 6.10 with 3.1 ns, 5.1 ns, 4.1 ns, and 6.1 ns, respectively. In a test bench, instantiate this module and use the *generate_data* task of Prob. 6.5 to test it.

6.14 Use the configured adder of Prob. 6.11 in a test bench to test it for its worst-case delay. You must apply test data such that the worst case occurs.

6.15 Design an 8-bit odd-parity checker using the XOR built-in gate. Use 0 for all its delay parameters. Write a top-level module for assigning 5- and 6-ns delay values to rise and fall delay parameters of all **xor** gates. Write a test bench to test this circuit for all possible input combinations.

6.16 Use the 4-bit adder of Prob. 6.13 in a top-level module and assign path delay values from its inputs to its outputs using the following values:
From bits of the operands to sum output bits: 8 ns
From carry-in to sum output bits: 12 ns
From carry-in to carry-out: 13 ns
From bits of either operand to carry-out: 11 ns

Write a test bench to test this circuit for all the above cases.

6.17 Use the full-adder of Prob. 6.4 in an upper-level full-adder module that will assign path delays to it as follows:
If a5b, sum output delay5(5, 7 ns) and carry output delay5(4, 6 ns)
If a > b, all delay values56 ns
If a < b, all delay values58 ns

6.18 Develop a 4-bit adder using *full_adder* of Prob. 6.17 and instantiate it in a test bench to test all possible input combinations. Take advantage of *generate_data* tasks developed in the text or in Prob. 6.5.

Chapter

7

Utilities for High-Level Descriptions

The previous two chapters discussed issues related to interconnecting, configuring, and testing hardware structures. In parallel with that, the Verilog constructs that support these tasks were also presented. For higher-level hardware descriptions, however, more advanced utilities than those introduced thus far are needed. This chapter is devoted to the presentation of such issues. Major topics covered in this chapter are types, arrays, operators, and system utilities. Section 7.1 discusses type declaration and usage. We will then discuss Verilog operators used at various levels of abstraction. A brief description of compiler directives and the way they are used follows this. The last section outlines system tasks and functions.

Unlike the two previous chapters, this chapter does not develop complete and evolutionary examples. Instead, it presents isolated examples or code fragments.

7.1 Verilog Types and Usage

Chapter 4 discussed data type declarations and the four-value logic system in Verilog. Because most data types are regarded as a collection of bits of this value system, various data types can easily be mixed in operations and assignments.

Register and **net** declarations are done in Verilog modules. Variable declaration statements can be placed anywhere in a Verilog module, although they must appear before they are used in assignments or instantiations. However, it is always better to place all declarations at the beginning of a module code, after input and output declarations.

Undeclared variables used in an instantiation statement or the right-hand side of a continuous assignment are said to be implicitly declared as **wire nets**.

In spite of the relaxed type rules in Verilog, there are certain assumptions based on language conventions that users should be aware of. This section is intended to discuss such type-related issues.

7.1.1 Module ports

Inputs and outputs of a model must be declared as **input, output,** or **inout.** By default, all declared ports are regarded as **net**s, and the default **net** type is used for the ports. For example, if defaults are not changed, an input or an output automatically assumes the **wire** type **net.** Ports declared as **output** may be declared as **reg.** This way they can be assigned values in procedural blocks. However, an **inout** port can be used only as a **net.** To assign values to an **inout** port in procedural bodies, a **reg** corresponding to the port must be declared and used. Values of this **reg** type variable can then be assigned to the **inout** port using continuous assignments.

7.1.2 Type conversions

Type conversions are important only when real numbers are used. Verilog provides several system tasks for type conversions to and from real data types. However, implicit type conversions are also done when real numbers are mixed with other data types.

A real number assigned to an integer variable is rounded to the nearest integer. Assigning a **net** or **reg** data type to a **real** data type is possible. Where **X** or **Z** values appear in the **net** or **reg** value, **0** is used for conversion to **real** type.

7.1.3 Parameters

In addition to variables that are used for modeling signals or hardware variables, parameters can also be used in Verilog. Parameters model hardware properties such as time, power, size, and other physical and nonsignaling properties of hardware. Parameters are treated as constants and cannot be modified at runtime. However, using **defparam** statements, parameter values can be modified at compile time.

7.1.4 Memory and vector declarations

Declaring a **net** or **reg** without a range specification declares corresponding data types as scalars. Verilog also provides formats for declaring vectors or memories.

Vectors. The vector declaration format is used for declaring multibit busses or registers. The statements shown below declare *abus* and *bbus* as an 8-bit wired-or bus and *breg* as a 16-bit register:

```
wor [7:0] abus, bbus;
reg [15:0] breg;
```

In this notation, the left-hand constant value specifies the most-significant bit and the right-hand constant is the least-significant bit of the vector. In the examples shown above, the least-significant bit of all three variables is bit number 0. The most-significant bit of *abus* and *bbus* is bit 7, and that of *breg* is 15.

A vector range can be descending, as in the above examples, or ascending, as in *areg,* declared below:

```
reg [0:15] areg;
```

In either case, the left range value is the most-significant bit position in an unsigned quantity. Figure 7.1 shows a graphical representation of the variables declared above.

Bit select. In operations involving vectors, all, part, or a certain bit of a vector may be selected. The operation selecting a bit of a vector is referred to as *bit_select.*

A set of square brackets bracketing a bit number selects that bit of a vector. An out-of-range bit returns an **X** value. For example, *areg[5]* selects bit 5 of our *areg* 16-bit vector declared above. A bit-select oper-

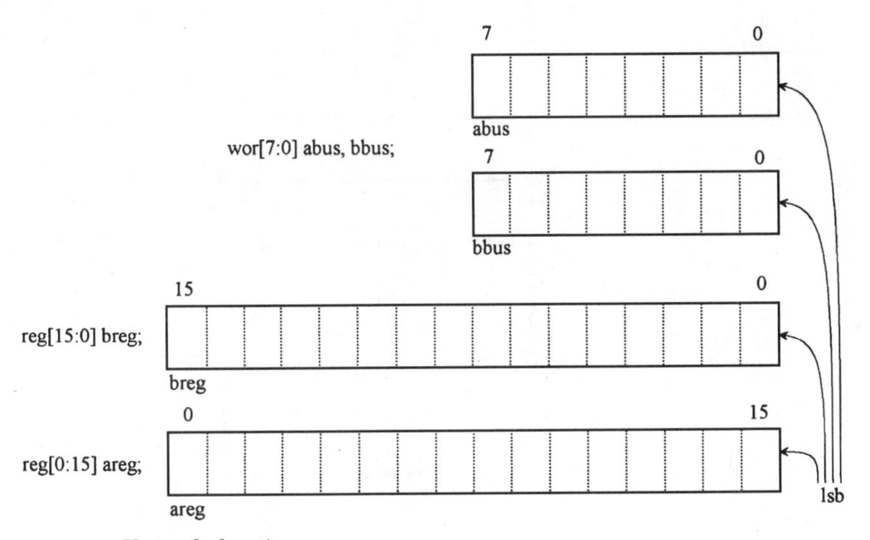

Figure 7.1 Vector declarations.

ation may be used on the right- or the left-hand side of assignments. Figure 7.2 shows examples of bit-select operation.

Part select. In order to select a set or a slice of a vector, the *part-select* operation may be used. For this purpose, a valid range of a vector must be specified within a set of square brackets. An out-of-range *part-select* specification returns **X.** As shown in Fig. 7.3, this operation may be used on the right- or the left-hand side of an assignment statement.

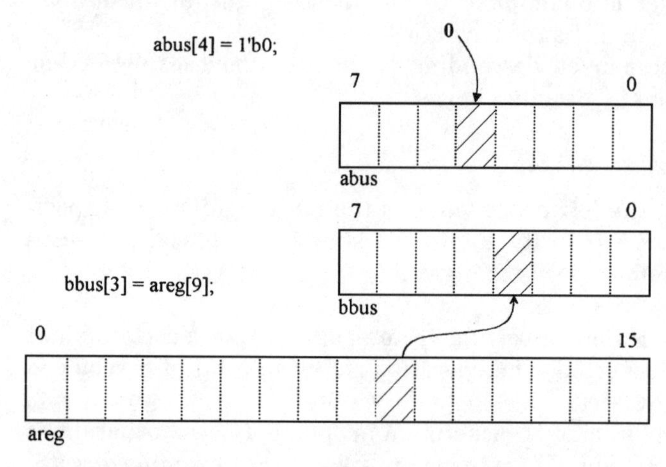

Figure 7.2 Bit-select operation on vectors.

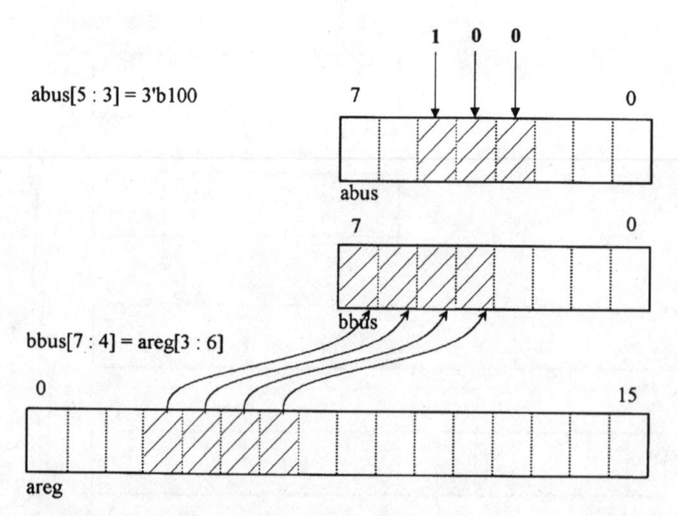

Figure 7.3 Part-select operation on vectors.

Memory. A memory array may be declared using the following format:

```
reg [word range] mem [memory range];
```

Using this format, a 1K array of 1-bit memory can be declared as

```
reg mem1k [0: 1023];
```

Graphically, we will represent this memory as a vertical array of 1-bit words, as shown in Fig. 7.4.

Memory arrays may be indexed, but a range of words cannot be selected at the same time. For example, *mem1k[25]* is a valid operation on the memory of Fig. 7.4, but *mem1k[20:25]* is not. This makes the memory declaration format of Fig. 7.4 different from that of a vector declaration when brackets precede the memory name. *Bit_select* and *part_select* operations apply only to vectors.

As another example, consider the 1K memory, *mema,* shown in Fig. 7.5. The word size of this memory is 8 and is declared by the following statement:

```
reg [7:0] mema[0:1023];
```

The operation shown in this figure indexes the memory, reads memory contents at location 25, and places them on the *bbus* declared in Fig. 7.1. Once they are placed on *bbus,* bit-select and part-select operations can be used to read individual bits or slices of the received word. Because of the selection limitation, a single memory word may be set to a specific data item, but a group of words or the entire memory cannot be. Verilog provides system tasks for memory in initialization. These tasks will be introduced later in this chapter.

String declaration. As previously discussed, most Verilog variables are declared as a bit or a group of bits. This is also true for string declarations. A string of *n* characters must be declared as a **reg** type vector of *8*n* bits. Therefore, to store the string "Memory Overflow," a **reg** vector of 120 bits providing room for 15 ASCII characters is needed. A convention often used in Verilog for string declarations is to use 8 multiplied by the number of characters for the size of a string variable. Using this convention, a 15-ASCII-character string must be declared as

```
reg [8*15:1] mem_status;
```

Because *mem_status* is a vector, bit-select or part-select operations can be performed to select a bit or a contiguous group of its bits.

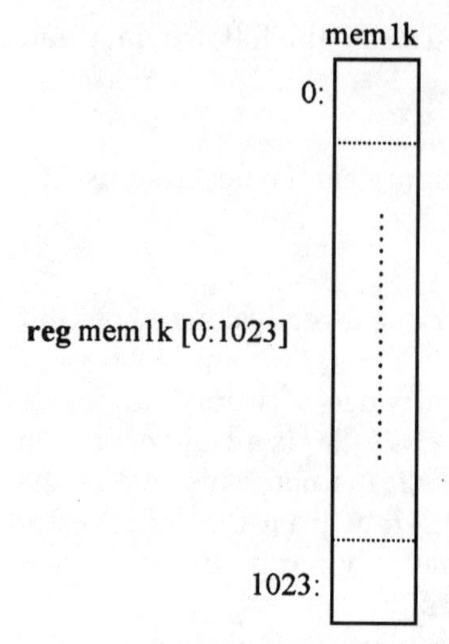

Figure 7.4 A 1K memory of 1-bit words.

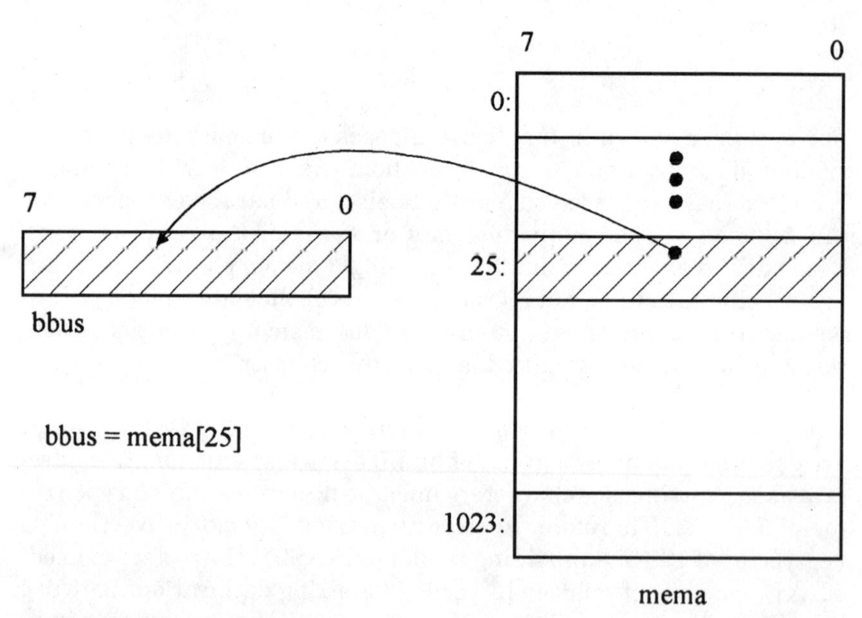

Figure 7.5 Reading a memory word.

7.1.5 Vector size mismatch

Operations or assignments involving vectors may have operands of different sizes. Adjustment of the size of an operand is done according to the operation in which it is used. In most cases, this adjustment is natural and is easily understood and justified. However, a brief discussion here formalizes this process.

Adjusting the size of operands in Verilog operations depends on the specific operation. For logical bitwise operations, a smaller vector operand is zero left-filled to match the size of the larger operand. For arithmetic operations, zero left filling is automatically done for the operands, and the size of the result is adjusted according to the variable the result is being assigned to.

Size adjustment in assignments is done by truncation or padding. If the left-hand side of a continuous or procedural assignment is smaller than the size of the right-hand side operation result or right-hand-side vector, the right-hand side is always truncated. In this case, as many rightmost bits of the right-hand side as can be used on the left are used. As an example, consider the following declarations:

```
wire [3:0]a;
wire [5:0]aa;
```

If the value of *aa* is 001110 and

```
assign a = aa;
```

takes place, the value of *a* becomes 1110, which is the contents of the rightmost 4 bits of *aa*.

Assigning values to a left-hand-side **reg** or **net** vector that is larger than the size of the right-hand side of the assignment is also allowed. In this case, padding is done to fill the extra left-hand-side bits. An unsigned number on the right-hand side of an assignment is padded with zeros to fit the left-hand-side size. For example, if *a* and *aa* are declared as above, and *a* is 1010, and

```
assign aa = a;
```

takes place, then the value of *aa* becomes 001010.

On the other hand, if a signed quantity is assigned to a larger left-hand-side vector, sign extension takes place to fill the extra left bits. For example,

```
assign a = 2'd2;
```

puts 0010 in *a,* and

```
assign aa = 2'd3;
```

puts 111101 in *aa.*

String vectors follow the rules for unsigned quantities. Assigning a larger string to a left-hand-side string vector causes the right-hand side to be truncated. If a string on the right is assigned to a larger string on the left, the assignment is done by padding the leftmost bits of the right-hand string with zeros or *null* characters.

7.2 Verilog Operators

As compared with those of many strongly typed languages, Verilog rules for using operators with various operand types are very relaxed. Also, as in any computer language, operators and their scope must be well understood if the language is to be used efficiently. This section presents Verilog operators categorized according to their applications.

7.2.1 Boolean operators

Boolean operators consist of bitwise operators, logical operators, and reduction operators.

Bitwise operators. Bitwise operators consist of &, |, ^, ^~, and ~ for bitwise AND, OR, XOR, XNOR, and NOT, respectively. For the XNOR operation, ~^ is also a valid symbol. All bitwise operators operate on scalars and vectors. In operations on vectors of differing sizes, the result becomes the larger of the two vectors. All four values of the standard Verilog logic value system are valid bit values for bitwise operators. Figure 7.6 shows the results of bitwise operators operating on two 1-bit operands for four bitwise operators. The complement operator ~, not shown in this figure, results in **1** and **0** for **0** and **1** inputs and **X** for **X** and **Z** inputs.

Logical operators. Logical operators are &&, ||, and ! for logical AND, OR, and NOT. Logical operators produce 1-bit results. The result of && is **1** if none of the operands is **0**. The result of || is 1 if at least one of the operands is nonzero, and the ! operator complements its operand and produces a **1** or a **0**. If an **X** or **Z** appears in an operand of the logical operators, an **X** will be the result. The ! operator has the highest precedence, followed by && and ||.

Reduction operators. A reduction operation is referred to as one that performs a certain bitwise operation on bits of a vector and reduces it by a single bit. There are six reduction operations in Verilog. These are &, ~&, |, ~|, ^, and ~^ (or ^~) for AND, NAND, OR, NOR, XOR, and XNOR. Applying a reduction operator to a vector performs bitwise operations on a pair of bits, starting in the least-significant position and repeating until all bits have been covered. For every pair of bits, the table in Fig. 7.6 determines the result of the operation. Complement reduction operations (~&,~|, and ~^) perform reduction first and then complement the result.

7.2.2 Compare operators

Comparing scalars or vectors can be achieved by relational or equality operators.

<table>
<tr><td>&</td><td>|</td><td colspan="2" rowspan="2">0</td><td colspan="2" rowspan="2">1</td><td colspan="2" rowspan="2">X</td><td colspan="2" rowspan="2">Z</td></tr>
<tr><td>^</td><td>^~</td></tr>
<tr><td colspan="2" rowspan="2">0</td><td>0</td><td>0</td><td>0</td><td>1</td><td>0</td><td>X</td><td>0</td><td>X</td></tr>
<tr><td>0</td><td>1</td><td>1</td><td>0</td><td>X</td><td>X</td><td>X</td><td>X</td></tr>
<tr><td colspan="2" rowspan="2">1</td><td>0</td><td>1</td><td>1</td><td>1</td><td>X</td><td>1</td><td>X</td><td>1</td></tr>
<tr><td>1</td><td>0</td><td>0</td><td>1</td><td>X</td><td>X</td><td>X</td><td>X</td></tr>
<tr><td colspan="2" rowspan="2">X</td><td>0</td><td>X</td><td>X</td><td>1</td><td>X</td><td>X</td><td>X</td><td>X</td></tr>
<tr><td>X</td><td>X</td><td>X</td><td>X</td><td>X</td><td>X</td><td>X</td><td>X</td></tr>
<tr><td colspan="2" rowspan="2">Z</td><td>0</td><td>X</td><td>X</td><td>1</td><td>X</td><td>X</td><td>X</td><td>X</td></tr>
<tr><td>X</td><td>X</td><td>X</td><td>X</td><td>X</td><td>X</td><td>X</td><td>X</td></tr>
</table>

Figure 7.6 Reduction operations.

Relational operators. Relational operators compare their operands and give the result as **0** or **1**. The **0** value represents false and **1** true for the result of a comparison. The relational operators are <, >, < =, and >= for less than, greater than, less than or equal, and greater than or equal. An **X** results if **X** or **Z** values in any of the operands make the comparison result uncertain.

Equality operators. Equality operators compare their scalar or vector operands for equality. Logical equality operators are == and != for equal and not equal. The == operator produces a **1** if only **0**s and **1**s appear in the operands and they are equal. The != operator is the complement of the == operator. An **X** or **Z** in any operand of these operators makes the result of the operation **X**.

Case equality operators are === and !== for equal and not equal. These operators are similar in function to logical equality operators, but also consider **X** and **Z** as actual bit values and include these values in calculating compare results.

7.2.3 Shift operators

A shift right and a shift left operator are provided in Verilog for shifting vector contents. The operators used are >> and <<. The >> operator right-shifts its left vector operand the number of positions given by its right operand. Right-hand-side bits of the shifted vector are ignored, and **0**s are moved in from the left.

7.2.4 Arithmetic operators

Verilog arithmetic operators are +, − , *, /, and % for plus, minus, multiply, divide, and modulo. These operators are the standard arithmetic operators found in most programming languages and have their usual meaning. The treatment of vectors and **X,** and **Z** values and the size of the result vector, which are particular to Verilog, are discussed here.

Arithmetic operations with **reg** data type operands treat the operands as unsigned quantities. For the multiply operation, the size of the result is determined by adding the bit lengths of the two operands. For the other arithmetic operations, the size of the result is the maximum of the bit lengths of the operands.

An **X** or a **Z** value in a bit of either of the operands causes the entire result of the multiply operation to become unknown **X.**

Unary plus (+) and minus (−) are allowed in Verilog. These operators take precedence over other arithmetic operators. Finally, the modulo operator (%) returns the remainder of division of its left-hand operand by its right-hand operand.

7.2.5 Concatenation operators

An important operation in hardware modeling is concatenation. This operation is used for formation of vectored sources or destinations from smaller vectors or scalars. The notation used for this operator is a pair of curly brackets ({...}) enclosing all scalars and vectors that are being concatenated.

The concatenation operator may be used on the left-hand side of an assignment. For example, if *a* is a 4-bit **reg** and *aa* is a 6-bit **reg** type variable, the following assignment places 1101 in *a* and 001001 in *aa:*

```
{a, aa} = 10'b1101001001;
```

A concatenation operation on the right-hand side forms a vector of the size of all the variables that are being concatenated. This vector may be used in other operations or can be assigned to a left-hand-side target.

A repetition multiplier can be used to form a vector. This vector may be used in another concatenation operation or may be used as a vector in other operations or as a right-hand-side assignment. If the *a* and *aa* registers have the values assigned to them above, and *aaa* is a 16-bit **reg** data type, then the assignment

```
aaa = {aa, {2{a}}, 2'b11}
```

puts 001001_1101_1101_11 in *aaa.* The leftmost 6 bits came from *aa,* the next 8 bits are two times repetition of *a,* and the least-significant 2 bits are the 2-bit constant **11.**

7.2.6 Other operators

The conditional operator in Verilog uses the ?: notation, the general format of this operation is

```
expression1 ? expression2 : expression3
```

If *expression1* is true, then *expression2* is selected as the result of the operation; otherwise *expression3* is selected. This operation provides a compact if–then–else type of construct for in-line continuous assignment statements.

7.3 Compiler Directives

Certain hardware modeling requirements in Verilog are provided as compiler directives instead of being incorporated in the main syntax of the language. Compiler directives provide facilities for file inclusion, timing specification, default settings, and string substitution. This section provides a brief description of these language facilities. Directives are presented in the order of their importance in hardware modeling.

7.3.1 `timescale

The **`timescale** directive sets the time unit and time precision in a module. Including the

```
`timescale 1ns/100 ps
```

directive before a module header causes all time-related numbers to be interpreted as having a 1-ns time unit. When expressions manipulating timing values are performed, 100-ps precision will be used.

7.3.2 `default_nettype

Undeclared **net**s are implicitly declared as **wire** type **net**s. The default **wire** type can be changed by the **`default_nettype.** For example,

```
`default-nettype wor
```

at the beginning of a module causes undeclared **net**s in constructs such as the terminal list of a module instance to be assumed to be **wor** type **net**s.

This default setting stays in effect for all future compilations until it is set to another value or reset by use of the **`resetall** directive.

7.3.3 `include

To include a parameter definition file or a section of shared code in a module, the **`include** directive may be used. Because Verilog does not provide a common library of parts and utilities, a shared code must be explicitly inserted in modules that use the code. This can be achieved using the **`include** directive.

7.3.4 `define

For better code readability, a meaningful string (referred to as a text macro) can be defined to represent a number or an expression. The **`define** directives shown below define *word_length* of 32 and assign the **101** binary code to state *begin_fetch_state* of a state machine.

```
`define word_length 32
`define begin_fetch_state 3'b101
```

Defined strings may be used in Verilog code text by preceding their defined names by a back quote. For example, if *`begin_fetch_state* is used anywhere in a Verilog code, it is replaced by 3'b101.

The **`undef** directive undefines a previously defined text macro.

7.3.5 `ifdef, `else, `endif

The **`ifdef** directive tests whether a text macro that immediately follows the **`ifdef** keyword has been defined using the **`define** directive. If the next macro has been defined, the group of lines bracketed between **`ifdef** and **`else** is compiled. If the text macro has not been defined, the group of lines bracketed between **`else** and **`endif** is compiled.

These directives can appear anywhere in a Verilog source code. The two groups of lines bracketed in an **`ifdef, `else, `endif** structure must independently have correct syntax.

7.3.6 `unconnected_drive

A port value left open in the connection list of a module instantiation assumes the default **net** value. To change this, and to force **pull0** or **pull1** values on unconnected ports, the **`unconnected_drive** directive can be used. The only arguments allowed with this directive are **pull0** or **pull1** for unconnected values **0** and **1,** respectively. The effect of this directive may be turned off by the **`nounconnected_drive** direction.

7.3.7 `celldefine, `endcelldefine

The **`celldefine** and **`endcelldefine** directives bracket modules that are to be considered as cells. The Verilog Programming Language Interface (PLI) uses cell modules.

7.3.8 `resetall

The **`resetall** directive turns off the effect of all compiler directives. Using this directive at the beginning of every module guarantees that no previous setting affects compilation of modules and that all defaults are set.

7.4 System Tasks and Functions

For test bench generation, data input and output, timing check, simulation flow control, data conversion, and memory initialization and specification, Verilog provides a number of system tasks and functions categorized into ten groups. The names of system tasks and functions begin with a dollar sign, **$,** followed by a task specifier. The name of the task or function usually contains characters and names that describe its functionality. A brief description of these language utilities will be given here. Details of several of these tasks have already been presented in the previous two chapters, and more tasks and functions will be presented in the examples that use them in the chapters that follow.

7.4.1 Display tasks

Display tasks are used for outputting to the standard output device. The most basic display task is the **$display** task, which writes its string argument to the display device. Other tasks include those for monitoring and outputting variable values as they change (the **$monitor** group of tasks) and those for displaying variables at a selected time (the **$display** tasks). Display tasks can display in binary, hexadecimal, or octal formats. The character **b, h,** or **o** at the end of the task name specifies the data type a task handles. For all display tasks, a generic task can be used to display data with specified formats and data types. Chapter 9 presents examples that take advantage of many of these tasks.

7.4.2 File I/O tasks

File I/O tasks begin with a dollar sign followed by the letter **f** and then by the same task names as those of the display tasks. These tasks perform the same functionalities as their display task counterparts, except that their output is to a file instead of to the display terminal. The **$fopen** function opens a file and assigns an integer file descriptor. The file descriptor will be used as an argument for all file I/O tasks. Chapter 9 presents examples that take advantage of many of these tasks.

7.4.3 Timescale tasks

Timescale tasks are **$printtimescale** and **$timeformat.** The **$printtimescale** task displays the timescale and precision of the module whose hierarchical name is being passed to it as its argument. The **$timeformat** task formats time for display by file IO and display tasks.

7.4.4 Simulation control tasks

Simulation control tasks are **$finish** and **$stop.** The **$finish** task ends the simulation and exits. Usually, simulation environments require a confirmation before the action of exiting the environment is taken. The **$stop** task suspends the simulation and does not exit the simulation environment.

7.4.5 Timing check tasks

Timing check tasks are used for checking timings, such as pulse width duration and setup and hold times. In general, timing check tasks check the timing on one signal or the relative timing of several signals for certain conditions to hold. If a violation is detected, a message will be issued in the user simulation environment display area. For example, the statement shown below uses the **$nochange** timing check task to report a violation if *d_input* changes in the period of 3 time units before and 5 time units after the positive edge of the *clock*.

```
$nochange (posedge clock, d_input, 3, 5);
```

7.4.6 PLA modeling tasks

PLA modeling tools use a declared memory as their first argument, configure it as synchronous or asynchronous, assign inputs and outputs to the PLA array, and configure the logical function of the PLA. Names used for PLA modeling tasks consist of three fields, *sync_async, and_or,* and *array_plane.* Tasks are named using these three fields separated by dollar signs. The general format that we use for describing these tasks is

```
$sync_async$and_or$array_plane
```

where *sync_async* can be either **sync** or **async**; *and_or* can be **and, or, nand,** or **nor**; and in place of *array_plane,* **array** or **plane** can be used.

Shown below is a PLA system task that defines an asynchronous PLA with a **nand** logical function, *a1* to *a7* inputs, and *b1* to *b4* outputs. PLA nand-plane fuses are determined by the contents of the *mem8by4* declared memory, as shown in Fig. 7.7:

```
$async$nand$array (mem8by4,{a1,a2,a3,a4,a5,a6,a7,a8},{b1,b2,b3,b4});
```

with the task and memory contents shown in Fig. 7.7. PLA outputs will be assigned values as shown in Fig. 7.8.

```
a1 a2 a3 a4 a5 a6 a7 a8

0  0  1  1  0  0  0  1  ──▶  b1
1  1  0  0  1  0  0  0  ──▶  b2
0  0  0  0  1  1  0  0  ──▶  b3
0  1  0  1  0  0  0  0  ──▶  b4
```

Figure 7.7 Contents of *mem8by4*.

$$b1 = (\overline{a3 \ \& \ a4 \ \& \ a8})$$

$$b2 = (\overline{a1 \ \& \ a2 \ \& \ a5})$$

$$b3 = (\overline{a5 \ \& \ a6})$$

$$b4 = (\overline{a2 \ \& \ a4})$$

Figure 7.8 PLA output expressions fused by *mem8by4*.

7.4.7 Conversion functions for reals

Verilog provides four system functions for converting from real to integer or bit, and for converting between bit or integer and real. These functions are **$bitstoreal, $realtobits, $itor,** and **$rtoi.**

7.4.8 Other tasks and functions

In addition to the above system utilities, Verilog provides several stochastic and probabilistic tasks and functions. Among these utilities, **$random** is a useful function for random data generation. We will use this task in examples in the chapters that follow.

There are also three time functions, **$realtime, $time,** and **$stime,** that return the simulation time in various formats.

7.5 Summary

This chapter presented tools for high-level descriptions. We started with type declarations and explained the defaults used when explicit types are not declared. Following that, we showed how vectors and arrays are declared in Verilog. The section that followed the type declaration section included a complete presentation of Verilog operators. Although operators have been used in the previous chapters and will be used and explained in the chapters that follow, the presentation in this chapter can be used as a reference for this topic. The last two sections of this chapter presented compiler directives and system tasks and functions. The presentations here discussed only the main topics; examples for these system utilities will be presented in the chapters that follow. IEEE Std. 1364-1995 serves as a complete and formal reference for language issues discussed here.

Further Reading

IEEE Standard Hardware Description Language Based on the Verilog Hardware Description Language, IEEE Std. 1364-1995, Institute of Electrical and Electronic Engineers, New York, 1996.

Palnitkar, S., *Verilog HDL: A Guide to Digital Design and Synthesis,* Prentice-Hall, Upper Saddle River, N.J., 1996.

Smith, Douglas J., *A Practical Guide for Designing, Synthesizing and Simulating ASICs and FPGAs Using VHDL or Verilog,* Doone Publications, Madison, Ala., 1996.

Thomas, D. E., and P. R. Moorby, *The Verilog Hardware Description Language,* 3d ed., Kluwer Academic Publishers, Norwell, Mass., 1996.

Problems

7.1 Write a module for defining a PLA for **BCD** to seven-segment display conversion. Using the **$readmemh** system task, read a memory array, and using a PLA task, configure a PLA as an asynchronous and array.

7.2 Write a test bench for applying random data to the PLA inputs of Prob. 7.1 and displaying the output. Use the **$random** and **$display** tasks.

Chapter

8

Dataflow Descriptions in Verilog

The middle ground between structural and behavioral descriptions is the dataflow or register transfer level of abstraction as we defined it in Chap. 1. Descriptions at this level specify the flow of data through the registers and busses of a system. This flow is controlled by external signals that can be generated by other dataflow machines. Verilog continuous assignments provide mechanisms for describing selection logic and bus structures. These constructs enable specification of various bus types and controlled placement of multiple data on busses. Furthermore, Verilog procedural statements allow specification of clocking schemes and register structures. In addition to busses and registers, dataflow descriptions, require specification of state machines for control of the timing and flow of data between registers and busses. Such controlling hardware can also be described in Verilog using basic procedural constructs of this language.

This chapter discusses Verilog constructs for bussing, basic registers, and state machines. We will show the use of various forms of continuous assignments for data selection, procedural constructs for register specifications, and statements that facilitate description of controllers.

8.1 Multiplexing and Data Selection

In a digital system, various forms of hardware structures are used for the selection and placement of data into buses, logic units, or registers. The simplest form of data selection is the AND-OR logic shown in Fig. 8.1.

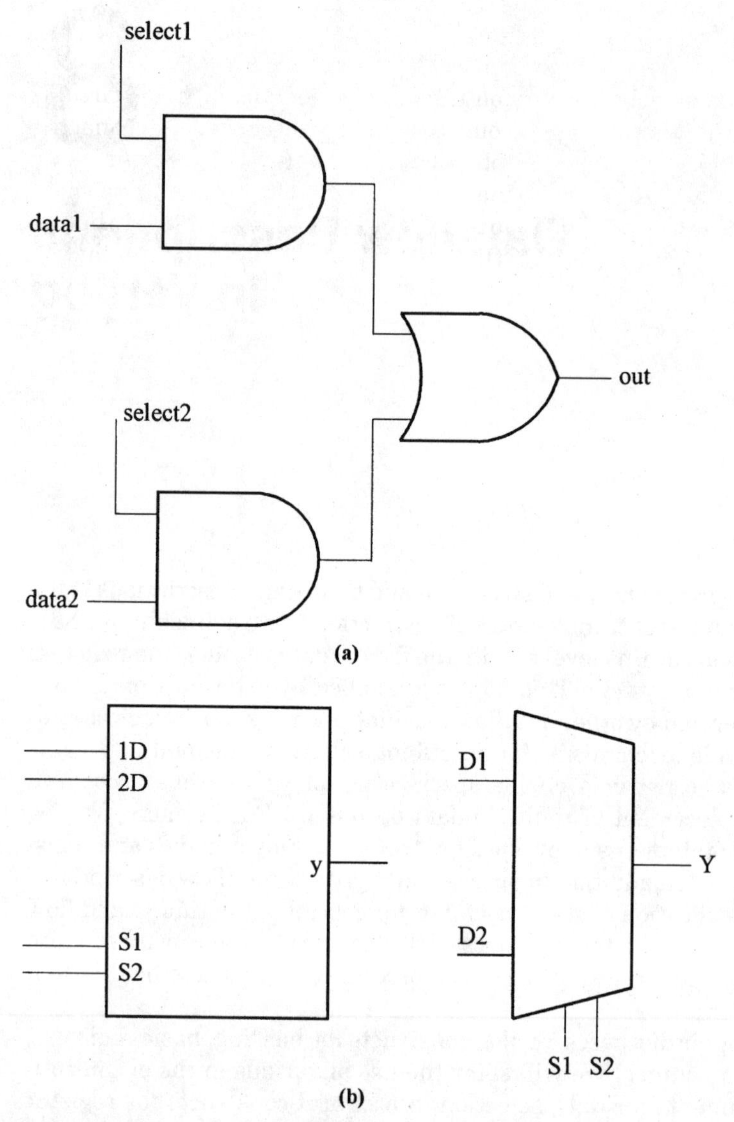

Figure 8.1 Basic data selection hardware. (*a*) Logic diagram; (*b*) symbols.

This structure selects *data1* or *data2,* depending on the values of *select1* and *select2.* Other forms of hardware for the selection of data may consist of a wired connection of three-state gate outputs or a parallel connection of MOS transmission gates. Data selection is also referred to as multiplexing, and the hardware that performs this task is called a multiplexer. Figure 8.1*b* shows two multiplexer symbols for the hardware of Fig. 8.1*a*. Another form of multiplexing may be achieved by use of three-state structures. Figure 8.2 shows data selection logic based on three-state noninverting buffers. If a select input is active, its corresponding data input appears on the output. Functionally this circuit is different from that of Fig. 8.1 only when both select inputs are inactive. Figure 8.2*b* shows symbolic notation for this multiplexer circuit.

In a dataflow-level design, hardware structures such as those shown above are used to provide inputs to logic units, registers, and other bus structures. In the Verilog HDL, various forms of continuous assignments may be used for describing this hardware.

8.1.1 Data selection

A 1-bit 8-to-1 multiplexer with eight select inputs is shown in Fig. 8.3*a*. The output of this structure becomes equal to one of the eight inputs when a corresponding select input is active. Several styles of coding for describing this hardware in Verilog will be presented.

One possible description for this multiplexer is shown in Fig. 8.3*b*. The statement shown in this figure is a continuous assignment statement.

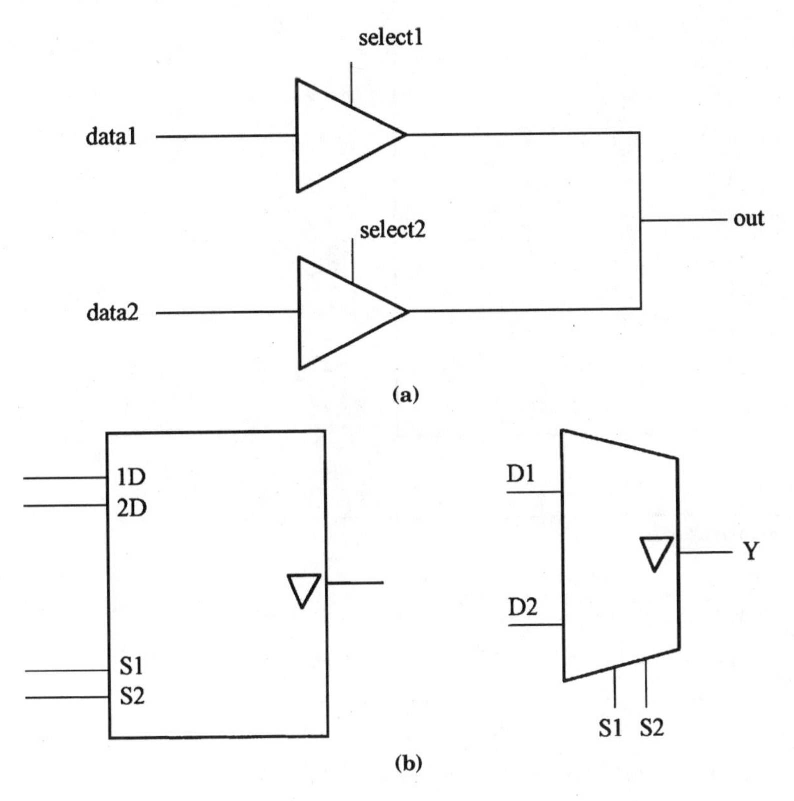

Figure 8.2 Three-state multiplexer. (*a*) Logic diagram; (*b*) symbols.

On the right-hand side an expression ANDs bits of select inputs with bits of the data input, ORs the AND results together, and assigns the result to *z*. Only **nets** (here using the **wire** type) may appear on the left-hand side of a continuous assignment. A **net** type is assumed by default for input and output ports of a module.

Figure 8.4 shows syntax details of a continuous assignment statement. Several *net_assignments* separated by commas constitute a *continuous_assign*. A *net_assignment* construct has a *net_lvalue* on the left-hand side of an equal sign and an expression on the right-hand

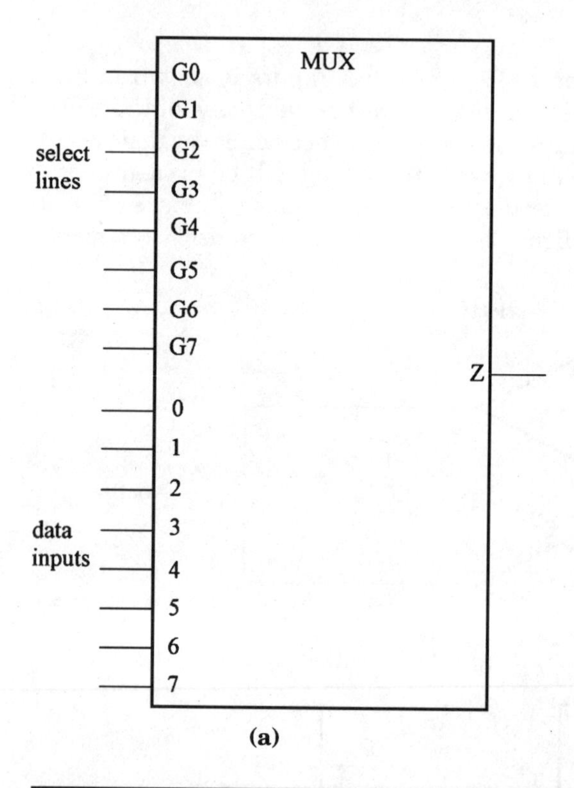


```
`timescale 1ns/100ps

module mux_8_to_1     (i7, i6, i5, i4, i3, i2, i1, i0,
                       s7, s6, s5, s4, s3, s2, s1, s0, z );
input i7, i6, i5, i4, i3, i2, i1, i0, s7, s6, s5, s4, s3, s2, s1, s0;
output z;
        assign z = i7 & s7 | i6 & s6 | i5 & s5 | i4 & s4 | i3 & s3 | i2 & s2 | i1 & s1 | i0 & s0;
endmodule
```

(b)

Figure 8.3 (*a*) An 8-to-1 multiplexer; (*b*) a simple 8-to-1 multiplexer description.

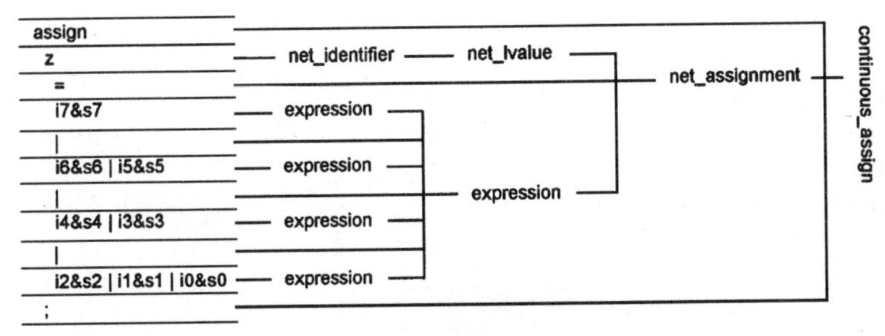

Figure 8.4 Syntax of a continuous assignment.

side. The *net_lvalue* indicates that a **net** or a bit or part select of a **net** may appear on the left-hand side of a *continuous_assign* construct.

A *continuous_assign* construct may include drive strength and delay values immediately after the **assign** keyword. The syntax for these specifications is the same as that used with gate instantiation in Chap. 5. Figure 8.5 shows another module description for our multiplexer example. In addition to the use of rise and fall parameters, another difference between the code of Fig. 8.3*b* and that of Fig. 8.5 is the right-hand expression. Instead of bit-by-bit ANDing of data and select inputs, the expression in Fig. 8.5 concatenates all select inputs together and all data inputs together using the {} operator. The vectors formed are ANDed, and an OR reduction ORs bits of the AND vector, resulting in 1 bit. This bit is assigned to **net** *z* using 5-ns and 7-ns rise and fall delay values.

Continuous assignment statements may be combined with **net** declarations. Figure 8.6 shows another module description for the multiplexer of Fig. 8.3 in which the **wire** *z* declaration also includes the expression for *z*. The expression used in this module is a conditional expression, the syntax of which is illustrated in Fig. 8.7. If any of the select inputs is **1,** the corresponding input is assigned to the output *z*. If none have the value **1, 0** will be assigned to *z*.

The operator **?:** forms a convenient and compact conditional expression. Nesting these expressions causes conditions to be examined from

```
`timescale 1ns/100ps

module mux_8_to_1    (i7, i6, i5, i4, i3, i2, i1, i0,
                      s7, s6, s5, s4, s3, s2, s1, s0, z );
input i7, i6, i5, i4, i3, i2, i1, i0, s7, s6, s5, s4, s3, s2, s1, s0;
output z;
     assign #(5,7) z = | ({i7, i6, i5, i4, i3, i2, i1, i0} & {s7, s6, s5, s4, s3, s2, s1, s0});
endmodule
```

Figure 8.5 Using delays and concatenation in a continuous assignment.

left to right, and the value of the first expression with a *true* condition will be used for the left-hand side. Using **?:** with three expressions *e1, e2,* and *e3* like the following:

```
e1_is_true ? take_e2 : take_e3
```

is interpreted as

```
IF e1_is_true THEN take_e2 ELSE take_e3.
```

In the expression of Fig. 8.7, the **else** part of the last condition is 0, and *s0*==1 ? i0:0 is itself the **else** condition for the conditional expression preceding it. Moving from the last 0 in Fig. 8.7 toward the first expression, expressions selecting the value **0** or *i0* through *i6* become the **else** part of the expression with the *s7* == 1 condition.

One last issue in the module of Fig. 8.6 that needs some explanation is the use of the relational operator == for comparing values appearing on select inputs of the multiplexer. Use of this operator instead of the case equality operator === causes **Z** or **X** values on the select

```
`timescale 1ns/100ps

module mux_8_to_1   (i7, i6, i5, i4, i3, i2, i1, i0,
                     s7, s6, s5, s4, s3, s2, s1, s0, z );
input i7, i6, i5, i4, i3, i2, i1, i0, s7, s6, s5, s4, s3, s2, s1, s0;
output z;
      wire z =
             s7 == 1 ? i7 :
             s6 == 1 ? i6 :
             s5 == 1 ? i5 :
             s4 == 1 ? i4 :
             s3 == 1 ? i3 :
             s2 == 1 ? i2 :
             s1 == 1 ? i1 :
             s0 == 1 ? i0 :
             0;
endmodule
```

Figure 8.6 Combining **wire** declaration and **assign.**

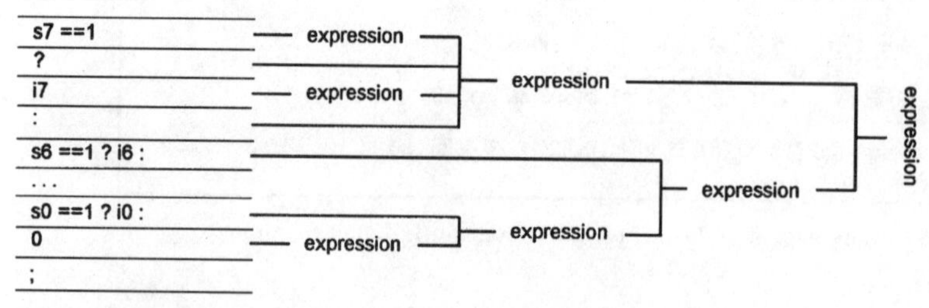

Figure 8.7 Syntax details of nested conditional expressions.

lines to be regarded as false. Case equality operators, however, do take unknown and high-impedance operators into account.

For another example of the use of conditional expressions, consider the 3-to-8 decoder in Fig. 8.8. This decoder has a 3-bit address input and eight output lines. An output *i* becomes active (high) when the decimal equivalent of the input address is equal to *Si*. A value of **Z** on an address line is treated as **1.**

The Verilog code for this unit is shown in Fig. 8.9. As in the multiplexer example, nested conditional expressions are used. Two 3-bit vectors with the case equality operator form the condition part of each of the conditional expressions. Values selected when conditions are satisfied are 8-bit vectors that are assigned to *so* 8-bit output lines with 5 ns and 7 ns rise and fall delays. The use of the case equality operator === causes **Z** or **X** values to be regarded as actual *adr* bit values which participate in comparing *adr* with the constants on the right-hand side of this operator.

8.1.2 Bussing

A bus in a digital system is a line or a vector of lines on which data from several sources may be placed. A bus extends beyond where its sources are located on a chip or a board to reach all its destinations. Reading from a bus by its destinations usually does not require special hardware, but writing into the bus by its sources requires a selection logic for each source and coordination between selection of various sources.

From the standpoint of providing hardware for selecting one of several data sources and making it available for various destinations to read its value, the multiplexer of the previous section qualifies as a bus. The code presented in the module of Fig. 8.3b may be regarded as

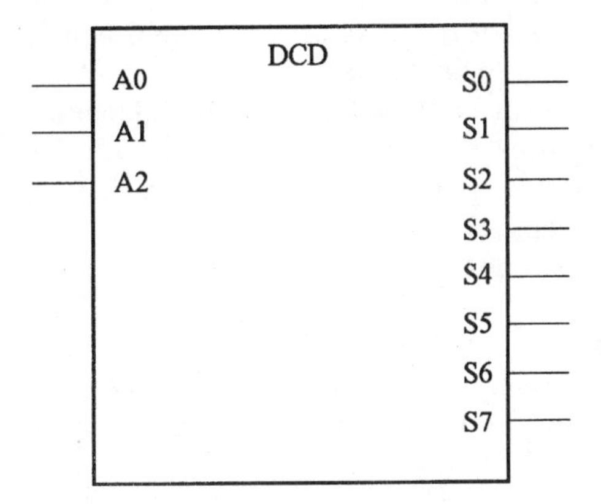

Figure 8.8 A 3-to-8 decoder.

```
`timescale 1ns/100ps

module dcd_3_to_8    (adr, so);
input [2:0] adr;
output [7:0] so;
    assign #(5,7)
            so = adr === 3'b000 ? 8'b00000001 :
              adr === 3'b001 ? 8'b00000010 :
              adr === 3'b00Z ? 8'b00000010 :
              adr === 3'b010 ? 8'b00000100 :
              adr === 3'b0Z0 ? 8'b00000100 :
              adr === 3'b011 ? 8'b00001000 :
              adr === 3'b01Z ? 8'b00001000 :
              adr === 3'b0Z1 ? 8'b00001000 :
              adr === 3'b0ZZ ? 8'b00001000 :
              adr === 3'b100 ? 8'b00010000 :
              adr === 3'b101 ? 8'b00100000 :
              adr === 3'b10Z ? 8'b00100000 :
              adr === 3'bZ01 ? 8'b00100000 :
              adr === 3'bZ0Z ? 8'b00100000 :
              adr === 3'b110 ? 8'b01000000 :
              adr === 3'b1Z0 ? 8'b01000000 :
              adr === 3'bZ10 ? 8'b01000000 :
              adr === 3'bZZ0 ? 8'b01000000 :
              adr === 3'b111 ? 8'b10000000 :
              adr === 3'b11Z ? 8'b10000000 :
              adr === 3'b1Z1 ? 8'b10000000 :
              adr === 3'b1ZZ ? 8'b10000000 :
              adr === 3'bZ11 ? 8'b10000000 :
              adr === 3'bZ1Z ? 8'b10000000 :
              adr === 3'bZZ1 ? 8'b10000000 :
              adr === 3'bZZZ ? 8'b10000000 :
              8'BXXXXXXXX;
endmodule
```

Figure 8.9 Conditional expressions for the decoder description.

a 1-bit AND-OR bus with eight sources. Changing the *ii* sources and bus output of Fig. 8.3*b* to *n*-bit vectors makes this module an *n*-bit bus.

Instead of routing bus inputs (sources) to a location on the chip for data selection, bus logic is often distributed to reach all its inputs and outputs, as in Fig. 8.10. The multiplexer modules presented in the previous section do not adequately model such distributed bussing.

The logical operation a bus performs on its selected sources depends on the bus pull configuration (pull-up, pull-down, or open), the source selection logic, and the driving source pull configuration. Figure 8.11 shows three bus types and their selection logic.

Figure 8.11*a* is an inverted AND-OR bus with *si* select lines that select *ii* input data and place them on the bus. The *bus_line* is in high-impedance state unless it is pulled down by at least one pull-down path, causing it to become 0. Because of the logic function that this type of bussing performs on its selected inputs, or its driving values, it is often referred to as a wired-nor or wired-or bus.

The value of *bus_line* in Fig. 8.11*b* is formed by wiring nodes *a, b, c,* and *d.* If an input *ii* is selected, its value appears on its three-state

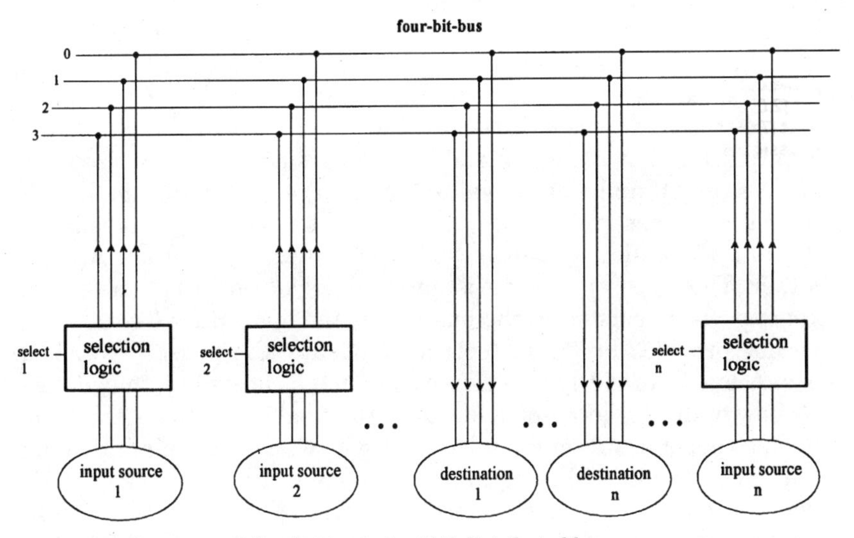

Figure 8.10 Sources and destinations of a 4-bit distributed bus.

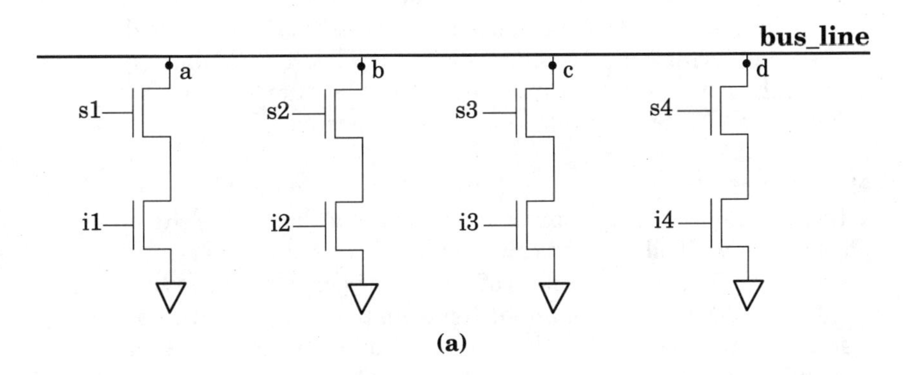

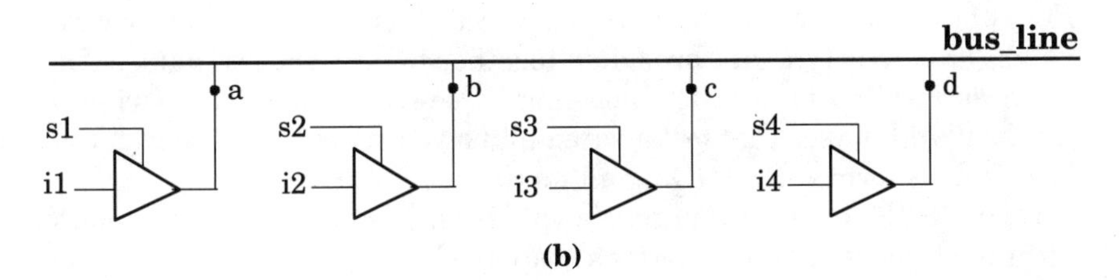

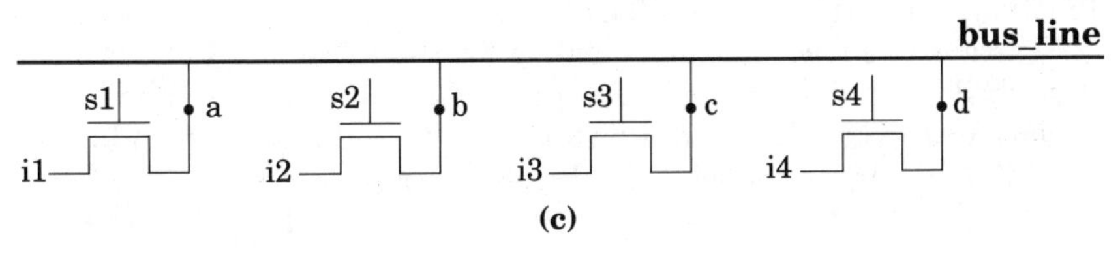

Figure 8.11 Bus types. (*a*) Wired-nor; (*b*) three-state; (*c*) wired-and.

gate output (node *a, b, c,* or *d*), which becomes the driving value for the bus. If all drivers are off, the bus will be in the high-impedance state. Conflicting values on active drivers will be wired together. This type of bussing is referred to as a three-state bus.

Figure 8.11*c* shows a wired-and bus. We assume that input sources provide strong **0** and weak **1** values. Without a pull-up on the bus, if none of the sources are selected, the bus will be at the high-impedance state, and if conflicting source values are selected, they are ANDed together. This type of bussing is generally referred to as wired-and. Using a pull-up resistor on the bus makes the bus values **0** or **1.**

Verilog provides language tools for modeling the busses described above. A bus is properly modeled by language constructs that determine how multiple values on a bus form the final bus value. A final bus value also depends on source selection logic, which determines driving values when the bus is being driven and when it is not. In Fig. 8.11, values on nodes *a, b, c,* and *d* are determined by the kind of selection logic used, i.e., controlled pull-down transistors, three-state gates, or pass gates. The declared bus type decides how these values contribute to the final bus value. What follows discusses final bus value determination and formation of bus driving values. These contributing factors to modeling a bus will be presented in subsections under bus value modeling and modeling selection logic, respectively.

Bus value modeling. The type declared for a bus determines how selected values from various sources affect the final bus value. Referring to Fig. 8.11, we will show how values on nodes *a, b, c,* and *d* form the final value on a *bus_line*, regardless of how values on these nodes are generated. For this discussion and for the examples used, we will externally generate data for these nodes, so that all possible cases are studied.

Verilog provides language tools for describing any of the bus types in Fig. 8.11. A bus declared as **wor** is a wired-or bus. Concurrent driving values on a **wor** type **net** are ORed together to form the bus value. The **wire** declaration is used for modeling three-state busses. Declaring a **net** as **wand** makes that **net** a wired-and bus. On a **wand** type **net,** the bus value is formed by ANDing all its concurrent drivers.

Figure 8.12 shows how multiple values on a bus line are resolved to form a bus value. The functions displayed in tabular form in these tables are recursively used for any number of active bus sources. The functions implied here are referred to as bus resolution functions.

Figure 8.13 shows Verilog modeling for the three types of bussings described. The *bus_line* module outputs in Fig. 8.13*a, b,* and *c* are declared as **wor, wire,** and **wand** type **net**s. Values driving a bus as a result of the bus selection logic (Fig. 8.11) are modeled in Fig. 8.13 as inputs *a, b, c,* and *d,* which correspond to the nodes shown in Fig. 8.11.

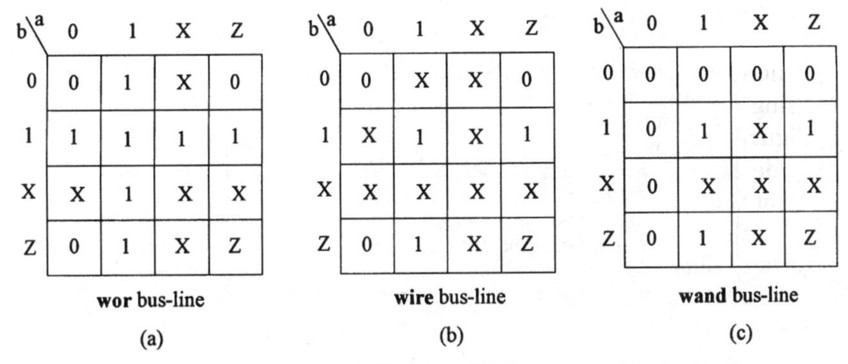

b\a	0	1	X	Z
0	0	1	X	0
1	1	1	1	1
X	X	1	X	X
Z	0	1	X	Z

wor bus-line

(a)

b\a	0	1	X	Z
0	0	X	X	0
1	X	1	X	1
X	X	X	X	X
Z	0	1	X	Z

wire bus-line

(b)

b\a	0	1	X	Z
0	0	0	0	0
1	0	1	X	1
X	0	X	X	X
Z	0	1	X	Z

wand bus-line

(c)

Figure 8.12 Bus resolution tables. (*a*) Wired-or; (*b*) three-state; (*c*) wired-and.

The test bench in Fig. 8.14 assigns **0, 1, X,** and **Z** values top 201 201 *a, b, c,* and *d* nodes. Using **always** statements in this test bench causes the sequence of values placed on each line to be repeated as long as the simulation runs. The outputs of the three bussing modules of Fig. 8.13 are shown in the listing in Fig. 8.15. In all three cases, bus values are **Z** if all inputs are **Z.** As shown in this listing, and as determined by the tables in Fig. 8.12, conflicting values are ORed, wired, or ANDed to generate the bus values of Fig. 8.13*a, b,* and *c,* respectively.

Modeling selection logic. The logic used for selecting or activating a source generates data on a bus line and is considered part of the bussing hardware. This logic determines the logical operation performed when multiple sources are active or when conflicts occur in bus source values. Appropriate modeling of this logic together with declared type of a bus, form Verilog modeling of a hardware bus.

Obviously, a straightforward and exact way to model bus selection logic for generating values on a bus line (hardware driving nodes *a, b, c,* and *d* in Fig. 8.11) is at the structural level. At this level of abstraction, bus selection hardware may be described by instantiating built-in gates. However, for convenience in coding and for better simulation speed performance, more abstract constructs are needed at the dataflow level of abstraction.

Figure 8.16 shows conditional signal assignments that select 8-bit sources and place them on the three bus types of Fig. 8.13. In each module, four continuous assignments represent placement of bus source inputs *i1, i2, i3,* and *i4* selected by *s1, s2, s3,* and *s4* on the 8-bit *bus_line.* Select conditions are checked by use of the equality operator (==). Therefore, values of **X** or **Z** on the select lines result in unknowns, causing the busses to become unknown. Using the case equality operator === instead of == would cause a condition to

(a)

```
`timescale 1ns/100ps

module wired_or_bussing (a, b, c, d, bus_line);
input a, b, c, d;
output bus_line;
wor bus_line;
        assign bus_line = a;
        assign bus_line = b;
        assign bus_line = c;
        assign bus_line = d;
endmodule
```

(b)

```
`timescale 1ns/100ps

module three_state_bussing (a, b, c, d, bus_line);
input a, b, c, d;
output bus_line;
wire bus_line;
        assign bus_line = a;
        assign bus_line = b;
        assign bus_line = c;
        assign bus_line = d;
endmodule
```

(c)

```
`timescale 1ns/100ps

module wired_and_bussing (a, b, c, d, bus_line);
input a, b, c, d;
output bus_line;
wand bus_line;
        assign bus_line = a;
        assign bus_line = b;
        assign bus_line = c;
        assign bus_line = d;
endmodule
```

Figure 8.13 Modeling three bus types. (*a*) Wired-or; (*b*) three-state; (*c*) wired-and.

become true if a corresponding select line is **1** and false otherwise. This way, unknown bus values are not generated.

A test bench for placing data on bus inputs and select lines is used to instantiate the modules of Fig. 8.16. The simulation run result is depicted in Fig. 8.17. As shown in this figure, when none of the sources is selected, all three buses—wired-nor, three-state, and wired-and—

are at the high-impedance state. Selecting multiple sources results in bits with conflicting **0** and **1** values that must be ORed, wired, or ANDed for the three modules in Fig. 8.16*a, b,* and *c,* respectively.

Combination of bus **net** types and assignments used for placing drivers on a bus enable a designer to model most forms of busses in Verilog. The examples in this section showed three forms of bussing used in digital systems. Busses using open-collector logic gates and pull-up or pull-down can also be modeled at the dataflow level in Verilog. In cases where an exact representation of a particular bus is

```
`timescale 10ns/1ns

module test_bus_circuit;
reg a, b, c, d;
wire wired_or, tri_state, wired_and;
     wired_or_bussing u1 (a, b, c, d, wired_or);
     three_state_bussing u3 (a, b, c, d, tri_state);
     wired_and_bussing u2 (a, b, c, d, wired_and);
     initial #55 $stop;
     always begin #08 a = 1'b0; #10 a = 1'bx; #10 a = 1'bz; #06 a = 1'b0; #10 a = 1'b1; end
     always begin #05 b = 1'bz; #10 b = 1'b1; #12 b = 1'bz; #07 b = 1'b0; #06 b = 1'b1; end
     always begin #06 c = 1'bz; #12 c = 1'b0; #11 c = 1'b1; #04 c = 1'bz; #09 c = 1'b1; end
     always begin #04 d = 1'bx; #16 d = 1'b0; #13 d = 1'bz; #08 d = 1'bx; #05 d = 1'b1; end
endmodule
```

Figure 8.14 Testing 1-bit busses.

ps	a	b	c	d	wired_or	tri_state	wired_and
0	x	x	x	x	x	x	x
50000	x	z	x	x	x	x	x
60000	x	z	z	x	x	x	x
80000	0	z	z	x	x	x	0
150000	0	1	z	x	1	x	0
180000	x	1	0	x	1	x	0
200000	x	1	0	0	1	x	0
270000	x	z	0	0	x	x	0
280000	z	z	0	0	0	0	0
290000	z	z	1	0	1	x	0
330000	z	z	z	z	z	z	z
340000	0	0	z	z	z	z	z
400000	0	1	z	z	1	x	0
410000	0	1	z	x	1	x	0
420000	0	1	1	x	1	x	0
440000	1	1	1	x	1	x	x
450000	1	z	1	x	1	x	x
460000	1	z	1	1	1	1	1
480000	1	z	z	1	1	1	1
500000	1	z	z	x	1	x	x
520000	0	z	z	x	x	x	0

Figure 8.15 Bus values that result from direct assignment of values to 1-bit busses.

```
`timescale 1ns/100ps

module n_wired_nor_bussing (i1, i2, i3, i4, s1, s2, s3, s4, bus_line);
input [7:0] i1, i2, i3, i4;
input s1, s2, s3, s4;
output [7:0] bus_line;
wor [7:0] bus_line;
        assign bus_line = s1 == 1 ? ~i1 : 8'bz;
        assign bus_line = s2 == 1 ? ~i2 : 8'bz;
        assign bus_line = s3 == 1 ? ~i3 : 8'bz;
        assign bus_line = s4 == 1 ? ~i4 : 8'bz;
endmodule
```

(a)

```
`timescale 1ns/100ps

module n_three_state_bussing (i1, i2, i3, i4, s1, s2, s3, s4, bus_line);
input [7:0] i1, i2, i3, i4;
input s1, s2, s3, s4;
output [7:0] bus_line;
wire [7:0] bus_line;
        assign bus_line = s1 == 1 ? i1 : 8'bz;
        assign bus_line = s2 == 1 ? i2 : 8'bz;
        assign bus_line = s3 == 1 ? i3 : 8'bz;
        assign bus_line = s4 == 1 ? i4 : 8'bz;
endmodule
```

(b)

```
`timescale 1ns/100ps

module n_wired_and_bussing (i1, i2, i3, i4, s1, s2, s3, s4, bus_line);
input [7:0] i1, i2, i3, i4;
input s1, s2, s3, s4;
output [7:0] bus_line;
wand [7:0] bus_line;
        assign bus_line = s1 == 1 ? i1 : 8'bz;
        assign bus_line = s2 == 1 ? i2 : 8'bz;
        assign bus_line = s3 == 1 ? i3 : 8'bz;
        assign bus_line = s4 == 1 ? i4 : 8'bz;
endmodule
```

(c)

Figure 8.16 Selecting inputs for three bus types. (a) Wired-nor; (b) three-state; and (c) wired-and.

ps	i1	i2	i3	i4	s1	s2	s3	s4	wired_nor	tri_state	wired_and
0	xxxxxxxx	xxxxxxxx	xxxxxxxx	xxxxxxxx	x	x	x	x	xxxxxxxx	xxxxxxxx	xxxxxxxx
10000	xxxxxxxx	xxxxxxxx	xxxxxxxx	00000100	x	x	x	x	xxxxxxxx	xxxxxxxx	xxxxxxxx
20000	xxxxxxxx	00001111	xxxxxxxx	00000100	x	x	x	x	xxxxxxxx	xxxxxxxx	xxxxxxxx
30000	00000101	00001111	00000111	00000100	x	x	x	x	xxxxxxxx	xxxxxxxx	xxxxxxxx
40000	00000101	00001111	00000111	00000100	0	x	x	x	xxxxxxxx	xxxxxxxx	xxxxxxxx
80000	00000101	00001111	00000111	00000100	0	0	x	x	xxxxxxxx	xxxxxxxx	xxxxxxxx
120000	00000101	00001100	00000111	00000100	0	0	0	x	xxxxxxxx	xxxxxxxx	xxxxxxxx
130000	00001111	00001100	00000111	00000100	0	0	0	x	xxxxxxxx	xxxxxxxx	xxxxxxxx
140000	00001111	00001111	00000111	00000100	0	0	0	0	zzzzzzzz	zzzzzzzz	zzzzzzzz
150000	00001111	00001111	00000010	00000100	0	0	0	0	zzzzzzzz	zzzzzzzz	zzzzzzzz
160000	00000101	00001111	00000010	00000100	1	0	0	0	11111010	00000101	00000101
170000	00000101	00001111	00000010	00001110	1	0	0	0	11111010	00000101	00000101
180000	00000101	00001111	00000111	00000100	1	0	0	0	11111010	00000101	00000101
200000	00000101	00001111	00000111	00000100	1	1	0	0	11111010	0000x1x1	00000101
240000	00000101	00001100	00000111	00000100	1	1	1	0	11111011	0000x1xx	00000100
260000	00001111	00001111	00000111	00000100	1	1	1	1	11111011	0000x1xx	00000100
280000	00001111	00001111	00000111	00000100	x	1	1	1	11111x11	xxxxxxxx	00000x00
290000	00000101	00001111	00000111	00000100	x	1	1	1	11111x11	xxxxxxxx	00000x00
300000	00000101	00001111	00000010	00000100	x	1	1	1	11111111	xxxxxxxx	00000000
320000	00000101	00001111	00000010	00000100	x	x	1	1	11111111	xxxxxxxx	00000000
330000	00000101	00001111	00000111	00000100	x	x	1	1	11111x11	xxxxxxxx	00000x00
340000	00000101	00001111	00000111	00001110	x	x	1	1	11111xx1	xxxxxxxx	00000xx0
350000	00000101	00001111	00000111	00000100	x	x	1	1	11111x11	xxxxxxxx	00000x00
360000	00000101	00001100	00000111	00000100	x	x	x	1	11111x11	xxxxxxxx	00000x00
380000	00000101	00001111	00000111	00000100	x	x	x	x	xxxxxxxx	xxxxxxxx	xxxxxxxx
390000	00001111	00001111	00000111	00000100	x	x	x	x	xxxxxxxx	xxxxxxxx	xxxxxxxx
400000	00001111	00001111	00000111	00000100	z	x	x	x	xxxxxxxx	xxxxxxxx	xxxxxxxx
420000	00000101	00001111	00000111	00000100	z	x	x	x	xxxxxxxx	xxxxxxxx	xxxxxxxx
440000	00000101	00001111	00000111	00000100	0	z	x	x	xxxxxxxx	xxxxxxxx	xxxxxxxx

Figure 8.17 Simulating wired-nor, three-state, and wired-and busses.

not possible at the dataflow level, built-in gates and the three **net** types, **wor, wire,** and **wand,** may be used for structural or gate-level detailed modeling of the bus.

8.1.3 Open-collector

A type of bus used in the bipolar technology uses open-collector gates. Here we will show how an open-collector structure may be described and utilized at the dataflow level, without having to instantiate built-in gates that are provided for this purpose.

Figure 8.18 shows four open-collector NAND gates described at the dataflow level. The continuous assignment statement performs bit-by-bit NANDing of the 4-bit *a* and *b* vectors. NANDing *a* and *b* results in output values of **0** and **1,** which are converted to open-collector **0** and **Z** values by the *oc* function. The function output is assigned to the **wire** output *y* of the *oc_nibble_nand* module.

The 2-bit *oc* function is defined in the *oc_nibble_nand* module. Bit 1 of the 2-bit *convert* argument is **Z,** and bit 0 has **0** value. A for loop in this function looks up bits of the *x* input vector and uses them as indexes for the *convert* parameter. A 0 index is converted to **0,** and a 1 index is converted to **Z.** Each bit that is looked up in *convert* is assigned to the function output *oc*. Bit values that appear on the *y* vector output of this module are either **0** or **Z,** as expected from an open-collector structure.

The circuit shown in Fig. 8.19 uses four open-collector NAND gates for generating an *xor* and an *xnor* output. The pull-ups are required to turn **Z** values at NAND outputs to logic **1** values.

Figure 8.20 shows a test bench for a module that corresponds to the circuit of Fig. 8.19. The *u1* instance uses concatenation for connecting the *{aa, ~aa, aa, ~aa}* vector to the 4-bit *a* input of the module of Fig. 8.18 and *{bb, ~bb, ~bb, bb}* vector to the *b* input. The 4-bit vector that connects to the output of *oc_nibble_nand* wires bits 3 and 2 of the *y* output and a pull-up named *pullup1* together to form the XOR output. It also wires bits 1 and 0 of *y* with *pullup2* to form the XNOR output, as shown in Fig. 8.19.

The test bench in Fig. 8.20 uses two for-loop statements and non-blocking assignments for generating bit data on the *aa* and *bb* inputs of the 4-bit open-collector NAND structure. Each loop generates 4-bit binary data between 0 and 14 on the *i* **reg.** The statement

```
aa <= #(i*30) i[0];
```

schedules bit 0 of *i* into *aa* at 300 ns (*i* times 30 times 10 ns **`timescale**) time intervals. A similar statement schedules ~i[0], as *i* is incremented, into the *bb* input at 400-ns time intervals.

```
`timescale 1ns/100ps

module oc_nibble_nand (a, b, y);
input [3:0] a, b;
output [3:0] y;
        function [3:0] oc;
        input [3:0] x;
        parameter [1:0] convert = 2'bz0;
        integer i;
                for (i=3; i>=0; i=i-1) oc[i] = convert [x[i]];
        endfunction
        assign y = oc(~a | ~ b);
endmodule
```

Figure 8.18 Dataflow description of open-collector NAND gates.

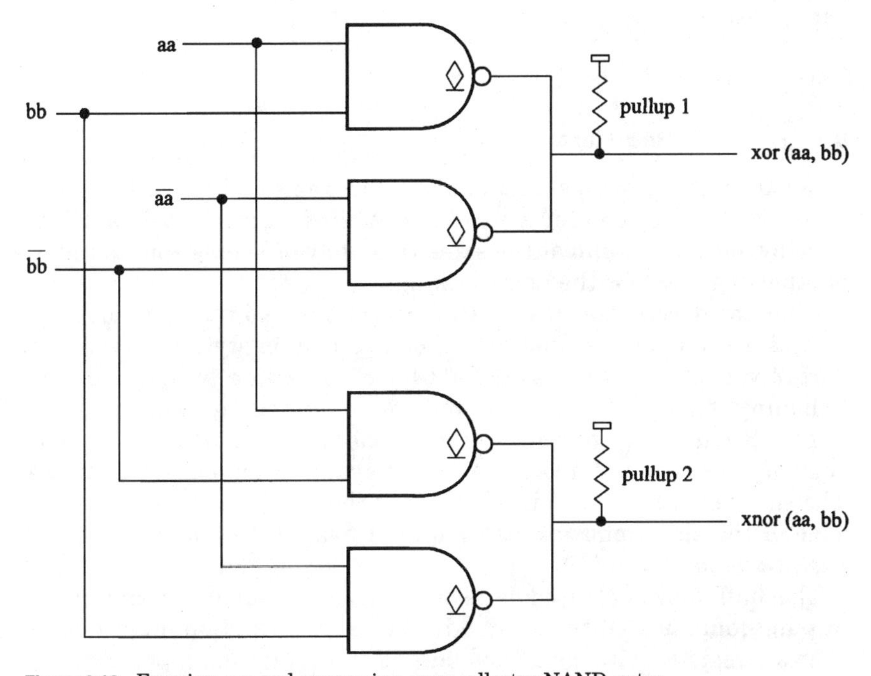

Figure 8.19 Forming *xor* and *xnor* using open-collector NAND gates.

Execution of a nonblocking procedural assignment is performed at zero time. Such a statement causes a value to be scheduled into the left-hand-side **reg** with a specified time delay. The statement that follows a nonblocking statement is processed at the same simulation time. Each of the for loops in Fig. 8.20 causes scheduling of 15 values into *aa* and *bb*. The **initial** block in this figure is useful for generating arbitrary data for test purposes.

```
`timescale 10ns/1ns

module test_oc_nibble_nand;
reg aa, bb;
reg [3:0] i;
    oc_nibble_nand u1 (
        {aa,        ~aa,        aa,        ~aa},
        {bb,        ~bb,        ~bb,        bb},
        {pullup1, pullup1, pullup2, pullup2});
    pullup
        (pullup1),
        (pullup2);
    initial begin
        for (i=0; i<=14; i=i+1) aa <= #(i*30) i[0];
        for (i=0; i<=14; i=i+1) bb <= #(i*40) ~i[0];
        #200 $stop;
    end
endmodule
```

Figure 8.20 Code for test bench for *xor-xnor* circuit.

8.2 Dataflow Registers

In a bus system, bus destinations may be register inputs. A register stores bus data when they are clocked. While in a register, data will be used by logic units, and at the same time they are being sent to inputs of other registers for the next clocking.

After the description of various forms of busses in the previous section, this section is devoted to Verilog description styles for registers.

Figure 8.21 shows the use of a 1-bit register in a bus-register system. Register inputs and outputs are connected to system busses. Figure 8.21*a* shows a register that is clocked with *abus* with every edge of the clock, and makes its output available on *bbus* at all times. The register in Fig. 8.21*b* is loaded with the contents of *abus* on every edge of the clock, and makes its output available on *bbus* when the *reg1_on_bbus* signal is **1.**

The buffer symbol used here is a generic symbol for any of the bussing configurations discussed in the previous section. Figure 8.21*c* shows a register with controlled output and controlled input. Data on *abus* will be clocked into *reg1* only on the rising edge of the clock when the *load_reg1* signal is active. Finally, in Fig. 8.21*d,* the *reg1* register may synchronously be reset if the *reset_reg1* signal is active.

The Verilog descriptions presented here will concentrate on various coding styles for the register in Fig. 8.21*a,* and will then extend them to cover other register types used in other forms of bussings.

8.2.1 Edge-trigger flip-flop with setup delay

Figure 8.22 shows a rising edge D-type flip-flop that suits the usage shown in Fig. 8.21*a.* Two concurrent **always** blocks are used for set-

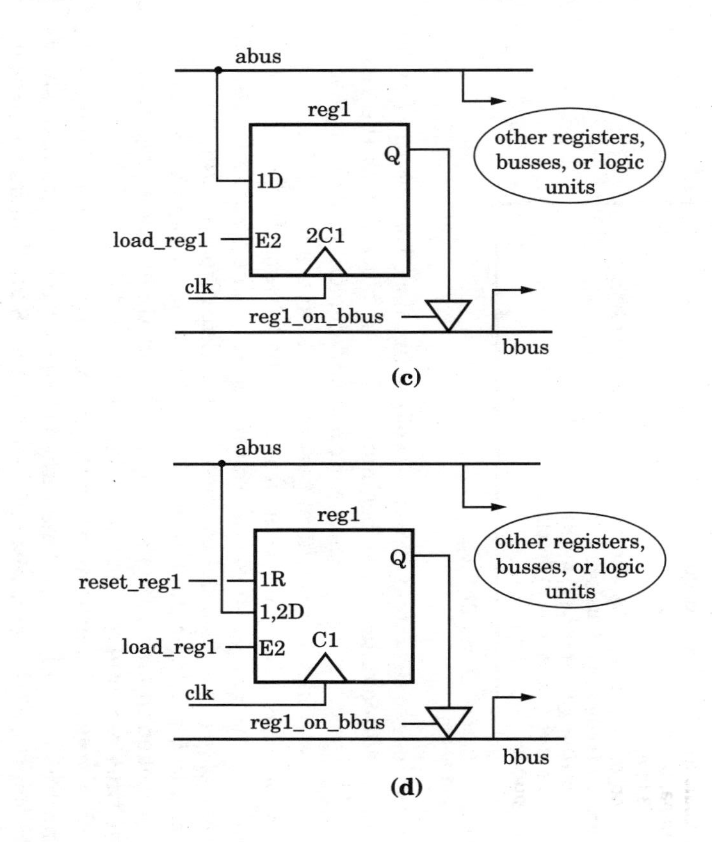

Figure 8.21 A 1-bit register in a dataflow description.

```
`timescale 1ns/100ps

module d_flipflop (d, c, q, qb);
input d, c;
output q, qb;
reg q, qb; // default is wire, use reg in concurrent statements
parameter delay1 = 4, delay2 = 5;
      always @(posedge c) #(delay1) q = d;
      always @(posedge c) #(delay2) qb = ~ d;
endmodule
```

Figure 8.22 Basic D-type flip-flop.

ting values to the *q* and *qb* outputs of the flip-flop. The *posedge c* event control statements at the beginning of each block detect the rising edge of the clock input. An event control statement controls the activity flow within the procedural part of an **always** statement.

Another activity flow control statement (*timing_control*) causes a wait of 4 ns (*delay1*) before *d* is assigned to *q*. The waiting for assignment of ~*d* to *qb* is determined by the *delay2* parameter, which is equal to 5 ns. After the rising edge of the clock, there is a wait before *d* is looked up and assigned to *q*. This wait time may be regarded as the time required for the clock edge to set up.

The value of *d* at 4 ns after the rising edge of *c* is assigned to *q*, and the value of ~*d* at 1 ns later is used for *qb*. Figure 8.23 shows the syntax of the first **always** block in Fig. 8.22. The *statement* construct that follows the **always** keyword consists of three statements. The first two statements are of the *procedural_timing_control_statement* type. The syntax details of the first of these statements are shown in Fig. 8.24. This statement consists of an *event_control* that begins with an at sign, @, and specifies an event for the enclosed expression.

Procedural timing control statements may be *event_control* or *delay_control*. Details of a *delay_control* statement, which begins with a sharp sign, #, are shown in Fig. 8.25. This figure shows that a complete expression may substitute for the *delay1* parameter that we used in Fig. 8.22.

8.2.2 Edge-trigger flip-flop with propagation delay

A major problem with the code in Fig. 8.22 is that after a 4-ns wait time, the *q* output immediately, with zero delay, receives the value of *d*. This arrangement does not account for gate delays between the flip-flop *d* input and its output.

Another model for a D-type flip-flop is shown in Fig. 8.26. In this code, two **always** statements are used for the two outputs of the flip-flop. Activity flow begins in each block with the rising edge of the clock. Expressions *d* and ~d are evaluated and, after the specified delay, the values are assigned to *q* and *qb*. Despite the similarities between the Verilog descriptions in Figs. 8.22 and 8.26, flow control in the two descriptions is very different. Unlike the description in Fig. 8.22, the description in Fig. 8.26 uses only one *procedural_timing_control_statement* in each block. The delay values in Fig. 8.26 are associated with the blocking assignment statements and do not constitute procedural timing control statements. The delay control in this figure is referred to as *intra_assignment* timing control.

Figure 8.27 clarifies the difference in timing and syntax. The *delay_or_event_control* in Fig. 8.26 is part of the *blocking_assignment* statement shown. The activity flow into this entire assignment statement is controlled only by the *@(posedge c) event_control* statement. On the other hand two procedural timing control statements control activity flow in the code of Fig. 8.22 (see also Figs. 8.23 and 8.25). As indicated in Fig. 8.23, in addition to the *event_control,* a *delay_control* statement also controls activity flow before the *q*=d blocking statement is processed.

After the rising edge of the clock, flows of activities reach the assignments in Fig. 8.26. At this time the values of the *d* and *~d* expressions are evaluated. The actual assignment of these values to *q* and *qb* is delayed by the *delay1* and *delay2* parameters.

Figure 8.28 shows a simulation run for the two flip-flops of Figs. 8.22 and 8.26 for the same test data. As shown in Fig. 8.28*a*, at

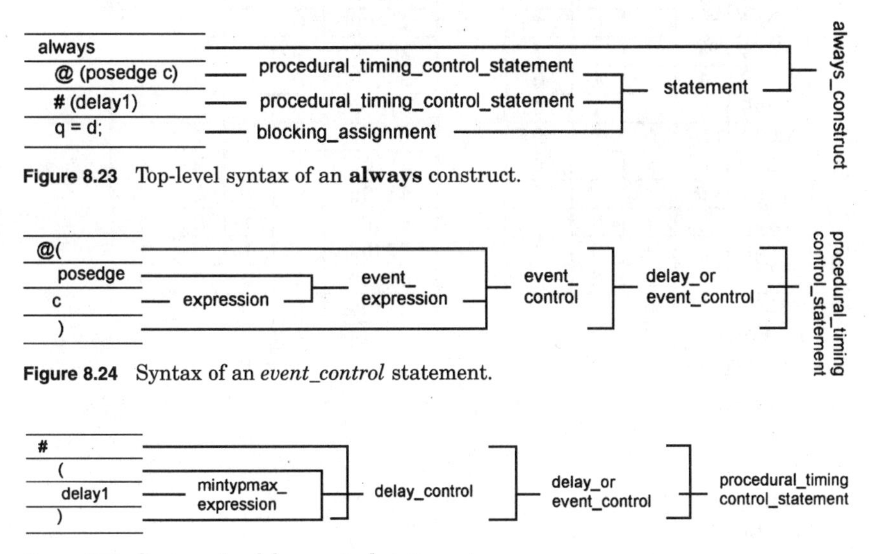

Figure 8.23 Top-level syntax of an **always** construct.

Figure 8.24 Syntax of an *event_control* statement.

Figure 8.25 Syntax of a *delay_control* statement.

```
`timescale 1ns/100ps

module d_flipflop (d, c, q, qb);
input d, c;
output q, qb;
reg q, qb;
parameter delay1 = 4, delay2 = 5;
    always @(posedge c) q = #delay1 d;
    always @(posedge c) qb = #delay2 ~ d;
endmodule
```

Figure 8.26 *d* input values are delayed.

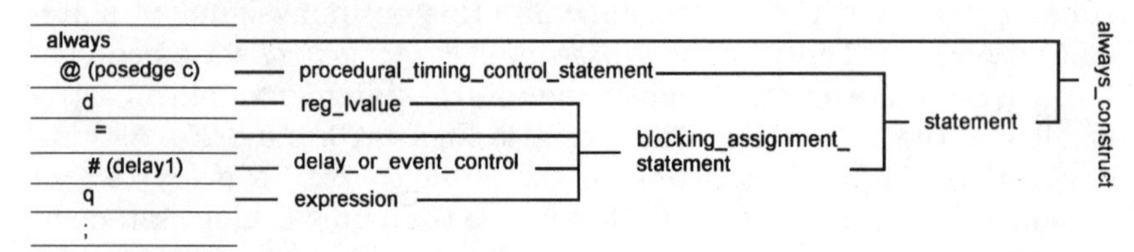

Figure 8.27 Syntax details of the **always** block in Fig. 8.26.

ps	d	c	q	qb
0	0	0	x	x
200000	0	1	x	x
204000	0	1	0	x
205000	0	1	0	1
300000	0	0	0	1
410000	*0*	*1*	*0*	*1*
413000	*1*	*1*	*0*	*1*
414000	*1*	*1*	*1*	*1*
415000	1	1	1	0
513000	1	0	1	0
533000	0	0	1	0
643000	*0*	*1*	*1*	*0*
647000	*0*	*1*	*0*	*0*
648000	0	1	0	1
733000	1	1	0	1
853000	1	0	0	1
873000	0	0	0	1

(a)

ps	d	c	q	qb
0	0	0	x	x
200000	0	1	x	x
204000	0	1	0	x
205000	0	1	0	1
300000	0	0	0	1
410000	*0*	*1*	*0*	*1*
413000	*1*	*1*	*0*	*1*
513000	1	0	0	1
533000	0	0	0	1
643000	0	1	0	1
733000	1	1	0	1
853000	1	0	0	1
873000	0	0	0	1

(b)

Figure 8.28 Comparing simulation of *d_flipflops* in (*a*) Fig. 8.22 and (*b*) Fig. 8.26.

410,000 ps, the clock edge starts the activity flow of **always** blocks of Fig. 8.22. The activity flow is then delayed for 4 ns in the block assigning *d* to *q*. After this time, the new value of *d* is **1,** which is read and immediately assigned to *q*. Figure 8.28*a* shows *q* changing at exactly 414 ns.

A simulation run for the flip-flop of Fig. 8.26 using the same test data yields the listing in Fig. 8.28*b,* which is significantly different from that of Fig. 8.28*a*. At 410,000 ps, after the positive edge of the clock, activity flow in both blocks in this code begins and immediately reaches statements assigning *d* and *~d* to *q* and *qb*. The *d* and *~d* expressions are evaluated and assigned to *q* and *qb* after 4 ns and 5 ns. Since the value of *d* at 410 ns time is **0,** this value will be clocked into the *q* output of the flip-flop of Fig. 8.26. Therefore, unlike the listing in Fig. 8.28*a,* the *q* and *qb* outputs in Fig. 8.28*b* are not affected by the **1** value on *d* at 413,000 ps.

This and the previous section presented two delay mechanisms that may be used in dataflow description of registers. These schemes may be combined for more accurate representation of delays caused by internal gates of flip-flops. Examples in the sections that follow will illustrate this.

8.2.3 Alternative flip-flop coding styles

The flip-flop descriptions in the previous two sections used an **always** block for each of the *q* and *qb* flip-flop outputs. The code shown in Fig. 8.29 shows an alternative description for a D-type flip-flop. This description uses only one **always** statement that encloses two blocking procedural assignments. The *d_flipflop* module of Fig. 8.29 has only one activity flow, which starts executing each time the positive edge of the clock is detected.

```
`timescale 1ns/100ps

module d_flipflop (d, c, q, qb);
input d, c;
output q, qb;
reg q, qb;
parameter delay1 = 4, delay2 = 5;
always @(posedge c) begin
    q = #delay1 d;
    qb = #(delay2 - delay1) ~ d;
end
endmodule
```

Figure 8.29 *d_flipflop* using a single **always** statement.

The first assignment waits for the edge of the clock, reads the value of *d,* and assigns this value to *q* after *delay1.* This blocking assignment blocks the execution of what follows it for *delay1* time period. The second blocking statement, which is executed *delay1* time period after the edge of the clock, evaluates *~d* and waits *delay2 - delay1* time period before assigning it to *qb.* As a result, *delay1* is used for propagation of *d* to *q,* and at the same time it is used for clock setup for *qb* clocking. The value *delay2 - delay1* is used for propagation of *~d* to *qb.*

Unlike the descriptions in Secs. 8.2.1 and 8.2.2, the code in this section causes the delay of *qb* to be dependent on that of the *q* output. For more independent specification of delay values, the coding styles in Figs. 8.22 and 8.26 are preferred. These flip-flop descriptions consist of concurrent independent activity flow constructs for each of the flip-flop outputs.

8.2.4 Transparent D-latch

The event control statements shown in the flip-flop descriptions in the previous section included the **posedge** keyword for rising edge detection on *c.* Verilog allows the use of other event expressions for describing various clocking schemes and flip-flop set and reset mechanisms.

Figure 8.30 shows a *d_latch* using an **always** statement. An event control statement, *@(c or d),* blocks the execution flow until an event occurs on *c* or *d.* When such an event happens, a delay control statement blocks the flow for *setup_delay,* which is equal to 5 ns. After this time elapses, the procedural blocking assignment assigning values to *q* executes.

The syntax of the event control statement is shown in Fig. 8.31. The event expression enclosed in this statement uses the **or** operator to form the ORing of events on *c* or *d* as events that allow activity flow of the **always** block.

When flow reaches the blocking assignment statement in Fig. 8.30, the right-hand-side operation is evaluated, and after *prop_delay,* it is

```
`timescale 1ns/100ps

module d_latch (d, c, q);
input d, c;
output q;
reg q, qb;
parameter prop_delay = 4, setup_delay = 5;
    always @(c or d) #(setup_delay) q = #(prop_delay) c===1 ? d : q;
endmodule
```

Figure 8.30 D-latch Verilog code.

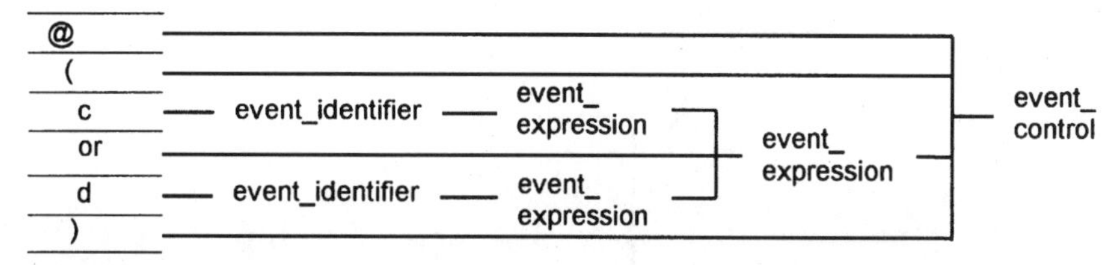

Figure 8.31 Using **or** in an event expression.

assigned to *q*. The condition operator in the expression on the right-hand side of *q* evaluates to *d* if *c* is **1**; if *c* is **0, Z,** or **X,** it evaluates to *q*. This value becomes the next value for *q*.

When values of *c* or *d* change, after a delay, the value of *c* is examined, and if *c* === 1 is true, the value of *q* will be updated with the *d* input. Because this **always** statement is sensitive to *c* and *d* levels, it implements a level-sensitive transparent D-type latch.

8.2.5 D flip-flop with asynchronous control

Often flip-flops use asynchronous set and reset inputs. When such an input becomes active, it affects the state of the flip-flop directly, without having to wait for a clock edge. A notation often used for these flip-flops is shown in Fig. 8.32.

The notation used here indicates dependence of the *D* input on *C* by use of number *1* on the right-hand side of *C,* the controlling input, and the left-hand side of *D,* the controlled input. Based on this notation, input *R* acts independently of the clock. Figure 8.33 shows Verilog code that corresponds to the flip-flop in Fig. 8.32.

The code in Fig. 8.33 uses two concurrent activities starting with **always** statements. The event control statements, @(***posedge*** *c* ***or posedge*** *r*), used in these **always** statements block the execution of blocking assignment statements until *c* **or** *r* changes to **1.** At any time, if *c* changes to **1,** activity flow in the first **always** statement reaches the assignment to the *q* output. The condition operator on the right-hand side of *q* checks for an active *r.* If *r* is **0,** *d* will be clocked into *q*; otherwise, the flip-flop will be reset by assigning a **0** to *q*.

Although edges on both *r* and *c* are detected in the event control statement, the way *r* and *c* affect the output are different. The *q* output will be held at **0** while *r* is **1**; however, when *q* is set to *d* on the positive edge of *c,* it does not necessarily hold to its value while the clock is **1.** This behavior is achieved by checking the value of *r* any time a value is being assigned to *q*. In effect, the *r* input has higher priority than the clock, and active *r* values override values determined by *c* and *d.*

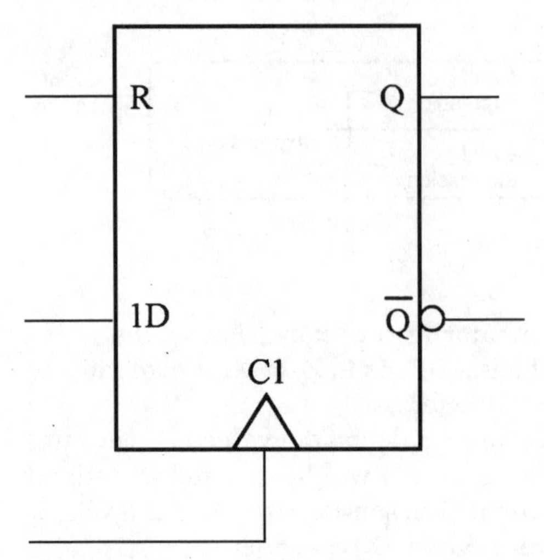

Figure 8.32 D-type flip-flop with asynchronous reset.

```
`timescale 1ns/100ps

module d_ar_flipflop (d, r, c, q, qb);
input d, r, c;
output q, qb;
reg q, qb;
parameter delay1 = 4, delay2 = 5;
    always @(posedge c or posedge r) q = #delay1 r ? 0 : d;
    always @(posedge c or posedge r) qb = #delay2 r ? 1 : ~d;
endmodule
```

Figure 8.33 Verilog code for a D-type flip-flop with asynchronous reset.

8.2.6 Flip-flop with asynchronous enable input

Figure 8.21*c* shows a flip-flop with an active high clock-enabling input. The clock edge is effective only if enable is **1.** Verilog code corresponding to this flip-flop is shown in Fig. 8.34.

An **always** statement sets *q* on the positive edge of *c*. Because all flip-flop activities are synchronized with the clock edge, only *posedge c* is used as an event expression in the event control statement that controls activity flow to reach the assignment that sets values to *q*. However, when this assignment statement executes, it will change the output only if *e* is active. Assigning *q* to *q* when *e* is not **1** implements clock disabling when *e* is not active.

```
`timescale 1ns/100ps

module de_flipflop (d, e, c, q, qb);
input d, e, c;
output q, qb;
reg q, qb;
parameter delay1 = 4, delay2 = 5;
    always @(posedge c) q = #delay1 e ? d : q;
    always @(posedge c) qb = #delay2 e ? ~d : qb;
endmodule
```

Figure 8.34 Verilog code for D-type flip-flop with enable input.

Before ending this section, a test bench for the flip-flop of Fig. 8.34 that can also be modified to test other structures presented in this section will be discussed. Figure 8.35 shows a test bench that instantiates *de_flipflop.* Three concurrent statements in this module determine test values for the *c, d,* and *e* inputs. The first such statement, **initial,** initializes *c* and *d,* puts a sequence of **1**s and **0**s on *e,* and ceases with a **$stop** system task.

The two **always** statements in the module of Fig. 8.35 execute concurrently, repeat forever, and generate periodic data on the *c* and *d* inputs. A delay control statement, *#10,* in the first **always** statement causes a wait of 100 ns before the blocking assignment that follows it is executed. When activity flow reaches this statement, the value of *c* is complemented and assigned back to it in zero time. Since this is the last procedural statement in this **always** block, activity flow returns to the delay control statement after completion of the $c = \sim c$ assignment. This process continues until activity flow reaches the **$stop** task in the **initial** block, which is running concurrently with the **always** statements.

8.3 State Machine Descriptions

In a dataflow description, the flow of data between busses and registers is controlled by signals coming from a controller. Generally the same clock used for register clocking is also used for generating control signals. Controllers are state machines with outputs fanning out to control inputs of registers and busses.

State diagrams are used to graphically represent state machines. An important part of digital systems, state machines can appear explicitly in a digital system for the control and sequencing of events, or they can be embedded in sequential components, such as counters and shift registers. At the dataflow level, where we separate control and data of a hardware system, the design and description of state machines for implementing the control unit become important. Verilog provides

```
`timescale 10ns/1ns

module test_de_flipflop;
reg d, e, c;
wire q, qb;
de_flipflop u1 (d, e, c, q, qb);
initial
  begin
     c = 1'b0; d= 1'b0;
     #10 e = 1; #91 e = 0;
     #60 e = 1; #91 $stop;
  end
  always #10 c = ~ c;
  always #13 d = ~ d;
endmodule
```

Figure 8.35 A test bench for *de_flipflop*.

convenient constructs to describe various forms of state machines at various levels of abstraction. At the dataflow level, a description of a state machine has a close correspondence to the state diagram of the machine. We will use **always** statements and condition operators in this section to accurately describe state machines. Other simpler state machine representations will be shown in the next chapter.

8.3.1 A sequence detector

A sequence detector is a classical example of an application of state machines in hardware. Figure 8.36 shows the state diagram for a sequence detector that continuously searches for the 1011 sequence on its x input. This diagram is a Mealy machine, which means that while the machine is in a stable state, input changes can affect the circuit output.

The states of this machine are labeled according to the significant input sequences they detect. For example, in the reset state, if a **1** followed by a **0** appears on the x input, the machine moves to the *10* state.

The Verilog description in Fig. 8.37 corresponds to the state diagram of the 1011 detector. The declarations specify inputs and outputs of the circuit as 1-bit signals. The **reg** declaration declares the 2-bit *current* variable for two state variables. Two state variables are enough for representing four states of the machine.

A parameter declaration declares the state names as 2-bit parameters with 0, 1, 2, and 3 values. The *detector* module in Fig. 8.37 uses an **initial** statement for initializing *current* to *reset* state, an **always** statement for state transitions, and a continuous assignment for setting the circuit output.

Activity flow that starts with the **always** statement is halted by the event control statement waiting for the positive edge of *clk*. After this

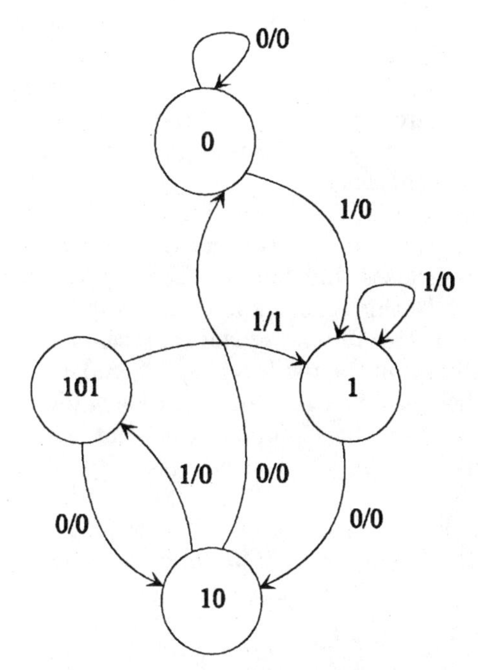

Figure 8.36 A 1011 Mealy sequence detector. State names indicate detected sequences.

```
`timescale 1ns/100ps

module detector (x, clk, z);
input x, clk;
output z;
reg [1:0] current;
parameter [1:0]     reset = 0,
                    got1 = 1,
                    got10 = 2,
                    got101 =3;
initial current = reset;
always @(posedge clk) current =
            current == reset  ? (x == 1 ? got1 : reset) :
            current == got1   ? (x == 0 ? got10 : got1) :
            current == got10  ? (x == 1 ? got101 : reset) :
            current == got101 ? (x == 1 ? got1 : got10) : reset;
assign z = current == got101 ? (x == 1 ? 1 : 0) : 0;
endmodule
```

Figure 8.37 Verilog description for 1011 detector.

occurs, activity flow reaches the only blocking assignment statement in this **always** block. The left-hand side of this statement is the *current* variable, and its right-hand side consists of an expression with nested use of the conditional operator. Evaluation of the right-hand side of *current* and assignment of a new value to this **reg** type variable repeats with every positive edge of the clock.

Here, we will describe the operation of the expression on the right-hand side of the blocking statement in the **always** block. If *current* is equal to 0 (*reset*), the set of parentheses that follows the *current* == *reset* expression will be evaluated. In this case, if x is **1,** the expression evaluates to *got1*; otherwise it evaluates to *reset.* If the expression *current* == *reset* is not true, the else part of the top-level conditional operator will be evaluated. This else part is itself an expression using nested conditional operators. This part checks for *current* being equal to *got1.* If none of the conditions are true, the last else part is taken, which sets *current* to *reset.*

Using a continuous assignment statement, the circuit output z is set to **1** if the current state of the machine is *got101* and the x input is **1.** As seen in Fig. 8.37, the right-hand side of z uses nested conditional operators.

8.3.2 Altering state flow

The flow of states of a state machine may be externally altered by the use of event control and the conditions described in relation to flip-flops in Sec. 8.2.

For adding an asynchronous reset input that puts the state machine of Fig. 8.37 in an initial state, the style used in Fig. 8.33 for the *d_ar_flipflop* should be used. For this purpose, the event expression of the **always** block must be modified to include the reset input, and a top-level conditional operator must be added to the right-hand side of *current.* The blocking assignment in the *detector* module should be modified as follows:

```
current = asynch_reset == 0 ? (present rhs) : reset;
```

8.3.3 Outputs of Mealy and Moore machines

Mealy and Moore machines make their state transitions in the same manner. The coding style used in the **always** statement that corresponds to state transitions applies to both Mealy and Moore machines. The main difference between these machines is the way they assign values to the outputs.

In a Mealy machine, inputs and states of the machine participate in the formation of conditions for assigning values to the outputs. In a Moore machine, the states alone are used in conditional signal assignments to the output signals. In either machine, several states can provide values for the output signals. This either requires **wor** type outputs with multiple continuous assignments or requires incorporating various conditions into the expression on the right-hand side of the circuit output. The advantage of using **wor** type outputs is that assignments to the outputs can be placed in the description next to the state making the assignment, instead of combining all conditions into a condition for a single continuous assignment.

Conditions for assigning values to outputs of a Moore machine do not include input values. Only states of a state machine are needed in output expressions for a Moore machine. For example if the *z* output of a Moore machine is to become **1** if the machine is in *state_a* or *state_b,* the following assignment may be used:

```
assign z = current == state_a | current == state_b;
```

8.4 General Dataflow Circuits

Application of dataflow descriptions is not limited to specification of bussing structures, simple register structures, and state machines, described in the previous sections. Hardware descriptions at this level can be used to describe a complete sequential circuit consisting of registers, combinational units, busses, and complex control algorithms. Using several examples, we show how word specification of a clocked sequential circuit can be translated into dataflow hardware descriptions.

The circuit to design is a sequential comparator that keeps a modulo-16 count of matching consecutive data set pairs. The circuit uses an 8-bit *data,* a *clk,* and a *reset* input. The 4-bit output is called *matches.* If on any two consecutive rising edges of the clock, the same data appear on *data,* then the output will be incremented by 1. The synchronous reset of the circuit resets the output count to zero.

The hardware implementation of this circuit, using standard parts, requires a register for holding the old data, a comparator for comparing new and old data, a counter for keeping the count, and perhaps a few logic gates used as glue logic. At the dataflow level, however, there is no need to be concerned with the component-level details of this circuit; rather, the flow of data into circuit registers can be captured directly in a Verilog description of this unit.

The Verilog code for the *sequential_comparator* is shown in Fig. 8.38. The circuit uses two registers that are described using **always** statements. One register clocks new *data* into *buff,* and another register keeps the count of matching data on *data* and *buff.*

```
`timescale 1ns/100ps

module sequential_comparator (data, clk, reset, matches);
input [7:0] data;
input clk, reset;
output [3:0] matches;
reg [3:0] matches;
reg [7:0] buff;
    always @(negedge clk) buff = data;
    always @(negedge clk)
        matches = (reset == 1) ? 0 : (data == buff) ? matches + 1 : matches;
endmodule
```

Figure 8.38 Dataflow description of the sequential comparator circuit.

The **always** statement that implements the counter uses an event control statement and a blocking assignment. The event control statement checks for the negative edge of *clk,* which makes operations in this **always** and the **always** statement setting *buff* synchronous. The right-hand side of the *matches* **reg,** which keeps the count, is a nested conditional expression. Synchronous with *clk,* if *reset* is **1,** *matches* becomes **0.** Otherwise, *matches* is incremented if *data* and *buff* are equal and holds its old value if they are not.

If at simulation time *t,* the clock makes a 1 to 0 transition, the data on the *data* input lines are assigned to *buff.* The new data in *buff* will not be available until the next simulation time. At time *t,* the old data in *buff* are compared with what appears on *data* at time *t.* If these data sets are equal, the *matches* variable is incremented. The result of this incrementing becomes available on *matches* during the next simulation time, which is the same time that the new data appear on *buff.*

Figure 8.39 shows the sequence of events in the *sequential_comparator.* Understanding timing and clocking is essential to understanding dataflow and to being able to use this level of abstraction for the description of systems.

A second dataflow example will demonstrate bussing, registers, and controlled flow. The example system has three registers, an input bus, and an output bus. It receives data on its input, performs several add and subtract operations on these data, and makes the result available on the system output. An RTL description for this system in the form of a flowchart is shown in Fig. 8.40.

We will implement this system by designing a bussing structure and a controller that controls the flow of data into busses and registers. Bus, register, and state machine styles presented in Secs. 8.1, 8.2, and 8.3 will be used for describing various parts of this system. The data path shown in Fig. 8.41 uses two busses and provides all paths for the transfers shown in Fig. 8.40 to take place.

ps	data	clk	reset	count
0	00000000	0	1	xxxx
0	00000000	0	1	0000
200000	11110101	0	0	0000
500000	11110101	1	0	0000
1000000	11110101	0	0	0000
1200000	01010110	0	0	0000
1500000	01010110	1	0	0000
1700000	11111110	1	0	0000
2000000	11111110	0	0	0000
2500000	11111110	1	0	0000
3000000	11111110	0	0	0001
3200000	01010100	0	0	0001
3500000	01010100	1	0	0001
3700000	00010001	1	0	0001
4000000	00010001	0	0	0001
4200000	10010110	0	0	0001
4500000	10010110	1	0	0001
5000000	10010110	0	0	0001
5500000	10010110	1	0	0001
6000000	10010110	0	0	0010
6500000	10010110	1	0	0010
7000000	10010110	0	0	0010
7000000	10010110	0	0	0011
7500000	10010110	1	0	0011
8000000	10010110	0	0	0011
8000000	10010110	0	0	0100
8500000	10010110	1	0	0100

Figure 8.39 Sequence of events in *sequential_comparator.*

The system has three 8-bit registers, *reg1, reg2,* and *reg3.* An adder-subtractor circuit performs logical add or subtract operations on its two 8-bit inputs. Two system busses, *abus* and *bbus,* are used for the flow of data into and out of registers and the adder-subtractor circuit.

Connections to busses are controlled or direct. For controlled bus connections, a triangle symbol with a signal name identifying its control is used. For direct control, an arrow specifies the connection.

All bus and register control signals are named according to their functionality. For example, the *enable* input of *reg1,* which will be referred to as *enable_reg1,* enables *reg1* clocking. The signal *inc_reg3* enables synchronous incrementing of *reg3* when it is active. When the control signal *reg2_on_abus* is active, *reg2* outputs will drive *abus.* If at the same time *abus_on_bbus* is also active, the contents of *reg2* will become available on *bbus* through *abus.* Data on *bbus* may be clocked into either of the two system registers, *reg1* and *reg2,* or into the *reg3* counter/register if a corresponding *enable* signal is active.

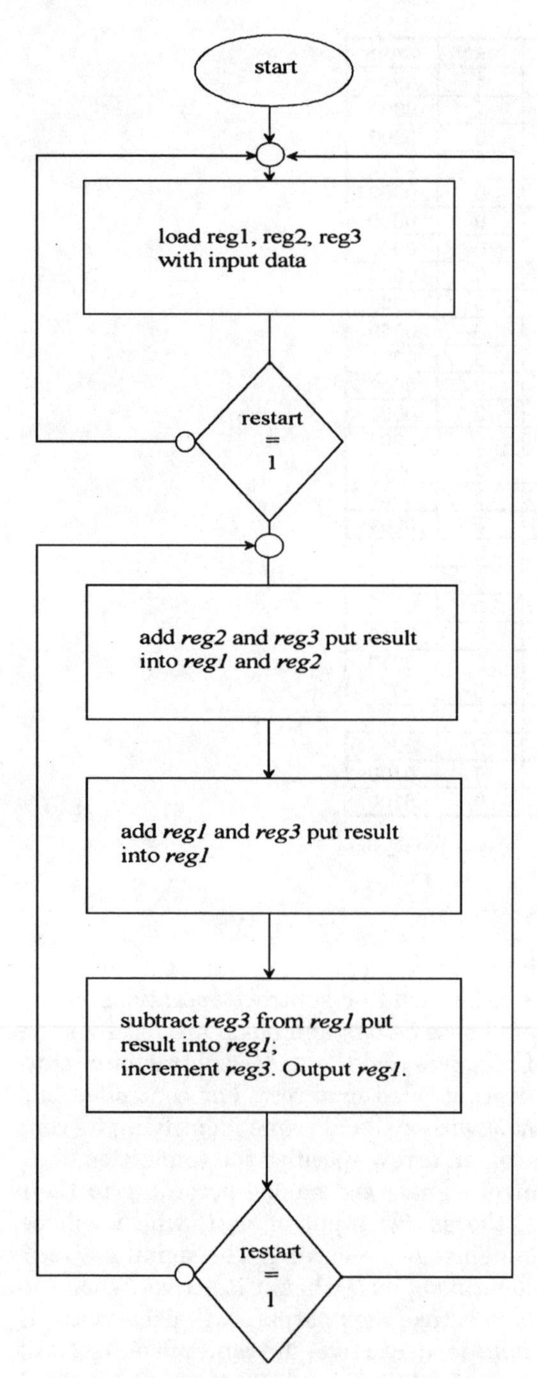

Figure 8.40 Dataflow example flowchart.

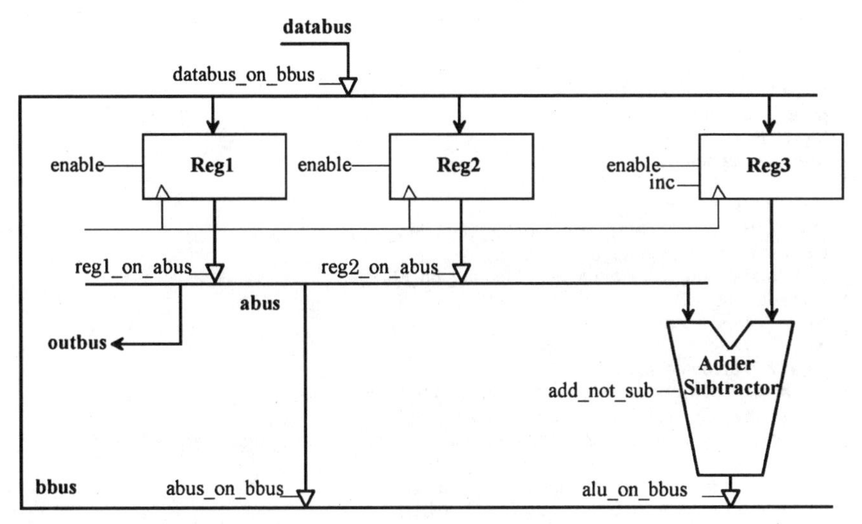

Figure 8.41 An example datapath with busses, registers, and controller.

Figure 8.42 shows the Verilog code that corresponds to the data path of Fig. 8.41. In the declarations, registers are declared as **reg,** since they will be assigned values in Verilog procedural blocks. All other variables are declared as three-state or wired-or **net**s. System busses are declared as three-state busses, **tri,** to correspond to the bus structure shown in Fig. 8.11*b*. Control lines are declared as **wor** so that they can be activated by multiple control states. The last item in the declarations declares *outbus* as an 8-bit bus and wires it to the *abus* system bus.

The second half of Fig. 8.42 shows register specifications and bus connections. Registers are specified with a synchronous enable input and use the register format described in Sec. 8.2.6, Fig. 8.34. The **always** statement for *reg3* uses a right-hand-side conditional expression to incorporate a synchronous increment input into this register. Following register specifications, the last part of Fig. 8.42 shows conditional continuous assignments for bus input connections. Bus connections follow the format shown in Fig. 8.16*b,* where a control signal, e.g., *reg1_on_abus,* constitutes the condition for the right-hand-side conditional expression. The last statement in Fig. 8.42 shows *alu* output receiving the add or subtract result of its inputs, conditioned by the *add_not_sub* control signal.

Control signals in Fig. 8.41 or the corresponding Verilog code of Fig. 8.42 are generated from a control circuit represented by the state machine shown in Fig. 8.43. Vertical lines designating state outputs connect to OR gates to generate control signals. The control signals

```
`timescale 1ns/100ps

module dataflow_structure (databus, outbus, clk, restart);
input [7:0] databus;
input clk, restart;
output [7:0] outbus;

reg [7:0] reg1, reg2, reg3;
wor enable_reg1, enable_reg2, enable_reg3, inc_reg3;
wor reg1_on_abus, reg2_on_abus;
wor databus_on_bbus, abus_on_bbus, alu_on_bbus;
wor add_not_sub;
tri [7:0] abus, bbus;
wire [7:0] alu;
wire [7:0] outbus = abus;
    always @(negedge clk) reg1 = enable_reg1 ? bbus : reg1;
    always @(negedge clk) reg2 = enable_reg2 ? bbus : reg2;
    always @(negedge clk) reg3 = enable_reg3 ? bbus : (inc_reg3 ? reg3 + 1 : reg3);

    assign abus = reg1_on_abus ? reg1 : 8'bz;
    assign abus = reg2_on_abus ? reg2 : 8'bz;
    assign bbus = databus_on_bbus ? databus : 8'bz;
    assign bbus = abus_on_bbus ? abus : 8'bz;
    assign bbus = alu_on_bbus ? alu : 8'bz;

    assign alu = add_not_sub ? abus + reg3 : abus - reg3;
    . . .
endmodule
```

Figure 8.42 Verilog code corresponding to the data path of Fig. 8.41.

issued by this controller connect to signals with the same names in Fig. 8.41. The flow of data in Fig. 8.41 is controlled by signals issued by the controller.

Verilog code that corresponds to Fig. 8.43 is shown in Fig. 8.44. The style used for this code is that of Sec. 8.3, Fig. 8.37. Four states of the machine are named *a, b, c,* and *d,* according to Fig. 8.43. After the declarations and the initialization of *current,* the first part of Fig. 8.44 describes state transitions shown in the right-hand side of Fig. 8.43. Continuous assignment statements in the lower part of Fig. 8.44 issue control signals based on states of the state machine. Four such statements are used for control signals issued in each state. State *a* activates *enable_reg1, enable_reg2, enable_reg3,* and *databus_on_bbus.* The continuous assignment shown concatenates these four signals and assigns 1111 to them if *current* is *a.* Other activities are done in the same fashion.

The complete system corresponding to the flowchart of Fig. 8.40 consists of the wiring diagrams in Figs. 8.41 and 8.43. The complete Verilog code, which includes the code shown in Figs. 8.42 and 8.44, is shown in Fig. 8.45.

8.5 Summary

A complete dataflow description consists of busses, registers, and state machines, as well as logic units performing operations on their inputs connected to registers and busses. This chapter presented basic coding styles for various types of busses, registers, and state machines. The chapter demonstrated how each component may be described and how these components are interconnected into a complete system at the dataflow level. In order to emphasize language constructs at the dataflow level, we limited our descriptions to concurrent statements with a minimum procedural coding. Procedural

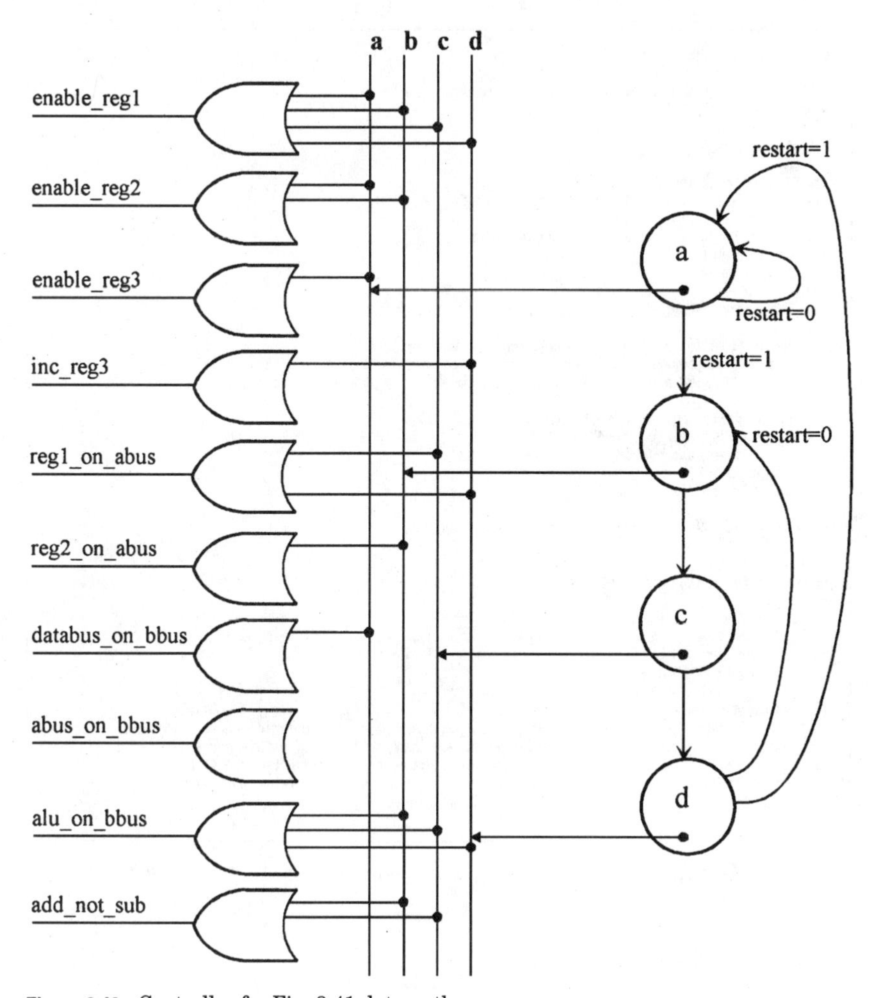

Figure 8.43 Controller for Fig. 8.41 data path.

```
reg [1:0] current;
parameter [1:0] a = 0, b = 1, c = 2, d = 3;
initial current = a;
     always @(negedge clk) current =
          current == a ? (restart == 1 ? b : a) :
          current == b ? c :
          current == c ? d :
          current == d ? (restart == 1 ? a : b) : a;
     assign abus_on_bbus = 0;
     assign {databus_on_bbus, enable_reg1, enable_reg2, enable_reg3} = (current == a) ? 4'b1111 : 0;
     assign {reg2_on_abus, add_not_sub, alu_on_bbus, enable_reg1, enable_reg2} = (current == b) ?  5'b11111 : 0;
     assign {reg1_on_abus, add_not_sub, alu_on_bbus, enable_reg1} = (current == c) ? 4'b1111 : 0;
     assign {reg1_on_abus, alu_on_bbus, enable_reg1, inc_reg3} = (current == d) ? 4'b1111 : 0;
```

Figure 8.44 Controller description for the design described in Fig. 8.40.

```
`timescale 1ns/100ps

module dataflow_structure (databus, outbus, clk, restart);
input [7:0] databus;
input clk, restart;
output [7:0] outbus;

reg [7:0] reg1, reg2, reg3;
wor enable_reg1, enable_reg2, enable_reg3, inc_reg3;
wor reg1_on_abus, reg2_on_abus;
wor databus_on_bbus, abus_on_bbus, alu_on_bbus;
wor add_not_sub;
tri [7:0] abus, bbus;
wire [7:0] alu;
wire [7:0] outbus = abus;

     always @(negedge clk) reg1 = enable_reg1 ? bbus : reg1;
     always @(negedge clk) reg2 = enable_reg2 ? bbus : reg2;
     always @(negedge clk) reg3 = enable_reg3 ? bbus : (inc_reg3 ? reg3 + 1 : reg3);

     assign abus = reg1_on_abus ? reg1 : 8'bz;
     assign abus = reg2_on_abus ? reg2 : 8'bz;
     assign bbus = databus_on_bbus ? databus : 8'bz;
     assign bbus = abus_on_bbus ? abus : 8'bz;
     assign bbus = alu_on_bbus ? alu : 8'bz;

     assign alu = add_not_sub ? abus + reg3 : abus - reg3;

reg [1:0] current;
parameter [1:0] a = 0, b = 1, c = 2, d = 3;
initial current = a;
     always @(negedge clk) current =
          current == a ? (restart == 1 ? b : a) :
          current == b ? c :
          current == c ? d :
          current == d ? (restart == 1 ? a : b) : a;
     assign abus_on_bbus = 0;
     assign {databus_on_bbus, enable_reg1, enable_reg2, enable_reg3} = (current == a) ? 4'b1111 : 0;
     assign {reg2_on_abus, add_not_sub, alu_on_bbus, enable_reg1, enable_reg2} = (current == b) ? 5'b11111 : 0;
     assign {reg1_on_abus, add_not_sub, alu_on_bbus, enable_reg1} = (current == c) ? 4'b1111 : 0;
     assign {reg1_on_abus, alu_on_bbus, enable_reg1, inc_reg3} = (current == d) ? 4'b1111 : 0;

endmodule
```

Figure 8.45 Verilog code for the system described by the flowchart of Fig. 8.40.

statements enable description of more complex hardware structures, which can still be used at the dataflow level by being instantiated as components or by their corresponding code being inserted. The next chapter emphasizes the use of procedural statements for description of components at the behavioral level.

Further Reading

Hill, F. J., and G. R. Peterson, *Digital Systems: Hardware Organization and Design,* 3d ed., John Wiley, New York, 1987.

IEEE Standard Hardware Description Language Based on the Verilog Hardware Description Language, IEEE Std. 1364-1995, Institute of Electrical and Electronic Engineers, New York, 1996.

Navabi, Zainalabedin, *VHDL: Analysis and Modeling of Digital Systems,* McGraw-Hill, New York, 1998.

Palnitkar, S., *Verilog HDL: A Guide to Digital Design and Synthesis,* Prentice-Hall, Upper Saddle River, N.J., 1996.

Smith, Douglas J., *A Practical Guide for Designing, Synthesizing and Simulating ASICs and FPGAs Using VHDL or Verilog,* Doone Publications, Madison, Ala., 1996.

Thomas, D. E., and P. R. Moorby, *The Verilog Hardware Description Language,* 3d ed., Kluwer Academic Publishers, Norwell, Mass., 1996.

Wakerly, J. F., *Digital Design Principles and Practices,* 2d ed., Prentice-Hall, Englewood Cliffs, N.J., 1993.

Problems

8.1 Use a continuous assignment statement with nested use of condition operators to describe a BCD seven-segment decoder.

8.2 Write a description for an 8-to-1 multiplexer with a 3-bit decoded input.

8.3 Wire the multiplexer in Fig. 8.3 and the decoder of Fig. 8.9 to generate a multiplexer with decoded input. Write a test bench and verify the operation of this circuit.

8.4 A decoder with an enable input is easily cascadable. Write a Verilog description for a 3-to-8 decoder with an active low enable input and an active high enable input. When disabled, all outputs have to be 0.

8.5 Write a Verilog description for wiring two of the decoders in Prob. 8.4 to implement a 4-to-16 decoder.

8.6 Use continuous assignments to describe a simple SR latch with q and NOT q outputs that functions the same way as a latch formed by cross-coupled NOR gates with clocked inputs. Use reasonable delay values.

8.7 Use two of the latches in Prob. 8.6 and necessary logic operations to describe a master-slave JK flip-flop.

8.8 Use continuous assignments to describe an 8-bit shift register. The structure has a serial input for right-shifting the data and a single serial output. All activities are synchronized with the leading edge of the clock.

8.9 Write a description for a universal 8-bit shift register with a 2-bit mode select input, an 8-bit parallel data input, and an 8-bit data output. The unit performs a right shift if the mode is 01, a left shift if the mode is 10, and a parallel load of the 8-bit input if the mode is 11. All activities are synchronized with the leading edge of the clock. Use continuous assignments and conditional expressions.

8.10 Write a description for a clocked T-type flip-flop. If T is **1** on the rising edge of the clock, the output of the flip-flop toggles.

8.11 Write a Verilog description for a rising-edge-trigger D-type flip-flop with asynchronous set and reset inputs and two outputs. Label the data, clock, set, and reset inputs *d, c, s,* and *r,* respectively. An active *s* or *r* input overrides the clocked values on the *d* input; *s* and *r* cannot simultaneously be active. Changes on *d* without the rising edge of *c* have no effect on the *q* and *qb* outputs of the flip-flop. Use delay parameters *sq_delay, rq_delay,* and *cq_delay* for setting, resetting, and clocking the flip-flop, respectively. Develop a test bench for testing this flip-flop. Generate a periodic clock that dies out after *n* clock pulses. Use this clock in your test bench.

8.12 Write a test bench that will bring out differences in the code of Figs. 8.26 and 8.29.

8.13 Write a clocked D-type flip-flop with active low asynchronous set and reset inputs. Set has higher priority than reset.

8.14 Design a Mealy sequence detector and develop a tester for this circuit. The circuit monitors its *x* input for the 10110 sequence. When this sequence is found, the *z* output becomes **1.** A valid data bit is one that coincides with the rising edge of the clock *c.* Make sure that you understand the behavior of a Mealy machine output. (*a*) Write a Verilog dataflow description for this sequence detector. (*b*) Show a test bench that tests this circuit for the 10110110101 sequence on the *x* input. Use a periodic clock.

8.15 Write the complete Verilog description for a Moore machine detecting 10111 or 11001. The circuit continuously monitors its *x* input. When in five consecutive clock pulses either sequence is found, the *z* output becomes **1** and stays at this level for a complete clock pulse. Write a Verilog dataflow description for this sequence detector. Show a test bench that tests this circuit for the 1011001011100110100 sequence on the *x* input. Use a periodic clock.

8.16 Write the complete Verilog description for a circuit with input *x* and two outputs *z1* and *z2.* The circuit consists of two concurrent Mealy machines. The *z1* output becomes **1** when a 1011 sequence is found on the input, and the *z2* output becomes **1** when a 110 sequence is found on *x*.

8.17 Write a Verilog description for a Moore state machine with resetting capability. While continuously searching for 1011 on the data input *x*, if the reset input *r* becomes **1**, the circuit returns to a reset state. In this state, all previously received data will be ignored and a complete 1011 is required before the output becomes **1**. While not reset, the circuit responds to overlapping valid sequences.

Chapter

9

Behavioral Description of Hardware

Most hardware characteristics can be described by the methods and techniques presented in the previous chapters. Although the emphasis has been on the structural and dataflow descriptions, we have also shown simple forms of procedural constructs that are used for behavioral description of hardware. These constructs were presented when discussing functions, tasks, test benches, registers, and state machines. This chapter gets more into the use of procedural constructs for describing complex hardware components. A component described partially or fully by procedural statements may be instantiated in a structural Verilog description, it may be used in a dataflow code, or it may completely specify a system at the behavioral level.

The chapter begins by presenting key Verilog constructs that can contain procedural statements for behavioral descriptions. We will describe constructs for activity control used in handshaking, timing, and asynchronous communication. Several examples will be presented that illustrate the use of procedural constructs for describing a complete system or for describing subcomponents that will be wired in a structural Verilog body.

9.1 Procedural Bodies

Verilog procedural bodies are those constructs within which procedural statements may be used. Between the header part of a module and the **endmodule** keyword, any number of *module_items* may be used.

Module items that are regarded as Verilog procedural bodies are **initial** and **always** block.

9.1.1 General format of procedural bodies

As shown in Fig. 9.1, an **always** block, which begins with the **always** keyword, marks the beginning of a procedural statement or a body of procedural statements. If an **always** statement is to apply to several procedural statements, these statements must be bracketed by **begin** and **end** keywords, forming a single sequential block. This format of an **always** statement is shown in Fig. 9.1*a*. Statements inside the **begin–end** block are procedural, and they execute one after another as the program flow reaches them. Figure 9.1*b* shows an **always** block that is followed by only one procedural statement. This format does not require **begin** and **end** keywords.

Figure 9.2 shows two formats for using the **initial** block. As with the **always** block, several grouped procedural statements must be bracketed by **begin** and **end** to form a sequential block which itself is a procedural statement. Execution of all procedural bodies of a model starts together at time 0 and runs concurrently. An **always** block repeats forever, but an **initial** block runs only once. Thcsc issues will be discussed in the following sections.

9.1.2 Activity flow

When simulation begins at time 0, the execution of procedural statements in **always** and **initial** blocks begins. Procedural statements in these bodies execute in the order in which they appear. As depicted in Fig. 9.3, when the last procedural statement in an

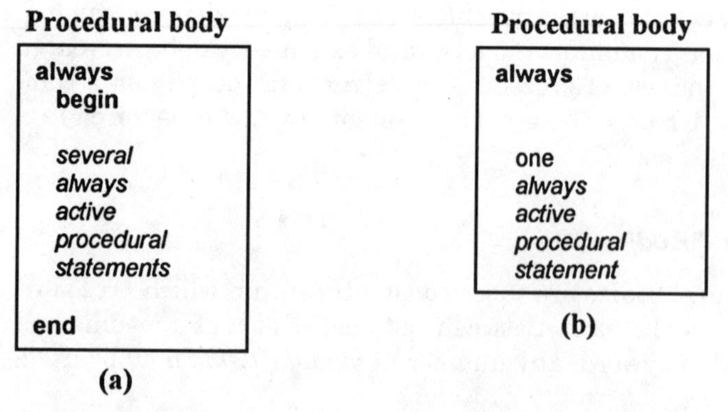

Figure 9.1 **always** procedural body (*a*) enclosing several procedural statements, and (*b*) using one procedural statement.

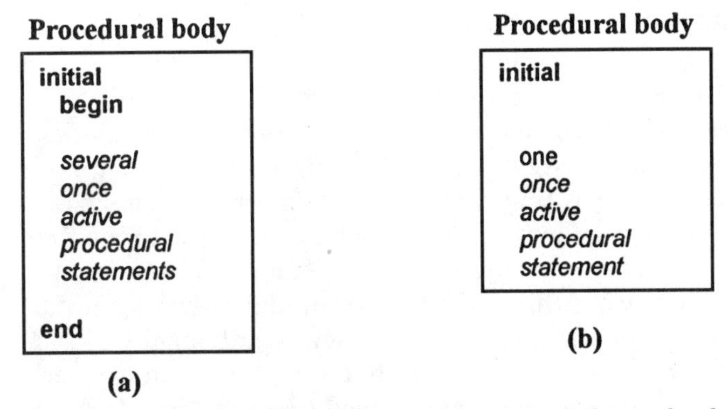

Figure 9.2 **initial** procedural body (*a*) enclosing several procedural statements, and (*b*) applying to one procedural statement.

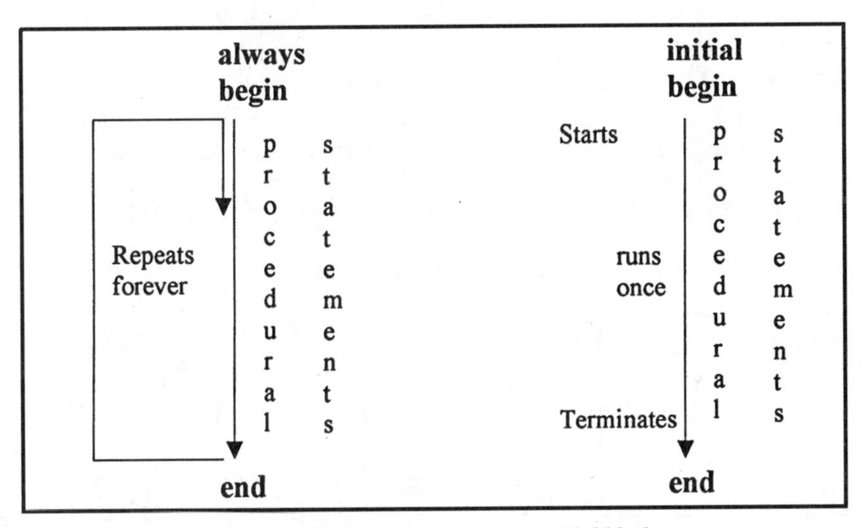

Figure 9.3 Activity flow in an **always** block and in an **initial** block.

always block is completed, activity flow returns to the first procedural statement. On the other hand, an **initial** statement runs once and terminates after the completion of its last statement.

Unless the flow of activities in a procedural body is halted or delayed by flow control statements, all procedural statements execute at time 0 in the order in which they appear. An **always** block without flow control runs at zero time forever. This means that once simulation begins, the simulation time never advances beyond time 0. This will be a case in which your wall-clock time will far exceed your Verilog simulation time, which should be avoided. A simple flow control statement is a procedural timing control statement, the simplest form of which is *#time_delay_value*. Before we discuss flow control in detail, we will use this simple format in our examples in this section. The statement will temporarily block the flow of activities in **initial** and **always** blocks.

Figure 9.4 shows an **always** block that waits for 5 time units (its activities are blocked for 5 time units) and then complements the *a* and *b* variables. Once this has completed, the flow returns to the first statement in this procedural body, and the waiting for 5 time units is executed again. Repeating this process forever, and assuming initial values of **0** and **1** on *a* and *b* variables, this **always** statement will generate two periodic waveforms on *a* and *b* that are complements of each other at all times. The same statements used in an **initial** block are executed only once. Assuming the same initial values for *a* and *b*, the code in Fig. 9.5*a* executes once and, at time 5, changes the values of *a* and *b* to their complements.

In addition to flow control statements, such as *#5*, that block the flow of execution of statements, assignments to *a* and *b* are also considered as blocking statements. A blocking assignment blocks the flow of activities in a procedural body until the assignment is completed. In our examples, this blocking is only for 0 amount of time, and therefore the new values of *a* and *b* are updated 0 time after they have been assigned. Figure 9.6 presents an example that illustrates this.

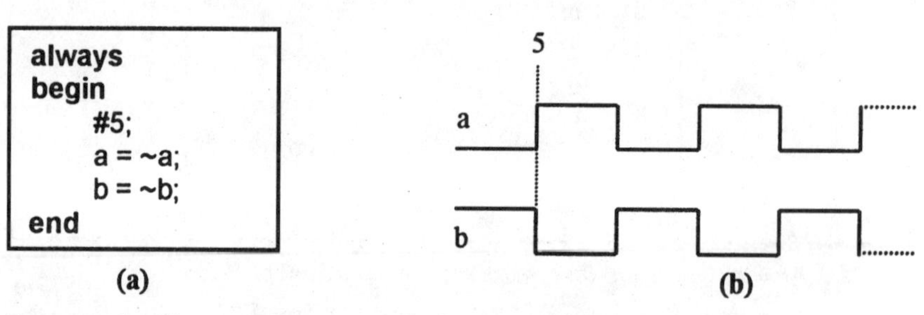

Figure 9.4 An **always** statement repeats forever. (*a*) Code; (*b*) waveform.

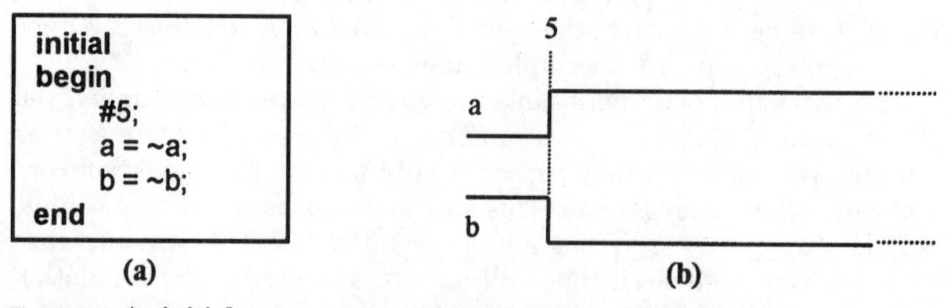

Figure 9.5 An **initial** statement runs only once. (*a*) Code; (*b*) waveform.

The **initial** statement in Fig. 9.6*a* *initializes* *a* to **0.** Since this **initial** statement encloses only one procedural statement, **begin** and **end** keywords are not used. This is in accordance with the **initial** statement format shown in Fig. 9.2*b*. While *a* is being initialized, an **always** statement that runs concurrently with this **initial** statement is blocked for 5 time units. At time 5, while *a* has its new value of **0,** the complement of *a* is assigned to *a,* which causes the new value of *a* to become **1.** While this assignment is taking place, the flow of execution within the **always** statement is blocked until *a* receives this new value. Therefore, the statement that follows this statement, which takes the complement of *a* and assigns it to *b,* uses the updated value of *a.* The waveform shown in Fig. 9.6*b* shows *b* changing to the complement of the new value of *a* at the same real time, but after a 0 time delay.

Procedural bodies may also contain nonblocking assignments. The assignment symbol used for nonblocking assignments is <=, as shown in Fig. 9.7.

In Fig. 9.7*a,* when activity flow reaches assignment to *a* at time 5, the complement of *a* is evaluated and scheduled for *a.* Before *a* is updated with its new value, activity flow, which is not blocked by this assignment, continues to the next assignment, which assigns ~*a* to *b.* Therefore, the assignment to *b* uses the old value of *a.* As a result, the same waveform that appears on *a* also appears on *b.* The waveforms on *a* and *b* are shown in Fig. 9.7*b.* Variables *a* and *b* change at exactly the same time, and, unlike the case with the waveforms in Fig. 9.6*b,* changes on *a* and *b* occur simultaneously, without zero delay. Nonblocking assignments may also be used in **initial** blocks and, like blocking assignments, may have an intra-assignment delay value. The **initial** block in Fig. 9.8 begins execution at time 0, and after a delay of 1 ns, it reaches the nonblocking assignment assigning **0** to *a.* At time 1 after the beginning of simulation, the **0** value will be scheduled for *a* for 4 ns later. Also at time 1, activity flow reaches the *b* <= #2 1 statement. Executing this statement causes the value **1** to be scheduled for *b* after 2 ns. At time 1, scheduling of values into *a* and *b* is done, and

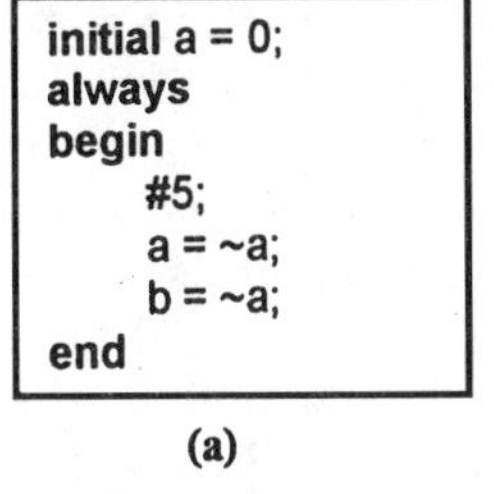

(a)

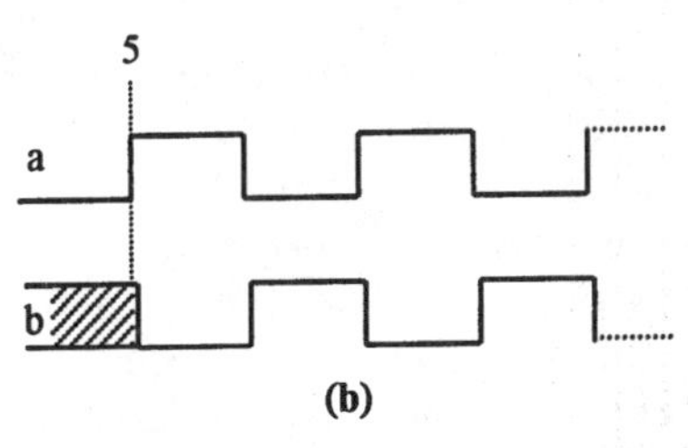

(b)

Figure 9.6 Blocking assignments block activity flow. (*a*) Code; (*b*) resulting waveform.

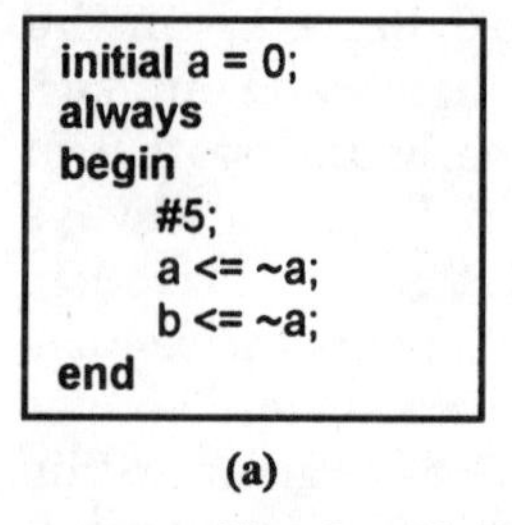

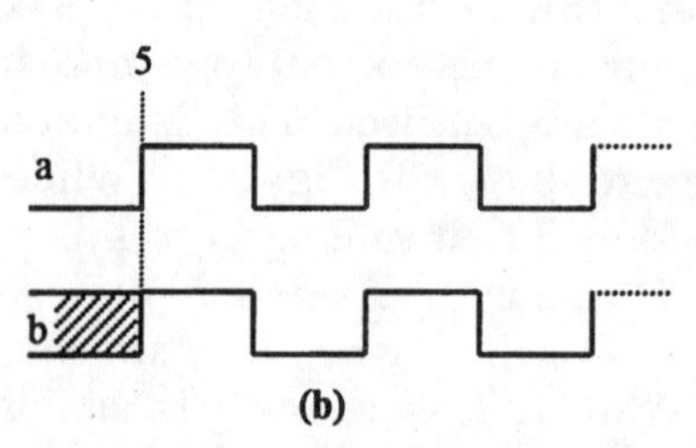

Figure 9.7 Use of nonblocking assignments in a procedural body. (*a*) Code; (*b*) waveform.

the **initial** block reaches its end. At time 3, 2 ns after termination of this block, *b* receives its initial value, and 2 ns later *a* receives its initial value.

Although scheduling **0** for *a* took place before scheduling **1** for *b*, variable *b* receives its value sooner than *a* receives its initial value. If blocking assignments were used, *a* would still receive its value at 5 ns, but *b* would receive its value 2 ns later. In this case, the **initial** block would terminate at time 7 ns.

9.1.3 Multiple procedural bodies

All procedural bodies in a module start at time 0, and they all run concurrently. Depending on the timing or flow control statements that appear in them, when procedural bodies terminate or loop back may be different, but all such blocks execute in parallel. The relative position of procedural bodies in a module does not affect when they are executed. Figure 9.9 shows a module with an **initial** and two **always** procedural bodies. Each body has several procedural statements. All three blocks begin at time 0.

At time 0, *a* and *b* are initialized to **0** and **1** by the **initial** block. After this initialization, the **initial** statement terminates. Also at time 0, execution of the two **always** blocks begins; these are blocked by 5 ns and 3 ns, respectively. These initial blocking times give enough time for *a* and *b* to be updated with their initial values. At time 3, activity flow reaches the *b*=~b statement in the second **always** statement. This blocking procedural statement blocks activity flow for 0 time until *b* is updated, which causes its value to become **0** after 3 ns into the simulation. Then 2 ns later, at time 5, activity flow reaches the *a*=~a blocking statement in the first **always** statement. This occurs while the execution of *b*=~b in the second **always** statement is being blocked by the *#3* statement. At time 5, *a*=~a is executed. For 0 time until *a* is updated, this procedural block is blocked. Following this, the **always** statement repeats, which causes *#5* to execute, which again blocks the flow for another 5 ns.

Waveforms resulting from running the Verilog code of Fig. 9.9 are shown in Fig. 9.10. Initially *a* and *b* are **X.** At time 0, they both receive their initial values. For the rest of the simulation run, *a* and *b* toggle every 5 ns and 3 ns, respectively.

9.1.4 Flow control

Anywhere in the body of an **initial** or an **always** statement, a timing control statement may be placed. When activity flow reaches such a statement, the procedural body is blocked until the condition or conditions required by the timing control statement are satisfied. As shown below, a timing control statement may be attached to a block of statements or a single statement, or it may appear as an independent statement.

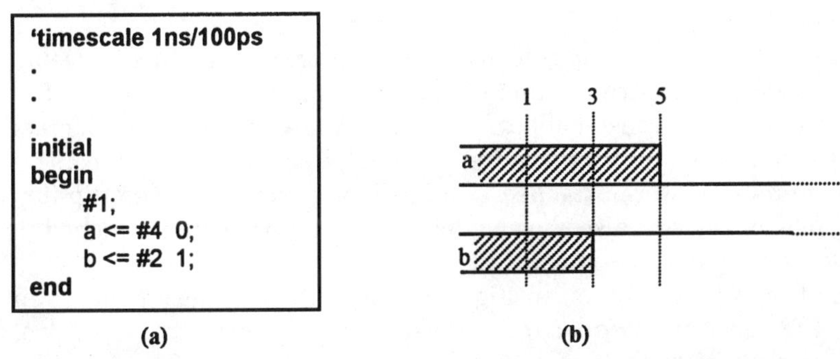

Figure 9.8 Nonblocking assignments with intra-assignment delays. (*a*) Initial block; (*b*) waveform.

```
'timescale 1ns/100ps
module try;
reg a, b;
    initial begin
        a = 0;
        b = 1;
    end
    always begin
        #5;
        a = ~a;
    end;
    always begin
        #3;
        b = ~b;
    end
endmodule
```

Figure 9.9 Concurrent procedural blocks.

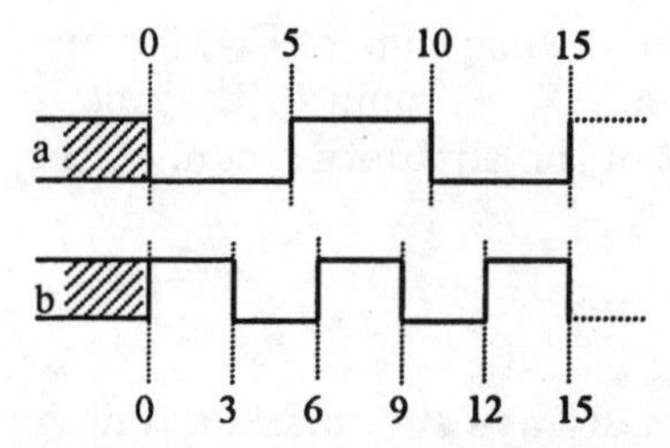

Figure 9.10 Waveforms resulting from concurrent execution of procedural bodies.

Timing_control **begin** ... `statements` ... **end**
Timing_control `statement`
Timing_control

Any of the combinations shown here forms a complete procedural statement.

Figure 9.11 shows various forms of timing control statements. Delay control, #, event control, **@,** and wait statements are shown in this figure. Any of these constructs can be used in any of the three formats described above. In Chap. 8 we used edge-sensitive event control statements that were attached to a block of statements. The flip-flop examples in Chap. 8 also showed delay control statements attached to single assignment statements.

The flip-flop description in Fig. 8.29, part of which also appears in Fig. 9.12, uses the ***@(posedge** c)* event control statement to control activity flow into the **begin–end** block that follows it. Together, the timing control statement and the **begin–end** block form the statement part of the **always** block.

The flip-flop example in Fig. 8.22, part of which is shown in Fig. 9.13, attaches delay control to a single procedural assignment. Flow into the delayed assignment in Fig. 9.13 is controlled by another timing control statement that detects the edge of the clock.

Examples for the third alternative for using timing control statements are those in Fig. 9.9. Delay control statements in this figure, i.e., *#5* and *#3,* appear as stand-alone procedural statements.

All the statements of Fig. 9.11 except the **wait** statement can be used on the right-hand side of an assignment after the = or <= operator for blocking and nonblocking assignments. This use of timing control is referred to as *intra-assignment* timing control. A timing control that precedes an assignment waits for a delay or event, then it evaluates the right-hand side and immediately assigns it to the left-hand side. In an *intra-assignment* timing control, the right-hand side is evaluated first, then waiting for the delay or event is performed, and then assignment to the left-hand side is done. In addition to delay and event controls, a repeat control event is also allowed in an intra-

#*time*	wait for *time*
#(*time_expression*)	wait for *time_expression*
@*variable*	wait for *variable* to change value
@*variable_or_expression*	wait for *variable_or_expression* to change value
@(posedge *variable_or_expression*)	wait for *variable_or_expression* to change to 1 or change from 0
@(negedge *variable_or_expression*)	wait for *variable_or_expression* to change to 0 or change from 1
@(*event1* or *event2* or event3 . . .)	wait for any of the events to occur
wait (*expression*)	wait while *expression* is false

Figure 9.11 Timing control statements.

```
always
@(posedge c)
    begin
      q = #delay1 d;
      qb = #(delay2 – delay1) ~d;
    end
```

Figure 9.12 Edge-trigger timing control controlling a block of statements.

```
always
@(posedge c)
    #(delay1) q = d;
```

Figure 9.13 Delay control controlling a single procedural statement.

assignment statement. A repeat construct preceding an event control delays assignment of a right-hand-side value until the occurrence of the specified event has been repeated the number of times specified by the repeat expression. For example, in

```
a = repeat(5) @(posedge c) b;
```

the value of b is evaluated and is assigned to a after 5 positive edges on c take place. Examples using timing control statements for handshaking will be presented later in this chapter.

9.1.5 A first procedural example

The topics discussed above will be illustrated in several flip-flop examples in this section. Figure 9.14 shows the logic symbol for a D-type flip-flop with asynchronous active-high set and reset inputs. The Verilog code that corresponds to this flip-flop is shown in Fig. 9.15. The description demonstrates parallelism of procedural bodies, event control expressions, and *intra_assignment* timing control constructs. The behavioral description of *d_sr_flipflop* in Fig. 9.15 uses the **reg** *state* to record the internal state of the flip-flop.

Two concurrent **always** statements are used in this module. The first statement uses an event control statement that controls the flow of activity into the sequential block to which it is attached.

This sequential block is entered when a positive edge is detected on the *clk* or *rst* or *set* input of the flip-flop. Once the block is entered, an if statement checks for *set* to be active. If this condition is satisfied, the internal state of the flip-flop, *state,* will be set to **1** after *sq_delay.* If *set* is not active, *rst* is examined, and if that is not active, the *clk* input is checked for being **1.** If activity flow reaches the if statement with *clk* condition and if *clk* is **1,** it means that the positive edge detected by the event control statement that precedes this procedural statement is that of *clk.* In this case, the value of *d* is looked up, and after a delay of *cq_delay,* it is assigned to *state.* The *state* variable, being a **reg,** retains its value until the next time this block is entered. The coding shown here gives higher priority to the *set* and *rst* inputs of the flip-flop by examining their values before examining the value of *clk.* This implies that if *set* or *rst* is **1,** the positive edges of the clock will be ignored and *set* and *rst* will determine

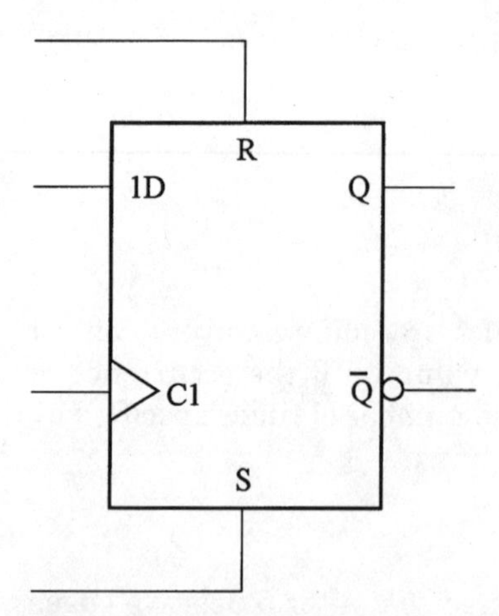

Figure 9.14 A positive-edge-trigger D-type flip-flop with asynchronous set and reset inputs.

```
`timescale 1ns/100ps

module d_sr_flipflop (d, set, rst, clk, q, qb);
input d, set, rst, clk;
output q, qb;
reg q, qb, state;
parameter sq_delay = 6, rq_delay = 6, cq_delay = 6;
    always @(posedge clk or posedge rst or posedge set) begin
        if (set) begin
            state = #sq_delay 1;
        end else if (rst) begin
            state = #rq_delay 0;
        end else if (clk) begin
            state = #cq_delay d;
        end
    end
    always @(state) begin
        q = state;
        qb = ~state;
    end
endmodule
```

Figure 9.15 Verilog code for the flip-flop of Fig. 9.14.

the value of *state*. This causes *state,* which is a **reg,** to retain its **1** or **0** value while *set* or *rst* is active. The resulting behavior is that *state* is sensitive to levels of *set* and *rst* and sensitive to positive edges of the *clk* input only.

When the flip-flop is being set or reset, using intra-assignment delay control for assigning constants **1** or **0** to *state* does not produce different results from those produced by the use of procedural delay control statements. This is because constants remain the same whether they are read after waiting for delay or before. On the other hand, assigning *d* to *state* using intra-assignment delay control or a procedural timing control statement will produce different values on *state* if *d* changes during the *cq_delay* wait period.

While the first **always** statement in Fig. 9.15 is waiting for a positive edge on *clk, set,* or *rst* to change the value of *state,* in another **always** statement, values of *state* are being monitored. The second **always** statement in this figure attaches an event control statement to a sequential block that sets values to the *q* and *qb* outputs of the flip-flop. The event control statement blocks flow into this sequential code until it detects a change on *state.*

Anytime the first **always** block causes *state* to change, flow into the body of the second **always** block begins. The flow of activities in this **always** statement stops zero time after *qb* receives its value, and it will resume after another change on *state.*

This example illustrated the dependency of several procedural bodies on one another. Describing the flip-flop of Fig. 9.14 could more easily be done by using one **always** statement for checking active inputs and setting output values.

In the flip-flop description of Fig. 9.16, *state* receives the flip-flop state value in a nesting of if–then–else statements, as was done in the code of Fig. 9.15. The blocking statements in these procedural statements block the activity flow until *state* receives its assigned value. The value assigned and available on *state* is assigned to *q*. While the procedural body is blocked, *q* gets updated, and then *qb* is set to its complement. This works only because blocking statements block the flow of activity until they are completed.

Figure 9.17 shows another alternative Verilog code for the flip-flop of Fig. 9.14. This code uses nonblocking assignments. Zero time after a positive edge is detected on *clk, rst,* or *set,* all schedulings are done, and flow reaches the end of the **always** statement and waits for another positive edge event on the inputs. The values scheduled for *q* and *qb* will appear on these outputs after the execution of the **always** statement has completed. If an intermediate *state* was used, as in the code in Fig. 9.16, assignment of *state* to *q* and *qb* would use the old value of *state* before it is updated.

For further illustration of sequential execution of statements in a procedural body, as well as delay and event control statements, we have developed a test bench for testing the flip-flop descriptions in

```
`timescale 1ns/100ps

module d_sr_flipflop (d, set, rst, clk, q, qb);
input d, set, rst, clk;
output q, qb;
reg q, qb, state;
parameter sq_delay = 6, rq_delay = 6, cq_delay = 6;
    always @(posedge clk or posedge rst or posedge set) begin
        if (set) begin
            state = #sq_delay 1;
        end else if (rst) begin
            state = #rq_delay 0;
        end else if (clk) begin
            state = #cq_delay d;
        end
        q  = state;
        qb = ~q;
    end
endmodule
```

Figure 9.16 Using one **always** block for the flip-flop of Fig. 9.14.

```
`timescale 1ns/100ps

module d_sr_flipflop (d, set, rst, clk, q, qb);
input d, set, rst, clk;
output q, qb;
reg q, qb;
parameter sq_delay = 6, rq_delay = 6, cq_delay = 6;
    always @(posedge clk or posedge rst or posedge set) begin
        if (set) begin
            q <= #sq_delay 1;
            qb <= #sq_delay 0;
        end else if (rst) begin
            q <= #rq_delay 0;
            qb <= #rq_delay 1;
        end else if (clk) begin    //can remove if (clk)
            q <= #cq_delay d;
            qb <= #cq_delay ~d;
        end
    end
endmodule
```

Figure 9.17 Using nonblocking statements for describing the flip-flop of Fig. 9.14.

Figs. 9.15, 9.16, and 9.17. The test bench shown in Fig. 9.18 instantiates one of the *d_sr_flipflop* modules and assigns values to its *d, set, rst,* and *clk* inputs.

In an **always** block, the *clk* input is toggled every 500 ns. After an initial wait period of 500 ns, *clk* is toggled for as long as simulation continues. The initial value of *clk* is **X,** and unless it is set to **0** prior to being toggled, it remains **X** for the entire time of the simulation. Concurrent with this **always** statement, at time 0, when this statement starts, an **initial** block also starts. In this **initial** block, **clk** is set to **0** at time 0, which provides an appropriate value for the toggling that takes place in the **always** block. Had an intra-assignment delay control been used for toggling *clk* in the **always** block, the toggling would start with the uninitialized value of *clk* before it is set to **0** in the **initial** block, and would not produce correct results.

In addition to setting an initial value for *clk,* the **initial** block sets initial values and generates waveforms on the *d, set,* and *rst* inputs of the circuit. The waveforms are placed on these signals by nonblocking assignments. All schedulings in this **initial** block are done at time 0, and starting at time 0, the block waits for 4000 ns before it is terminated by the **$stop** system task. Because nonblocking assignments are used, all delay values are with respect to time 0. The result will be, for example, that if *d* is set to an initial value of **0,** it becomes **1** at 2400 ns and **0** at 3300 ns. Had we used blocking assignments, relative delay values should have been used.

```
`timescale 10ns/1ns

module test_d_sr_flipflop;
reg d, set, rst, clk;
wire q, qb;
    d_sr_flipflop u1 (d, set, rst, clk, q, qb);
    always #50 clk =~clk;
    initial begin
        clk <= 0; d <= 0; set <= 0; rst <= 0;
        d <= #240 1; d <= #330 0;
        set <= #20 1; set <= #120 0;
        rst <= #140 1; rst <= #220 0;
        #400 $stop;
    end
endmodule
```

Figure 9.18 Testing *d_sr_flipflop* descriptions.

Figure 9.19 shows simulation results obtained by running the test bench of Fig. 9.18. As a result of execution of the first line of the **initial** block of Fig. 9.18, all inputs change from **X** to **0** at time 0. Flip-flop outputs, which also start with the value **X,** do not receive any values until after the *set* input becomes active. The operation of the *rst* input is verified at time 1406 ns, when *q* becomes **0** as a result of *rst* being asserted at 1400 ns. While *set* or *rst* is active, all *clk* edges are ignored. When these inputs are both **0,** a value on *d* is clocked into the flip-flop on the rising edge of *clk*. In Fig. 9.19, this happens at times 2506 ns and 3506 ns, 6 ns after the positive edge of *clk*.

9.1.6 Procedural body syntax details

Figure 9.20 shows syntax details of an *always_construct* used in the module of Fig. 9.16. The **always** keyword and a *statement* that is of a *procedural_timing_control* type are the top-level constructs of the *always_construct*. The *procedural_timing_control* statement is formed by attaching an *event_control* construct to a *sequential_block*.

The *event_control* statement is formed by an @ sign and an *event_expression*. The *event_expression* itself is formed by **or** operations on smaller *event_expressions*. The *event_control* construct controls the flow of activity into the statement to which it is attached. In Fig. 9.20, the *statement* is a sequential block that begins and ends with **begin** and **end** keywords. These keywords bracket two *statements*, the first one of which is a *conditional_statement* and the second a *blocking_assignment*. The *blocking_assignment* is the *qb*=~q assignment, and the *conditional_statement* is a nesting of several if-else-if statements. Details for this statement are shown in Fig. 9.21.

As shown in Fig. 9.21, the general format of a *conditional_statement* is *if(condition)...else....* The **if** and **else** parts may be other conditional statements or any other form of a *statement* construct, including a *sequential_block.*

Figures 9.20 and 9.21 indicate that the main body for behavioral description of hardware is the *statement* construct. In addition to the statements discussed so far in this chapter, most procedural constructs used in software languages can be used in an **always** or an **initial**

ps	d	set	rst	clk	q	qb
0	x	x	x	x	x	x
0	0	0	0	0	x	x
200000	0	1	0	0	x	x
206000	0	1	0	0	1	0
500000	0	1	0	1	1	0
1000000	0	1	0	0	1	0
1200000	0	0	0	0	1	0
1400000	0	0	1	0	1	0
1406000	0	0	1	0	0	1
1500000	0	0	1	1	0	1
2000000	0	0	1	0	0	1
2200000	0	0	0	0	0	1
2400000	1	0	0	0	0	1
2500000	1	0	0	1	0	1
2506000	1	0	0	1	1	0
3000000	1	0	0	0	1	0
3300000	0	0	0	0	1	0
3500000	0	0	0	1	1	0
3506000	0	0	0	1	0	1

Figure 9.19 Simulation of *d_sr_flipflop.*

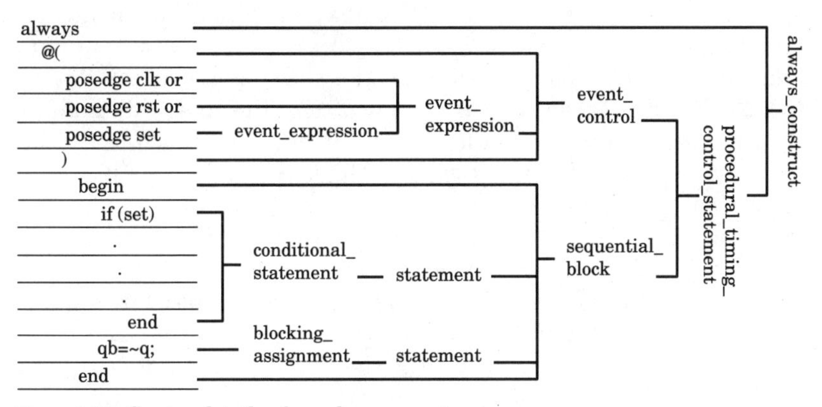

Figure 9.20 Syntax details of an *always_construct.*

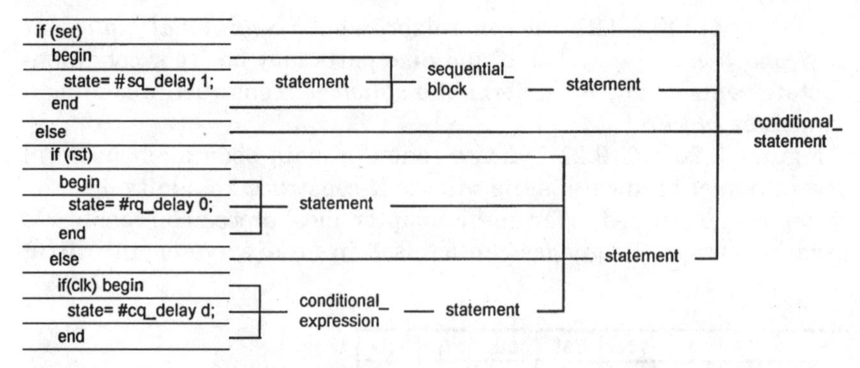

Figure 9.21 Syntax details of *conditional_statement* of *d_sr_flipflop* of Fig. 9.16.

block of Verilog. Examples in the sections that follow will show the use of other procedural statements not discussed so far. In a syntax tree, all such statements reduce to the *statement* syntax construct.

9.1.7 Statement-level concurrency

Verilog allows concurrency at several levels. Modules instantiated within a top-level module are treated concurrently. Also, within an instantiated module, all procedural bodies are concurrent. In a procedural body, nonblocking assignments have the effect of concurrent placement of values on left-hand-side **reg** type variables. Verilog allows still another concurrency at the level of procedural statements. Using a *parallel_block* instead of a *sequential_block* makes statements that are immediately contained in this block to execute concurrently. The bracketing used for a parallel block is the **fork** and **join** keywords instead of **begin** and **end** as for a sequential block.

For illustration of statement-level concurrency, we refer back to our D-type flip-flop example of Chap. 8. The **always** block of the module of Fig. 8.29 is shown here in Fig. 9.22.

After detection of a positive edge on the clock, *d* is evaluated, and after a block time of *delay1,* it is assigned to *q*. Because of this block time, assignment to *qb,* which follows assignment to *q,* uses only the difference of the *qb* and *q* delay values. This code works only if *delay2* is larger than *delay1.*

This problem is alleviated if assignments to *q* and *qb* are nonblocking, as shown in Fig. 9.23. Assignments to *q* and *qb* are still done in a sequential block (bracketed by **begin** and **end** keywords), but the use of nonblocking assignments causes scheduling ~d to *qb* to take place at the same time as scheduling *d* to *q*. In effect, assignments to *q* and *qb* appear as if they take place concurrently.

A true concurrency in executing statements is implemented when the sequential block in Fig. 9.22 is replaced by a parallel block, as depicted in Fig. 9.24. Except for the use of the **fork** and **join** keywords instead of **begin** and **end,** the syntax of statements bracketed by a parallel block is no different from that of statements of a sequential block. This can be seen by comparing the codes shown in Fig. 9.24 for a *parallel_block* with those of Fig. 9.22 for a *sequential_block.* Both blocks reduce to the *statement* syntax construct and can be used anywhere a *statement* is allowed, some cases of which are shown in Figs. 9.20 and 9.21.

Like a sequential block, a parallel block can enclose any number of separate statements. Each can be sequential or parallel. The code shown in Fig. 9.24 uses two *blocking_assignment* statements in the body part of the *parallel_block,* the flow into which is controlled by the positive edge of the clock. When activity flow reaches this parallel block, activity forks into all statements that appear immediately within the block. In our example, both blocking assignments are treated as separate statements and execute at the same time.

The first statement evaluates *d,* blocks the flow of activity to other sequential statements if there are any, and after *delay1* assigns the value of *d* before *delay1* to *q.* The second statement becomes active at exactly the same time as the first, and assigns ~d to *qb* after *delay2.*

```
always @(posedge c) begin
      q = #delay1 d;
      qb = (delay2 – delay1) ~d;
end
```

Figure 9.22 A sequential block using blocking assignments.

```
`timescale 1ns/100ps

module d_flipflop (d, c, q, qb);
input d, c;
output q, qb;
reg q, qb;
parameter delay1 = 4, delay2 = 5;
always @(posedge c) begin
      q <= #delay1 d;
      qb <= #delay2 ~ d;
end
endmodule
```

Figure 9.23 A sequential block using nonblocking assignments.

```
`timescale 1ns/100ps

module d_flipflop (d, c, q, qb);
input d, c;
output q, qb;
reg q, qb;
parameter delay1 = 4, delay2 = 5;
always @(posedge c) begin
     fork
          q = #delay1 d;
          qb = #delay2 ~d;
     join
end
endmodule
```

Figure 9.24 A parallel block using blocking assignments.

After both assignments complete, activity flow reaches the point where **join** appears in the code. The flip-flop description in Fig. 9.24 simulates the same as that in Fig. 8.22, in which two separate **always** statements are used. Sequential blocks are especially useful in handshaking where separate events are to occur in any order before a certain action is taken. Figure 9.25 shows an example in which an assignment that involves two sets of data has to wait until both are ready.

When activity flow reaches the **fork–join** statement, it splits into a composite statement that detects a positive pulse on *a_ready* and another statement that detects a positive pulse on *b_ready.* Positive pulses are detected by waiting for a positive edge to be followed by a negative edge. When pulses occur on both *a_ready* and *b_ready,* the split activity flows merge at **join.** At this time, *a_data* and *b_data* are read, the **&** operation is performed, and, when *clock* becomes active, the AND result is assigned to *data_register.*

9.1.8 Behavioral constructs

As stated previously, softwarelike procedural constructs such as if, loop, and case statements are allowed in **always** and **initial** blocks in Verilog. These constructs form a *statement* syntax construct and can be used everywhere statements are allowed. Syntax details in Figs. 9.20 and 9.21 illustrate several variations of using statements in procedural bodies. As with other *statement* constructs, flow of activity into an *if, loop,* or *case* statement may be controlled by a delay or event control construct.

Verilog offers four ways for specifying loop statements. The first is the forever loop, which is formed by preceding a *statement* with the **forever** keyword. When this is done, the statement repeats while the simulation continues.

Using **forever** with an **initial** construct makes the **initial** block repeat itself forever just like an **always** statement. Figure 9.26 shows a **forever** keyword attached to a sequential block that forms the statement part of an **initial** block.

Repeating the statement part of a loop statement may be limited by using **repeat** (*expression*) instead of the **forever** keyword. The expression that follows the **repeat** keyword specifies the number of times that the statement part of this construct repeats. The **always** block in Fig. 9.27 loads *areg* with *abus* after every 5 clock edges. In this example, the loop statement causes the flow of activity in the **always** block not to go beyond this statement until *5* positive edges are detected on *clock*. When this occurs, activity flow reaches the assignment statement that assigns *abus* to *areg*. The assignment is done immediately after detection of the fifth clock edge. After the assignment statement executes, flow returns to the beginning of the **always** block and the search for another 5 clock edges resumes. This process repeats forever.

Another form of a *loop_statement* starts with **while** (*expression*). The statement that follows the expression repeats for as long as

```
begin
    fork
      @(posedge a_ready) @(negedge a_ready);
      @(posedge b_ready) @(negedge b_ready);
    join
    data_register = @(clock) a_data & b_data;
end
```

Figure 9.25 Detecting positive pulses in any order.

```
initial forever begin

    sequential statements in this
    part repeat forever

end
```

Figure 9.26 An example use of **forever.**

```
always begin

    repeat (5) @(posedge clock);
    areg = abus;

end
```

Figure 9.27 An example use of **repeat.**

expression is true. Parameters that the expression uses may be assigned values within the **while** *loop_statement* or may be modified by statements that execute in parallel with the loop statement.

The last form of a loop statement is

for (*initial_assignment*; *expression*; *step_assignment*) statement

As in the **while** loop, looping continues while *expression* is true. The **for** loop also allows for a **reg** assignment before the loop begins (*initial_assignment*) and a **reg** assignment each time the *statement* part is entered (*step_assignment*). The *step_assignment* takes place after all procedural statements in the *statement* execute.

The Verilog *case_statement* is another behavioral statement used in functions, tasks, and **initial** and **always** blocks. Instead of using an **if–else–if** statement for repeated checking of an expression against several alternative values that it can take, a *case_statement* may be used, which is more readable and more compact. Figure 9.28 shows the general format of a *case_statement.*

The *case_statement,* as shown in this figure, is considered a *statement* construct in Verilog syntax. When activity flow reaches a *case_statement,* the *expression* that follows the **case** keyword is evaluated, and its value is compared with alternative values that appear in the body of this statement. The *statement* construct that follows the first alternative that matches the value of the case *expression* will execute. If no matches are found, the **default** alternative will be taken.

Instead of the **case** keyword, a *case_statement* may begin with **casez** or **casex.** When the **case** keyword is used, the *case_statement* requires an exact match between the case *expression* and its alternative values. To allow don't cares in bit values, **casez** and **casex** may be used. When **casez** is used, **Z** values that appear in bits of *expression* or any of its alternatives do not participate in comparing the values. The **casex** statement treats both **Z** and **X** as don't care, and neither will participate in finding a match with *expression* among the alternatives.

```
case (expression)
    alternative1 : statement;
    alternative2 : statement;
    .
    .
    .
    alternativen : statement;
    default : statement;
endcase
```

Figure 9.28 General format of a *case_statement.*

This section has presented small examples or just the general format of many behavioral constructs. Various forms of the *statement* construct and the syntax and semantics associated with this construct were discussed. In the sections that follow, complete examples will be used to demonstrate interaction among various forms of statements and program flow control constructs.

9.2 Behavioral Timing

The behavioral timing control constructs discussed in this chapter allow specification of the timing dependency of module outputs on the module's inputs or other variables. On the other hand, behavioral procedural statements allow for manipulation of timing variables for a very accurate representation of timing of a circuit. Alternatively, Verilog offers **specify** blocks for delay specification, as discussed in Chap. 6. This language construct is placed immediately within a module and acts concurrently with procedural bodies in the same module. Specification of module path delays, input-output timing, and timing checks can appear in a **specify** block.

9.2.1 Behavioral delay specification

Concurrency of a **specify** block and an **always** block is demonstrated in Fig. 9.29. This is a description for the flip-flop of Fig. 9.14 and behaves the same as the Verilog module in Fig. 9.15.

The **always** block in this figure specifies the behavior of module outputs with 0 delay values. While this block is looking for a positive edge on *clk, rst,* or *set* to assign values to the outputs, the **specify** block is watching the outputs to delay events on them if they are to occur. When a delay value in the **specify** block is found to apply to an input event and a resulting output event pair, the assignment to the output is delayed by the specified delay time.

9.2.2 Behavioral timing checks

A **specify** block can be set up to check for the relative timing of events on inputs of a module. Checking pulse width, clock, skew, setup, and hold times and reporting violations can be achieved by system tasks in a **specify** block. Figure 9.30*a* shows setup and hold times for a positive-edge-trigger clocked D-type flip-flop. The setup time is the minimum required time between changes on the data input and the triggering edge of the clock. A system task for checking this timing is

```
$setup (d, posedge clk, t_setup);
```

```
`timescale 1ns/100ps

module d_sr_flipflop (d, set, rst, clk, q, qb);
input d, set, rst, clk;
output q, qb;
reg q, qb;
parameter sq_delay = 6, rq_delay = 6, cq_delay = 6;
parameter t_setup = 4, t_hold = 4;
    specify
        (clk *> q, qb) = cq_delay;
        (set *> q, qb) = sq_delay;
        (rst *> q, qb) = rq_delay;
    endspecify
    always @(posedge clk or posedge rst or posedge set) begin
        if (set) begin
            q = 1; qb = 0;
        end else if (rst) begin
            q = 0; qb = 1;
        end else if (clk) begin    //if (clk) not essential
            q = d; qb = ~d;
        end
    end
endmodule
```

Figure 9.29 Using a **specify** block.

The hold time, also shown in Fig. 9.30*a,* is the minimum time that the data input of a flip-flop should stay stable after the effective edge of the clock. The **$hold** system task with the format shown below checks this timing:

```
$hold (posedge clk, d, t_hold);
```

Figure 9.30*b* shows a flip-flop description similar to that in Fig. 9.17 with the added system tasks for checking setup and hold violations. As with module path delay declarations, the order in which system timing check constructs appear in a **specify** block is not important. When a timing violation occurs, the Verilog simulator reports the time of the violation, the specified setup or hold time, and the measured time.

Other system tasks that may be used in a **specify** block are **$period, $skew, $nochange, $recovery, $setuphold,** and **$width.** Shown below is an example use of the **$period** system task:

```
$period (posedge clk, period);
```

A violation is reported if the time from one positive edge of the *clk* to another is less than the specified *period.* The **$width** system task is similar to **$period** except that the timing check is done from one edge to an opposite edge.

The **$skew** system task reports a violation if the time difference between events on its first argument and its second argument exceeds a specified limit. To report an error if a subsystem clock is more delayed from the main clock than can be tolerated by overall system clocking, the following timing check should be done:

```
$skew (main_clock, sub_system_clock, tolerated_skew);
```

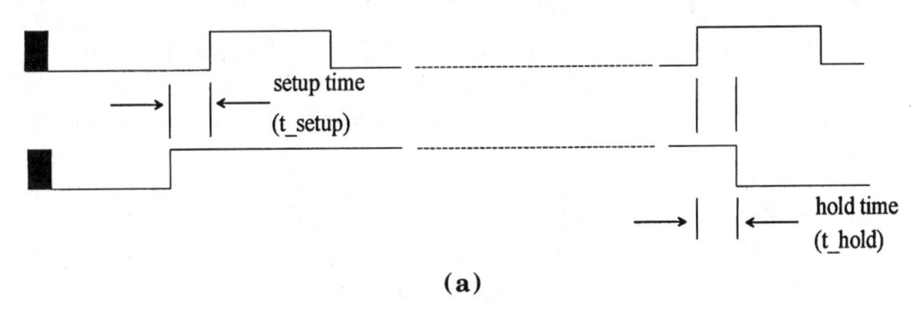

(a)

```
`timescale 1ns/100ps

module d_sr_flipflop (d, set, rst, clk, q, qb);
input d, set, rst, clk;
output q, qb;
reg q, qb;
parameter sq_delay = 6, rq_delay = 6, cq_delay = 6;
parameter t_setup = 4, t_hold = 4;
    specify
        (clk *> q, qb) = cq_delay;
        (set *> q, qb) = sq_delay;
        (rst *> q, qb) = rq_delay;
        $setup (d, posedge clk, t_setup);
        $hold (posedge clk, d, t_hold);
    endspecify
    always @(posedge clk or posedge rst or posedge set) begin
        if (set) begin
            q = 1; qb = 0;
        end else if (rst) begin
            q = 0; qb = 1;
        end else if (clk) begin    //if (clk) not essential
            q = d; qb = ~d;
        end
    end
endmodule
```

(b)

Figure 9.30 Setup and hold times for a positive-edge-trigger D-type flip-flop. (*a*) Diagram; (*b*)Verilog code.

9.3 Behavioral Event Control

Event control statements enable designers to describe a system with complex timing very accurately at the behavioral level. The *wait_statement* and *event_control* statement discussed in Sec. 9.1 provide such capabilities. An *event_control* statement is edge-sensitive, while a wait statement is level-sensitive. This section will use these language constructs for describing state machines, handshaking mechanisms, interfacing systems, and resource sharing examples.

9.3.1 A behavioral state machine

Chapter 8 presented two methods for describing state machines. A more behavioral method will be presented here. The example we use is the Moore implementation of the 1011 detector in the previous chapter, whose state diagram is shown in Fig. 9.31. As shown in this diagram, the output becomes **1** and stays **1** for a complete clock cycle when the 1011 sequence appears on the x input on four consecutive clock edges.

Figure 9.32 shows the Verilog description for this sequence detector. A parameter declaration specifies state designators by declaring

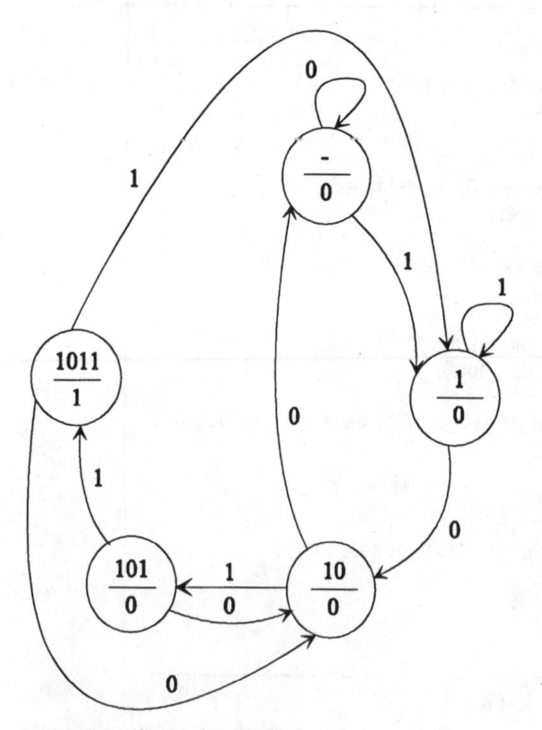

Figure 9.31 State diagram of a Moore machine for detecting 1011 sequence.

named parameters and assigning 3-bit values to them. As a result, the states of the machine become 000, 001, 010, 011, and 100, which are named *reset, got1, got10, got101,* and *got1011,* respectively. The *moore_detector* module declares *current* **reg,** which is initialized to *reset* in an **initial** block. The *current* **reg** type variable is used in the state transition part of this module for containing the present active state of the machine. An **always** block specifies state transitions and assignment of values to the state machine output. The **always** block has a statement part that consists of an event control statement attached to a case statement. The event control statement blocks activity flow into the case statement until a positive edge of the *clk* arrives. The case statement checks the value of *current* against the states of the machine. If *current* is found to match any of the 3-bit state values designated by state names, the sequential block that follows that alternative will execute. In each such block, the next value of *current* is set and the *z* output is assigned a value of **0** or **1.** After completion of this block, the remainder of the case that follows the matching state designator will be skipped. While *current* and *z* retain their values, the case statement will not be entered again (it is blocked by the event control statement) until the next edge of the clock is detected.

This description style is an accurate and convenient representation for state machines. It is easy to read, can be used for Mealy and Moore machine representations, and uses convenient behavioral constructs for state transitions. The use of a sequential block for the activities of each of the states makes this a very flexible coding scheme. More complex behavioral constructs for examining multiple inputs or assigning values to several outputs can be used in each of these blocks.

The syntax of the case statement of the *moore_detector* state machine is shown in Fig. 9.33. A case statement begins with the **case** keyword, followed by an expression and *case_item*s.

Each *case_item* consists of a statement and an expression that precedes it. Activity flows into a *case_item* statement if a match is found between its corresponding expression and the case expression.

Figure 9.34 shows an alternative description for the state machine of Fig. 9.31. In this description, a continuous assignment statement is concurrent with the **always** block. When *current* changes, the right-hand side of the assignment to *z* is evaluated, and a new value will be assigned to *z*. In a case, such as our example, with a simple output logic, the style shown in Fig. 9.34 offers a more compact form for describing state machines. The flexibility in assigning output values in Fig. 9.32 allows complex statements involving circuit inputs to participate in determining output values. Chapter 10 shows a description of a CPU using this general coding style.

```
`timescale 1ns/100ps

module moore_detector (x, clk, z);
input x, clk;
output z;
reg [2:0] current;
reg z;
parameter [2:0]
   reset = 0,
   got1 = 1,
   got10 = 2,
   got101 =3,
   got1011 =4;
   initial current = reset;
   always @(posedge clk)
   case (current)
      reset: begin
             if (x==1) current = got1;
             else current = reset;
             z = 0;
      end
      got1: begin
             if (x==0) current = got10;
             else current = got1;
             z = 0;
      end
      got10: begin
             if (x==1) current = got101;
             else current = reset;
             z = 0;
      end
      got101: begin
             if (x==1) begin
                current = got1011;  z = 1;
             end else begin
                current = got10;  z = 0;
             end
      end
      got1011: begin
             if (x==1) current = got1;
             else current = got10;
             z = 0;
      end
   endcase
endmodule
```

Figure 9.32 Verilog description of the 1011 sequence detector of Fig. 9.31.

Inclusion of a synchronous or asynchronous reset or preset is possible in any of the state machine styles presented thus far. However, a style often used for synthesis, which can easily be modified for any form of *set* or *reset,* will be presented for describing the Mealy machine shown in Fig. 9.35.

Like any other small or large digital system, a Mealy machine may be represented by the model shown in Fig. 9.36. This models a system as an interconnection of a combinational circuit and a register in its feedback path. For the combinational part, primary inputs and state variables feeding back from the outputs of the register are considered as inputs. The outputs of this part are the circuit outputs and state variables connected to register inputs. With each clock, the register in the feedback path makes the present state variable values become the same as those in the next state before the clock. The register is treated as a separate entity with its own inputs and outputs as shown in the first **always** block of Fig. 9.37. Presetting, resetting, or special register functions can be incorporated into the register part and do not influence the combinational part. We will use the general model in Fig. 9.36 to describe the state machine shown in Fig. 9.35.

The Verilog description in Fig. 9.37 corresponds to the Mealy machine in Fig. 9.35. A **reg** declaration declares 2-bit **reg** types *nxt* and *present* for the inputs and outputs of the register part of Fig. 9.36.

The *register* sequential block in Fig. 9.37 describes the register part. A block label is optional in Verilog and is used here for correspondence with the blocks of the diagram in Fig. 9.36. The flow into the *registering* block is controlled by the positive edge of *clk* and *r*. The if–else statement in this block prioritizes *r* over *clk* and implements a D-type register with asynchronous reset. This part assigns values to the *present* variable.

The *combinational* block in the Verilog code of Fig. 9.37 represents the combinational part in Fig. 9.36. This block is entered when an

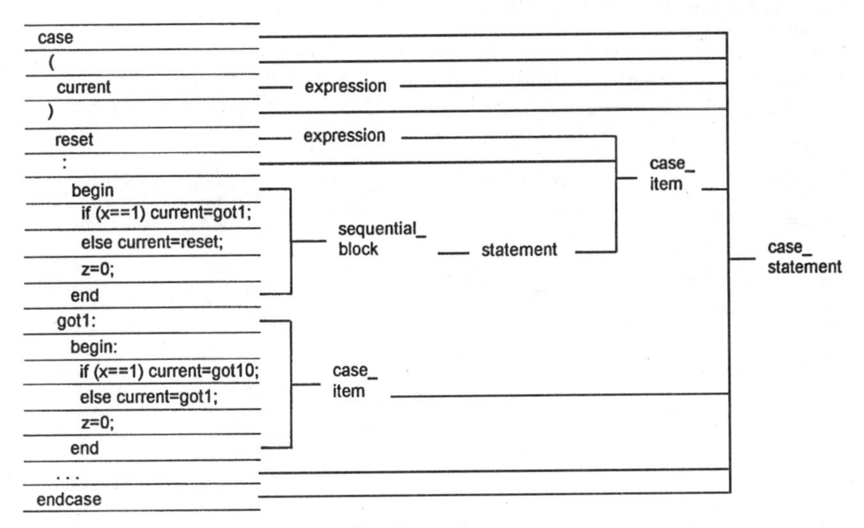

Figure 9.33 *case_statement* syntax details.

```
`timescale 1ns/100ps

module moore_detector (x, clk, z);
input x, clk;
output z;
reg [2:0] current;
parameter [1:0]
          reset = 0,
          got1 = 1,
          got10 = 2,
          got101 =3,
          got1011 =4;
    initial current = reset;
    always @(posedge clk)
        case (current)
            reset:
                if (x==1) current = got1;
                else current = reset;
            got1:
                if (x==0) current = got10;
                else current = got1;
            got10:
                if (x==1) current = got101;
                else current = reset;
            got101:
                if (x==1) current = got1011;
                else current = got10;
            got1011:
                begin
                    if (x==1) current = got1;
                    else current = got10;
                end
        endcase
    assign z = (current == got1011) ? 1 : 0;
endmodule
```

Figure 9.34 An alternative to the description of Fig. 9.32.

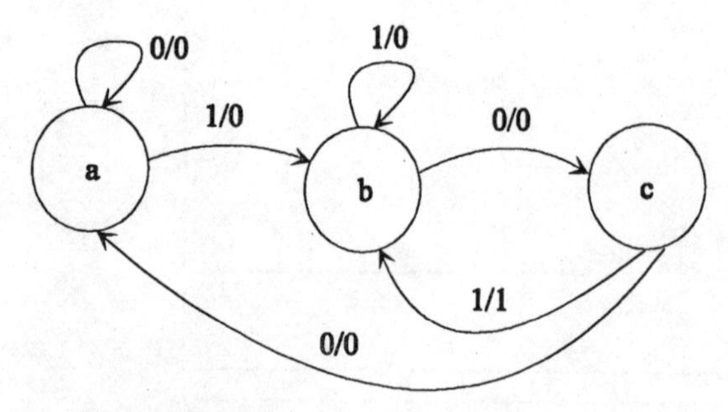

Figure 9.35 A Mealy machine detecting 101.

event occurs on *present* or on *x*. A case statement checks *present* against *a, b,* and *c* parameters and sets the next state according to transitions in the diagram of Fig. 9.35. Every time the *combinational* block is entered, *z* is set to **0.** When exiting the block, *z* may remain **0** or be set to **1** by the last if statement in this block. Assignment of **1** to *z* is conditioned by the *present* state being *c* and input *x* being **1.** Adding a preset or making the reset input synchronous affects only the *register* block, while the *combinational* block is responsible for state transitions. Separation of tasks makes it easy to adapt this coding style to most controller descriptions.

9.3.2 Two-phase clocking

A very common clocking scheme in MOS circuits is two-phase nonoverlapped clocking. This scheme ensures input-to-output isolation in master-slave registers and eliminates many charge-sharing problems. Figure 9.38 shows generation of a second phase, *c2,* from a periodic first phase, *c1.* We are assuming that the period of *c1* is 1 μs with a 500-ns duty cycle.

A one-line **always** block in the upper part of the code and a one-line **initial** block in the lower part of the code in Fig. 9.38 are responsible

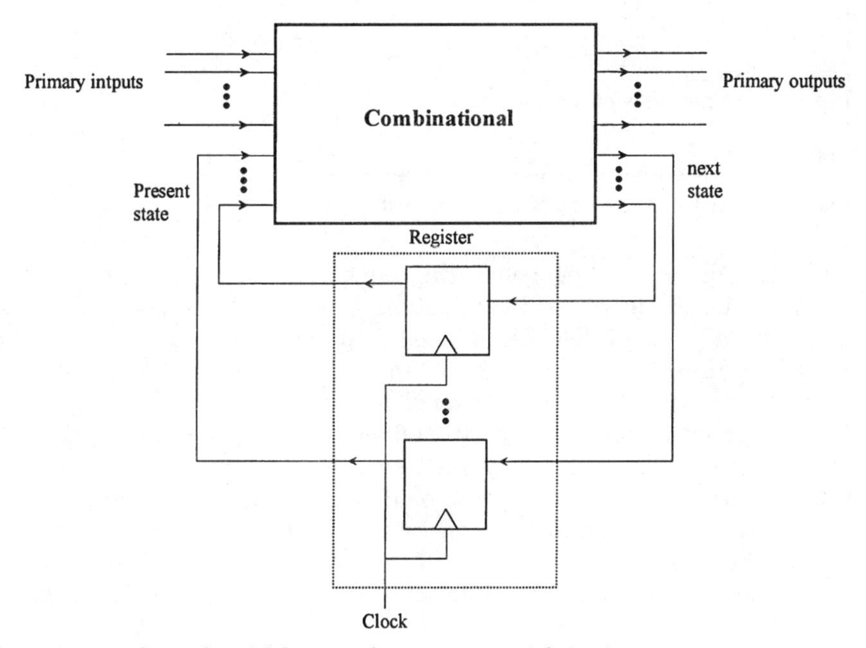

Figure 9.36 General model for a synchronous sequential circuit.

```
`timescale 1ns/100ps

module asynch_reset_detector (x, r, clk, z);
input x, r, clk;
output z;
reg [1:0] nxt, present;
reg z;
    parameter
        a = 0,
        b = 1,
        c = 2;
    initial nxt = a;
    always @(posedge clk or posedge r) begin : registering
        if (r==1) present = a;
        else present = nxt;
        end
    always @(present or x) begin : combinational
        z = 0;
        case (present)
            a:
                if (x==0) nxt = a;
                else nxt = b;
            b:
                if (x==0) nxt = c;
                else nxt = b;
            c:
                if (x==0) nxt = a;
                else nxt = b;
            default:
                nxt = a;
        endcase
        if (present == c && x == 1) z = 1;
    end
endmodule
```

Figure 9.37 Verilog code for the Mealy machine of Fig. 9.35.

for initializing *c1* and *c2,* generating a periodic waveform on *c1,* and stopping the simulation at time 4000 ns. The *phase2* sequential block generates *c2* from *c1.* The block label is optional and is used here for reference. Entering this block is not controlled, and because it is preceded by **always,** it repeats forever. The body of this sequential block waits for a level **1** followed by a level **0** on *c1.* Waiting for these values is performed by level-sensitive **wait** statements. Then, 10 ns after a complete positive pulse terminates on *c1, c2* becomes **1** and stays **1** for 480 ns. It then becomes **0,** and control returns to the beginning of the **always** block and waits for a **1** on *c1.* Figure 9.39 shows a timing diagram that results from the simulation run of the module of Fig. 9.38.

9.3.3 Implementing handshaking

A synchronous communication between systems is done by handshaking. Handshaking refers to the signaling that occurs between two systems as one transfers data to the other. When a system prepares data for transfer to another system, the sending system informs the receiving system that the data are ready. When the receiving system accepts the data, it informs the sending system that it has received them.

Handshaking can be fully responsive or partially responsive. In a fully responsive process, all events on the handshaking signals of one system occur in response to events on the signals of the other system as they communicate. Handshaking requires at least one signal for this specific purpose and can use as many as six for a two-way, fully responsive communication. Exchanging data without handshaking is called nonresponsive communication. Figure 9.40 shows a fully responsive two-line handshaking process for transferring data on *data_lines* from system A to system B.

System A places valid data on *data_lines* and informs system B of these new data by raising the *data_ready* line. When system B is ready to accept data, it does so, and it informs system A that it has accepted data by raising *accepted.* When system A sees that data on *data_lines* are no longer needed by system B, it removes the valid data from *data_lines* and lowers the *data_ready* line. System B acknowledges this and informs system A that it can accept new data by lowering its *accepted* signal.

A variation of this system can include a third handshaking line used by system B to inform system A that it is ready to accept new data.

```
`timescale 1ns/100ps

module two_phase;
reg c1, c2;
always #500 c1 = ~c1;
  always
  begin: phase2
    wait (c1==1)
    wait (c1==0)
    #10
    c2 = 1;
    #480;
    c2 = 0;
  end
  initial begin c1 = 1; c2 = 0; #4000 $stop; end
endmodule
```

Figure 9.38 Module generating *c1* and nonoverlapping *c2* clock phases.

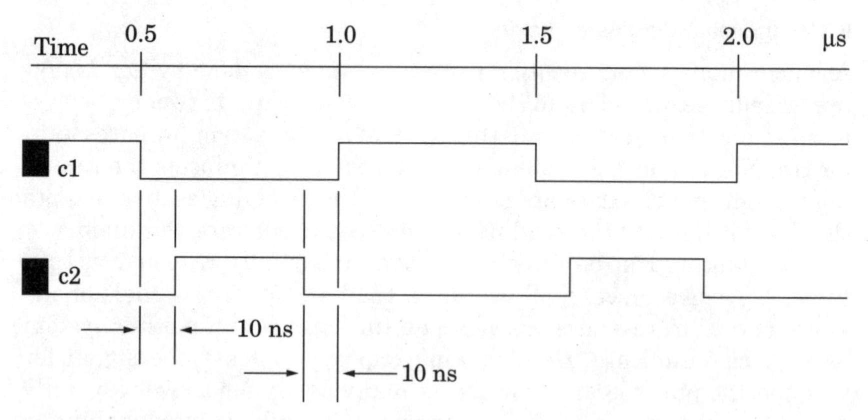

Figure 9.39 Two nonoverlapping phases of clock *c2* generated by the *phase2* block of Fig. 9.38 (timing not to scale).

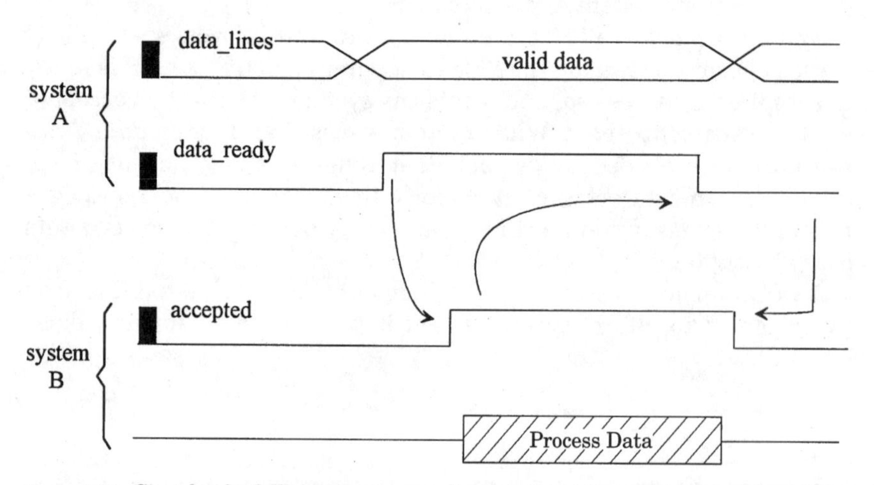

Figure 9.40 Signals of a fully responsive two-line handshaking.

Fully responsive handshaking is performed when no assumptions are possible as to the relative speed of the communicating systems. Other less responsive handshakings can be done in which the *data_ready* line returns to **0** after a fixed amount of time, instead of waiting for *accepted* to become **1.**

Figure 9.41 shows the corresponding Verilog pseudocode for the handshaking process in Fig. 9.40. Partial code sections in this figure are sequentially executed by systems A and B when they need to talk to each other. In all forms of handshaking, various forms of wait and event control statements are very useful and descriptive.

For a comprehensive example of modeling handshaking in Verilog, consider *system_i,* which works as an interface between *system_a* and *system_b,* as depicted in Fig. 9.42. *system_a* uses handshaking to provide 4-bit data, and *system_b* uses handshaking to receive 16-bit data. The interface *system_i* accumulates four data nibbles that it receives from *system_a* and makes 16-bit data available to *system_b*. The first nibble received from *system_a* forms the least significant 4 bits of the data that become available for *system_b*. *system_i* is capable of talking to *system_a* and *system_b* simultaneously. It should be possible for *system_i* to be involved in transmitting the previously accumulated data to *system_b* while accumulating new 16-bit data from *system_a*.

When *system_a* has a nibble ready on the *in_data* lines, it places a **1** on the *in_ready* line. The data and the *in_ready* line stay valid until this system sees a **1** on its *in_received* input. The interface *system_i*

```
System A:
        // start the following when ready to send
        data_lines = newly_prepared_data;
        data_ready = 1;
        wait for a positive edge on accepted
        data_ready = 0;
        // can use data_lines for other purposes

System B:
        // start the following when ready to accept data
        wait for a positive edge on data_ready
        accepted = 1;
        // start processing the newly received data
        wait for a negative edge on data_ready
        accepted = 0;
```

Figure 9.41 Verilog partial code for fully responsive two-line handshaking.

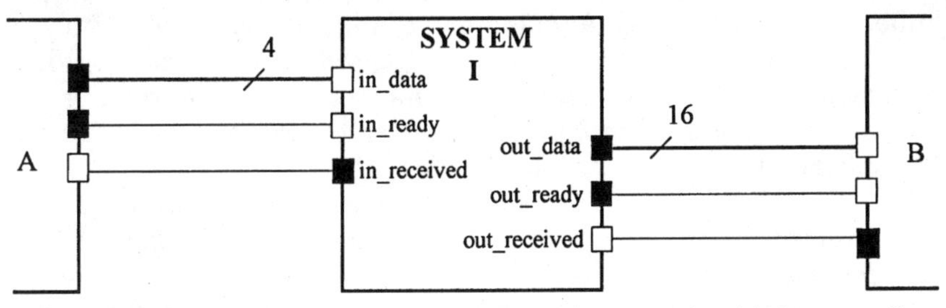

Figure 9.42 Interfacing *system_a* and *system_b*. *system_i* uses handshaking to talk to both systems.

waits in an idle state looking for *in_ready* to become **1.** When this happens, it receives data from *in_data* and acknowledges that it has received the data by placing a **1** on *in_received.* The interface holds *in_received* active until *in_ready* becomes **0.** On the other side, *system_i* talks to *system_b* by providing data on the *out_data* output bus and, by activating the *out_ready* line, informing the other system of the new data. When *system_b* receives the data, it places a **1** on the *out_received* line and holds this line active until *system_i* deactivates its *out_ready* output.

The Verilog description for *system_i* is shown in Fig. 9.43. This implementation has three handshaking involvements: one for talking to *system_a,* one for talking to *system_b,* and the third for communication between the transmitting and receiving parts of *system_i.*

The *a_talk* block waits for *in_ready* to become **1,** receives data, and places them in the part of the *word_buffer* indicated by *i.* When this is complete, it indicates that the data have been accepted by placing a **1** on *in_received* (the statement following **end else**). This line stays active until *system_a* deactivates the *in_ready* line. In accumulating data in the *word_buffer,* if the data received are the fourth nibble (that is, when *i* is 3, detected by the *i* == 3 condition of the if statement), the *a_talk* process asserts the *buffer_full* internal handshaking signal to indicate that data are ready to be transmitted to *system_b.* This line stays active until the buffer has been received by the *b_talk* process, as indicated by the *buffer_picked* signal issued by *b_talk.* When the buffer is picked, the nibble count *i* is set to 0 and *word_buffer* starts being refilled. While waiting for the buffer to be picked, *system_i* keeps *system_a* waiting by not issuing the *in_received* signal.

The *b_talk* block waits for a full buffer. This waiting is implemented by the wait *(buffer_full)* statement. When this statement is reached, if *buffer_full* is **1,** the flow continues, and if it is not **1,** it waits for it to become **1** as a result of its being set in the *a_talk* block. When this process finds a full buffer, it assigns it to the *out_data* output lines and causes the resumption of *a_talk* by setting *buffer_picked* to **1.** This line returns to zero only when *a_talk* acknowledges that it knows that a buffer has been received. At this time, *b_talk* communicates with *system_b* for sending the 16-bit data on *out_data* to this system. When *system_b* acknowledges the reception of data by raising its *out_received* line, the *b_talk* process deactivates *out_ready* and returns to the beginning of its statement part.

In the *a_talk* block, the *i* variable keeps a count of the number of nibbles received. When *i* becomes 3, flow is blocked until the positive edge of *buffer_picked* is observed.

Figure 9.44 shows a test bench for *system_i* of Fig. 9.43. An **initial** block schedules 15 values from 1 to 15 for the variable *seed* every 30 ns.

```
`timescale 1ns/100ps

module system_i (in_data, out_data, in_ready, out_received,
                                    in_received, out_ready);
input [3:0] in_data;
input in_ready, out_received;
output [15:0] out_data;
output out_ready, in_received;
reg [15:0] out_data;
reg out_ready, in_received;
reg [3:0] word_buffer [3:0];
reg buffer_full, buffer_picked;
reg [1:0] i;
  initial begin i = 0; end
  always begin: a-talk
          @(posedge in_ready);
          word_buffer [i] = in_data;
          if (i==3) begin
               buffer_full = 1;
               @(posedge buffer_picked);
               buffer_full = 0;
               i = 0;
          end else i = i + 1;
          in_received = 1;
          @(negedge in_ready);
          in_received = 0;
  end
     always begin: b-talk
          wait(buffer_full);
          out_data = {word_buffer [3],
                   word_buffer [2],
                   word_buffer [1],
                   word_buffer [0]};
          buffer_picked = 1;
          @(negedge buffer_full);
          buffer_picked = 0;
          out_ready = 1;
          @(posedge out_received);
          out_ready = 0;
  end
endmodule
```

Figure 9.43 Verilog model for the interface between systems A and B of Fig. 9.42.

This block also initializes the *a_ready* and *b_received* inputs of *system_i*. An **always** block modeling system *A* uses *seed* for calling the **$random** system function. Every 450 ns this *seed* is changing and **$random** uses its current value every time flow reaches it in the **always** block. A **repeat** loop statement in the block modeling system *A*, generates four sets of 4-bit data from the data generated by the **$random** system task and makes them available on *a_data*, which is connected to *in_data* of *system_i*. Another **always** block in Fig. 9.44 models the system *B* side of the interface. It waits for the positive edge of *b_ready* (*out_ready* of *system_i*), waits for 10 ns, and issues *b_received*. This **reg** type variable connects to *out_received* of *system_i* and is deasserted 20 ns after *out_ready* is set back to **0** by the interface circuit.

9.3.4 Interface handshaking

Several examples of handshaking and the use of process statements for high-level timing specifications will be presented here. The examples relate to components of a board-level interface design and are intended to demonstrate features of the language for abstract timing description.

Figure 9.45 shows a four-port bus arbiter. This component is used when several devices are sharing a bus. When a device requires the use of the system bus, it issues its *request* line and waits until the bus arbiter grants it permission to use the bus. The bus will not be granted to any other device until the *request* is deasserted by a device using the bus. In case of multiple bus requests, the arbiter grants the bus to the requesting device connected to the higher-number port.

As shown in the Verilog code in Fig. 9.46, the circuit has four *request* and four *grant* lines. The arbiter activities are synchronized with a system clock, *clock*. After a short delay (20 ns) after the falling edge of the clock, *request* lines are monitored, and a *grant* corresponding to the *request* with highest priority will become **1.** This will be repeated with each clock, and a *grant* will be set to **0** only when its corresponding request is not active. It is expected that a system that is granted the bus will hold its *request* active until it no longer needs access to the bus.

The description in Fig. 9.47 is written for testing the arbiter. The arbiter is instantiated here, and independent data are applied to each of its ports. The *delays* time array models the time between subsequent requests from a requesting device. The *usage* array specifies the time each device uses the bus.

For another interface example, exploring the capabilities of event control statements, consider the serial-to-parallel adapter shown in Fig. 9.48. The device has a serial input, an 8-bit parallel output, and several error and handshaking signals. Asynchronous data on the *serial* input are converted to parallel and become available on the *parallel_out, dataready* is issued, and the output remains valid while not *received* by the device reading the generated parallel data.

The format of the serial data on *serial* is according to the RS232 standard, as shown in Fig. 9.49. A frame consisting of 10 bits begins when the normally high input line drops to **0.** This and all other bits

```
`timescale 10ns/1ns

module test_system_interface;
reg [3:0] a_data;
reg a_ready;
wire a_received;
wire [15:0] b_data;
wire b_ready;
reg b_received;
integer i, seed;
system_i u1 (a_data, b_data, a_ready, b_received, a_received, b_ready);
  initial begin
     for (i=1; i <= 15; i = i+ 1) seed <= #(i*3) i;
     a_ready = 0; b_received = 0;
     #42 $stop;
  end

  // this models system_a
  always begin
     a_data = $random (seed);
     repeat (4) begin
       #2 a_data = {~a_data [0], a_data [3:1]};
       a_ready = 1;
       @(posedge a_received);
       #1 a_ready = 0;
     end
  end

  // this models system_b
  always begin
     @(posedge b_ready);
     #1 b_received = 1;
     @(negedge b_ready);
     #2 b_received = 0;
  end
endmodule
```

Figure 9.44 Test bench for Verilog code of handshaking problem.

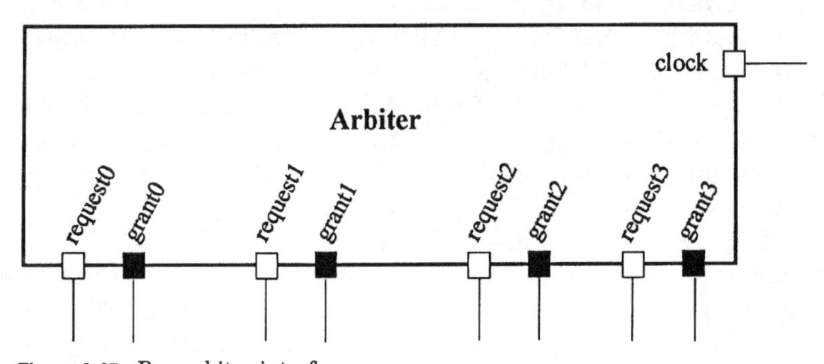

Figure 9.45 Bus arbiter interface.

```
`timescale 1ns/100ps

module arbiter (request, grant, clock);
input [3:0] request;
input clock;
output [3:0] grant;
reg [3:0] grant;
reg found;
integer i;
  initial i = 0;
  always begin
    @(negedge clock);
    #20;
    found = 0;
    for (i=3; i>=0; i=i-1)
      if (request[i]==0) grant[i] = 0;t[i] = 0;
      if (grant == 4'd0)
        for (i=3; i>=0; i=i-1)
          if (request[i] && found == 0) begin
            grant[i] = 1;
            found = 1;
          end
  end
endmodule
```

Figure 9.46 Bus arbiter description.

in a frame are equal in duration and depend on the speed of the line. For 9600 bits per second (bps), each bit duration is 1/9600 of a second.

This first **0** is the start bit, and it is followed by 8 bits of data. After completion of the serial data, the serial input line becomes **1** and stays at this level for a 1- or 2-bit duration. These 2 bits are referred to as stop bits. We will use only one stop bit in our interface description.

Because each frame begins with a start bit on the serial line, receiving data can be synchronized with each frame. Half a bit after the **1**-to-**0** transition, reading begins. The first reading reads the start bit, which must be **0.** The next eight readings read the data bits. The last reading reads the stop bit, which must be **1.** A framing error occurs if a **0** appears where a stop bit is expected. An overrun error occurs if the receiving system does not receive the parallel data before the next set of data is ready.

Figure 9.50 shows the Verilog code for the *serial2parallel* adapter. Two concurrent sequential blocks handle this conversion. The *collect* block receives serial data, collects 8-bit data, and handshakes them to the receiving device. The *too_fast* block issues *overrun* if handshaking is not complete when new data arrive.

A delay control in the *collect* block moves a time pointer to the middle of the start bit. At this point, in a repeat loop statement, *#full_bit* moves the time pointer to the middle of each data bit, reads data from the serial input, and places each data bit in *buff*. When reading of data bits is done, the time pointer is advanced to the middle of the next bit and the stop bit is read. The *frame_error* signal is set to **1** if the stop

```
`timescale 10ns/1ns

module test_arbiter;
reg clk;
reg [3:0] r;
wire [3:0] g;
integer delays [3:0], usage [3:0];
     arbiter u1 (r, g, clk);
  initial begin
          delays [3] = 400; delays [2] = 300;
          delays [1] = 1500; delays [0] = 800;
          usage [3] = 300; usage [2] = 400;
          usage [1] = 600; usage [0] = 100;
          clk = 0; r = 0;
          #2000 $stop;
     end
     always #50 clk = ~clk;
     always begin
          #(delays[3]) r[3] = 1;
          @(posedge g[3]);
          #(usage[3]);
          @(negedge clk) r[3] = 0;
     end
     always begin
          #(delays[2]) r[2] = 1;
          @(posedge g[2]);
          #(usage[2]);
          @(negedge clk) r[2] = 0;
     end
     always begin
          #(delays[1]) r[1] = 1;
          @(posedge g[1]);
          #(usage[1]);
          @(negedge clk) r[1] = 0;
     end
     always begin
          #(delays[0]) r[0] = 1;
          @(posedge g[0]);
          #(usage[0]);
          @(negedge clk) r[0] = 0;
     end
endmodule
```

Figure 9.47 Testing the arbiter.

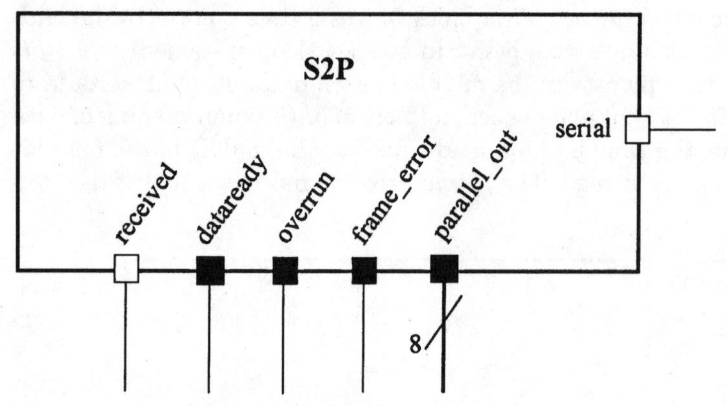

Figure 9.48 Serial-to-parallel interface.

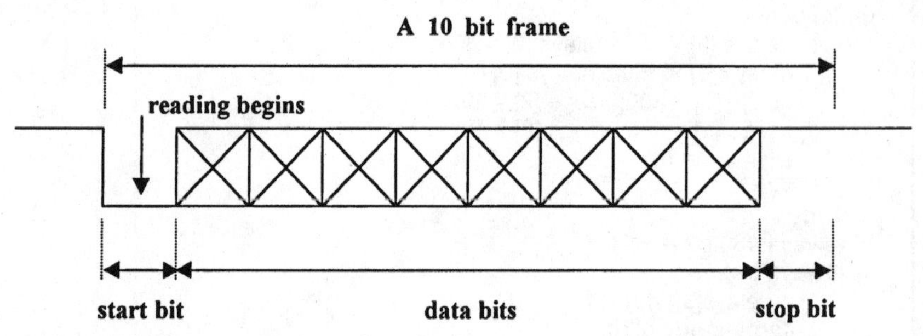

Figure 9.49 RS232 frame.

bit is not detected. Otherwise, *dataready* is issued, data are placed on *parallel_out*, and the flow of activity in the *collect* block is blocked until *received* becomes **1**.

While waiting for *received* to become **1**, the *too_fast* block, the flow into which started when *dataready* was issued, monitors the *serial* input. If a start bit appears on this line, i.e., if it becomes **0** (checked by @(**negedge** *serial*)) while *dataready* is still **1**, the *overrun* error flag will be raised.

In this example, two concurrent **always** blocks (*collect* and *too_fast*) are used. The need for this is due to the fact that the serial-to-parallel adapter is simultaneously talking to the receiving device and monitoring the serial input. This interfacing device will be used in Chap. 11 in a complete interface design example.

9.4 Display and File Output

Generally, simulation environments provide convenient ways of displaying **net** or **reg** values. For debugging purposes, or for generating outputs independent of a simulation environment, Verilog system tasks for display and file output may be used.

Generic system tasks for this purpose are **$display, $write, $strobe,** and **$monitor.** These tasks output to the standard output device using the format specified with task arguments when the task

```
`timescale 1ns/100ps
`define BPS 10000

module serial2parallel (serial, received, dataready,
                overrun, frame_error, parallel_out);
    input serial, received;
    output dataready, overrun, frame_error;
    output [7:0] parallel_out;
    reg dataready, overrun, frame_error;
    reg [7:0] parallel_out;
    reg [7:0] buff;
    initial begin
        frame_error = 0;
        dataready = 0;
        overrun = 0;
        parallel_out = 1'b0;
    end
    parameter
        half_bit = 1E9/(`BPS*2.0),
        full_bit = 1E9/`BPS;
    always begin: collect
        @(negedge serial);
        #half_bit;
        repeat(8) #full_bit buff = {serial, buff[7:1]};
        #full_bit;
        if (serial == 0) begin
            frame_error = 1;
            @(posedge serial);
        end else begin
            frame_error = 0;
            dataready = 1;
            parallel_out = buff;
            @(posedge received) @(negedge received);
            dataready = 0;
        end
    end
    always begin: too_fast
        @(dataready);
        if (dataready==1) begin
            @(negedge serial) if (dataready==1) overrun = 1;
        end else
            overrun = 0;
    end
endmodule
```

Figure 9.50 *Serial2parallel* Verilog description.

is enabled. The same task names preceded by the letter **f** become file output tasks. For these tasks, the first argument is the file descriptor to which the output is being done. Also, a **b, h,** or **o** at the end of these task names makes them specific for display or file output in binary, hexadecimal, or octal format.

9.4.1 Standard output

The **$display** task outputs a formatted line to the output device when it is enabled. This task is enabled in a sequential block when activity flow reaches it. Each time it is enabled, a complete line including a newline character is outputted. The **$write** task is similar to **$display** except that it does not output a newline character, causing the next output to appear on the same line. The **$strobe** system task is identical to **$display** except that it waits for all simulation activities to occur before it displays the values specified by its arguments.

The **$monitor** task is different from the tasks described above in that it is enabled once, and while it stays on, it displays a formatted line each time a value change occurs on one of the variables in its argument list. This task is turned on by the **$monitoron** task, and it is turned off by the **$monitoroff** task.

The examples that follow show ways of using the **$display** and **$monitor** tasks for displaying formatted data on the standard output. Features found in these tasks represent those found in most standard-output system tasks, and examples for the other forms of display routine are only slightly different from those being presented here.

The first example uses the **$display** task shown in Fig. 9.51. The line shown in bold is added to the *two_phase* module of Fig. 9.38 to display values of the *c1* and *c2* phases of clock. This part of the code is a sequential block enclosed in an **always** block which becomes active every 5 ns. Every 5 ns, for the entire time of simulation, the current time and the current values of *c1* and *c2* are displayed on the standard output device. The **$time** system task returns an integer representing the actual time of simulation when it is invoked. In addition to enabling the **$time** system task in its argument, three other arguments appear in the argument list of **$display** of Fig. 9.51. The quoted string specifies the output format, which is separated by commas from *c1* and *c2,* which are variables whose values will be displayed using the format in the quoted string. Characters in the quoted string argument are displayed as they are unless they are preceded by **** or **%.** The **\t** in Fig. 9.51 is the escape sequence for the tab character. Two **%b** strings in the quoted string are format specifications for the values of *c1* and *c2.* Format specifications and escape sequences used with the **$display** task are shown in Fig. 9.52.

```
`timescale 1ns/100ps

module two_phase_with_display;
reg c1, c2;
      always #500 c1 = ~c1;
      always
      begin : phase2
            wait (c1==1)
            wait (c1==0)
            #10
            c2 = 1;
            #480;
            c2 = 0;
      end
      initial begin c1 = 1; c2 = 0; #4000 $stop; end

      always #5 $display ($time, " ns \t %b %b", c1, c2);
endmodule
```

Figure 9.51 Using the **$display** task.

% h	Display value in Hexadecimal
% d	Display value in Decimal
% o	Display value in Octal
% b	Display value in Binary
% c	Display value in 8-bit ASCII
% m	Display hierarchical name
% s	Display in string format
% t	Display in current time format

(a)

\n	Display newline character
\t	Display tab character
\\	Display backslash
\"	Display double-quote
\ddd	Display character coded *ddd*
% %	Display percent sign

(b)

Figure 9.52 Escape sequences. (*a*) Format specification; (*b*) character display.

Simulating and running the module of Fig. 9.51 results in a display such as that in Fig. 9.53 appearing on the user's standard output screen. Usually this is where simulation commands are entered and simulator messages are displayed. This result is generated in addition to any variable waveform or list outputs that a Verilog simulation environment may generate.

Another example of the use of a display system routine is shown in Fig. 9.54. The part shown in bold is added to the *two_phase* module of Fig. 9.38. The **$monitor** task is used here for displaying values of *c1* and *c2*. Being in an **initial** block, this task is executed once at time 0.

Other system tasks in this **initial** block turn the monitor on only for 2000 ns from time 1000 ns. During the simulation, while monitoring is turned on, anytime an event occurs on *c1,* or *c2,* the **$monitor** task is executed. Execution of this task causes time, *c1* and *c2* values to appear on the standard output using the formatting specified by the quoted string argument. Figure 9.55 shows the display generated by running the module of Fig. 9.54.

```
# Loading work.two_phase_with_display
Run -All
#          5 ns    1 0
#         10 ns    1 0
#         15 ns    1 0
#         20 ns    1 0
#         25 ns    1 0
#         30 ns    1 0
#         35 ns    1 0
#         40 ns    1 0
..        .. ..    ...
#       3930 ns    0 1
#       3935 ns    0 1
#       3940 ns    0 1
#       3945 ns    0 1
#       3950 ns    0 1
#       3955 ns    0 1
#       3960 ns    0 1
#       3965 ns    0 1
#       3970 ns    0 1
#       3975 ns    0 1
#       3980 ns    0 1
#       3985 ns    0 1
#       3990 ns    0 0
#       3995 ns    0 0
# Break at c:/VERILOG/CHAP9/two_phase_p1.v line 18
```

Figure 9.53 Display result of the module of Fig. 9.51.

```
`timescale 1ns/100ps

module two_phase_with_display;
reg c1, c2;
      always #500 c1 = ~c1;
      always
      begin : phase2
            wait (c1==1)
            wait (c1==0)
            #10
            c2 = 1;
            #480;
            c2 = 0;
      end
      initial begin c1 = 1; c2 = 0; #4000 $stop; end

      initial begin
            $monitor ($time, " ns \t %b %b", c1, c2);
            $monitoroff; #1000 $monitoron; #2000 $monitoroff;
      end
endmodule
```

Figure 9.54 Using the **$monitor** system task.

```
# Loading work.two_phase_with_display
Run -All
#           0 ns     1 0
#        1000 ns     1 0
#        1500 ns     0 0
#        1510 ns     0 1
#        1990 ns     0 0
#        2000 ns     1 0
#        2500 ns     0 0
#        2510 ns     0 1
#        2990 ns     0 0
# Break at c:/VERILOG/Chap9/two_phase_p2.v line 18
```

Figure 9.55 Display generated by the **$monitor** task in Fig. 9.54.

Another example using **$display** system tasks is shown in Fig. 9.56. As in the other two examples in this section, this example inserts code in the test bench of the *two_phase* clock generation module. The part shown in bold is used for display purposes. In an **initial** block, a **$display** task is enabled to generate a header for data that will be displayed. Following the **initial** block, two **always** blocks are shown in the bold part of Fig. 9.56. Activity flow into these blocks is controlled by changes in the values of *c1* and *c2*, respectively. When *c1* changes, the corresponding **$display** task is enabled, which writes the time, the value of *c1,* and . . (two dots) to the standard output. Similarly, when *c2* changes, the current time is written, followed by two dots for the value of *c1* and the current value of *c2.* The simulation run of the module of Fig. 9.56 results in the display shown in Fig. 9.57. Simultaneous changes of *c1* and *c2* cause both **$display** tasks in the two **always** blocks to be enabled. The order in which these statements are executed is unknown.

9.4.2 File output

Functionally, file output system tasks are identical to standard output tasks except that they require a file descriptor for outputting to file. The letter **f** precedes file output task names to distinguish them from standard output tasks. The first argument of a file output task is an integer descriptor, which must be defined by use of the **$fopen** system function. This integer identifies the file for all writing purposes. When file output is completed, the integer file descriptor should be used in the **$fclose** system task. The following statements show opening the file *listing.out,* writing to it, and closing it:

```
`timescale 1ns/100ps

module two_phase_with_display;
reg c1, c2;
    always #500 c1 = ~c1;
    always
    begin : phase2
        wait (c1==1)
        wait (c1==0)
        #10
        c2 = 1;
        #480;
        c2 = 0;
    end
    initial begin
        c1 = 1; c2 = 0; #4000 $stop;
    end

    initial begin
        $display ("\t\t Time (ns) \t  c1  c2");
    end
    always @(c1) $display ($time, " ns \t   %b  ..", c1);
    always @(c2) $display ($time, " ns \t  ..   %b", c2);
endmodule
```

Figure 9.56 Displaying clock phases only when they change.

```
Run -All
#          Time (ns)      c1  c2
#           500 ns        0  ..
#           510 ns       ..   1
#           990 ns       ..   0
#          1000 ns        1  ..
#          1500 ns        0  ..
#          1510 ns       ..   1
#          1990 ns       ..   0
#          2000 ns        1  ..
#          2500 ns        0  ..
#          2510 ns       ..   1
#          2990 ns       ..   0
#          3000 ns        1  ..
#          3500 ns        0  ..
#          3510 ns       ..   1
#          3990 ns       ..   0
# Break at c:/VERILOG/Chap9/two_phase_p4.v line 19
```

Figure 9.57 Display result of the simulation run of the module of Fig. 9.56.

```
. . .
integer list1 = $fopen ("listing.out");
. . .
$fdisplay (list1, . . .);
. . .
$fclose (list1);
```

In these statements, *list1* is a descriptor used for referring to the disk file *listing.out*. The descriptor may be defined when declared, or it may be assigned a value in a sequential body using the **$fopen** system task. Writing to this file and closing it are done by **$fdisplay** and **$fclose** system tasks in a Verilog sequential block.

Figure 9.58 shows an example of using file output system tasks. In this example, code shown in lower part of the figure is added to our *two_phase* clock generation example to generate a waveform simulation report. Running the module shown in this figure generates an ASCII time plot with 5-ns time resolution for the *c1* and *c2* phases of the clock. High and low values and their transitions will be displayed using ASCII special characters.

Starting from the top portion of the code in Fig. 9.58 that is used for output file generation, a **reg** declaration declares *old_c1* and *old_c2*. These variables will be used for detecting transitions on *c1* and *c2*, respectively. An **integer** declaration follows this **reg** declaration. The integer *flush* is declared for the output file descriptor. Before any writing, an **$fopen** task must assign a value to this descriptor. The next statement in the lower part of Fig. 9.58 is an *event_declaration* that declares *print_resolution* as an **event** name. The **always** block that follows this declaration activates an event triggering statement every 5 ns, which causes events to occur on *print_resolution* every 5 ns while simulation continues. This **always** block is followed by an **initial** block.

The **initial** statement shown assigns a file descriptor to *flush* to represent the *clock4.out* disk file. It then initializes *old_c1* and *old_c2,* and writes a header that includes time and the *c1* and *c2* names to the *flush* logical file. Control of data written to this file is done in the **always** block that appears after the **initial** block in Fig. 9.58. Events on *print_resolution* every 5 ns make the sequential block in this statement active. Every 5 ns, the current time is written to *flush* followed by a wave slice for *c1* and a wave slice for *c2.* In order for the time, the *c1* slice, and the *c2* slice to appear on the same line, the **$fwrite** system task is used, which does not output a newline character. An output line is ended by enabling the **$fdisplay** task. The **$fdisplay** task with only a file descriptor argument writes a newline character to this file.

For writing the *c1* and *c2* portions of the waveform to the line that is being written to the *flush* file, the *append_wave_slice* user task is used. This task and the parameters used by it are shown in the lower part of Fig. 9.58. The parameter declaration declares three character-wide strings to be used for high, low, and transition values of *c1* and *c2.* When *append_wave_slice* is enabled, the *old* and *new* values in its argument are compared. If both these values are **1** or **0,** *hi_value* or *lo_value* will be written to the *flush* file. If the *old* and *new* values are different, corresponding strings representing low-to-high or high-to-low transitions are written into the *flush* file. The "| " and " |" strings represent low and high, respectively, while the ".−+" and "+−." strings are used for low-to-high and high-to-low transitions.

Figure 9.59 shows a portion of the waveform generated by running the simulation of the module of Fig. 9.58. If turned 90° in a counter-clockwise direction, this output makes a good substitute for graphic waveforms for ASCII terminals or data transmission.

9.5 MSI-Based Design

This section shows the design of a complete system using standard parts. The parts we use are from the 74LS00 logic family, while the example circuit is a variation of the sequential comparator in Chap. 8. This example demonstrates how generic size modules are defined, top-down design strategy, and top-level configuration of a design.

```
`timescale 1ns/100ps

module two_phase_with_display;
reg c1, c2;
    always #500 c1 = ~c1;
    always
    begin : phase2
        wait (c1==1)
        wait (c1==0)
        #10
        c2 = 1;
        #480;
        c2 = 0;
    end
    initial begin
        c1 = 1; c2 = 0; #4000 $fclose(flush); $stop;
    end

reg old_c1, old_c2;
integer flush;
event print_resolution;
    always #5 -> print_resolution;
    initial begin
        flush = $fopen ("clock4.out");
        old_c1 = 0;  old_c2 = 0;
        $fdisplay (flush, "\t\t Time (ns) \t  c1  c2");
    end
    always @(print_resolution) begin
        $fwrite (flush, $time, " ns \t ");
        append_wave_slice (old_c1, c1); old_c1 = c1;
        append_wave_slice (old_c2, c2); old_c2 = c2;
        $fdisplay (flush);
    end

parameter [8*3:1] lo_value = "|  ", hi_value = "  |", lo_to_hi = ".-+", hi_to_lo = "+-.";
    task append_wave_slice;
    input old, new;
        begin : append
            if (old == 1 & new == 1) $fwrite (flush, " %s ", hi_value); else
            if (old == 0 & new == 0) $fwrite (flush, " %s ", lo_value); else
            if (old == 0 & new == 1) $fwrite (flush, " %s ", lo_to_hi); else
            if (old == 1 & new == 0) $fwrite (flush, " %s ", hi_to_lo);
        end
    endtask
endmodule
```

Figure 9.58 Generating an ASCII plot file with 5-ns resolution.

```
 Time (ns)    c1 c2
 . . . .      . . . . .
 470 ns       |  |
 475 ns       |  |
 480 ns       |  |
 485 ns       |  |
 490 ns       |  |
 495 ns       |  |
 500 ns     +-.  |
 505 ns     |    |
 510 ns     |    .-+
 515 ns     |      |
 520 ns     |      |
 525 ns     |      |
 530 ns     |      |
 . . . .      . . . . .
1970 ns     |      |
1975 ns     |      |
1980 ns     |      |
1985 ns     |      |
1990 ns     |    +-.
1995 ns     |    |
2000 ns     .-+  |
2005 ns       |  |
2010 ns       |  |
2015 ns       |  |
```

Figure 9.59 Waveform simulation report written to a disk file.

The design strategy is as follows: After an initial understanding of the functionality of the circuit we are to design, we partition it into several functional components, and the components are mapped into standard parts. If such a mapping is not possible, more partitioning is done. Once the standard parts are chosen, they are assembled to perform the necessary functions and wired together to form the implementation of the system.

9.5.1 Top-level partitioning

The circuit shown in the block diagram in Fig. 9.60 keeps a modulo-16 count of consecutive equal data bytes on *data_in*. It has synchronous active low clear and load inputs, *clear_bar* and *load_bar*. The *clear_bar* input clears the output count, and the *load_bar* loads the *count_in* into the counter.

Figure 9.61 shows the partitioning of this circuit into a register, a comparator pair, and a counter. The register component keeps the most recent data byte. The comparator pair compares the incoming data on *data_in* with the old data in the register and asserts its output if the two data bytes are equal. This output enables the synchronous counting of the counter.

The MSI components that most closely correspond to this partitioning are the 74LS85 4-bit magnitude comparator, the 74LS377 8-bit register, and the 74LS163 4-bit binary counter, all shown in Fig. 9.62.

The 74LS85 package has three outputs that indicate the relation of values of the 4-bit inputs. The $P < Q$ outputs become **1** when input P is less than input Q; likewise, the $P > Q$ output becomes **1** when P is greater than Q. When the two inputs are equal, the values on the <, =, and > inputs appear on their corresponding output lines.

The 74LS377 package is a positive-edge-trigger 8-bit register with an active low enable input. If this input is low on the edge of the clock, data are clocked into the register.

The 74LS163 package is a 4-bit binary counter with a synchronous active low parallel load and reset inputs. The counting is enabled

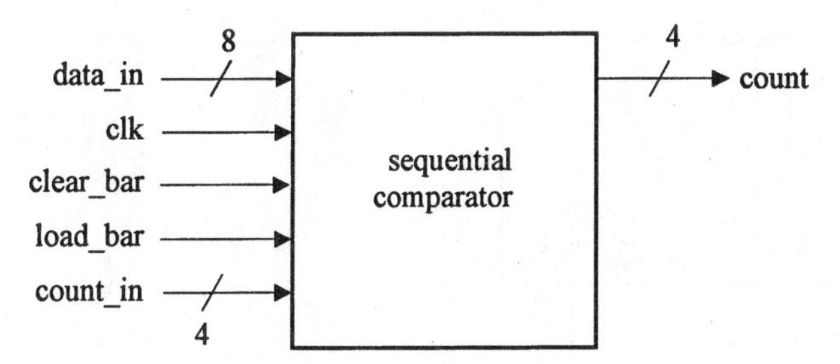

Figure 9.60 Block diagram of the sequential comparator circuit.

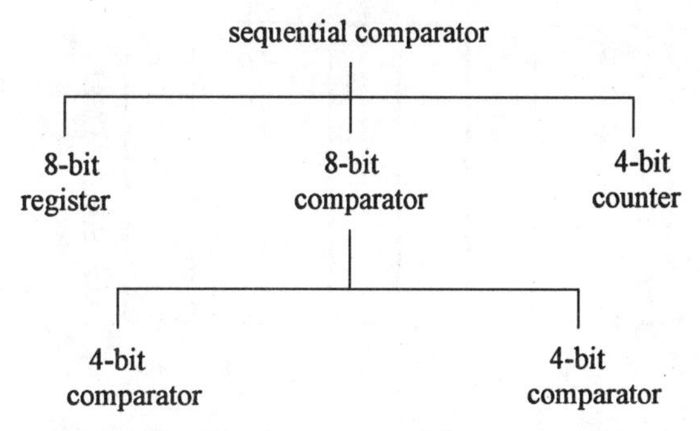

Figure 9.61 Partitioning a sequential comparator circuit into smaller functional components.

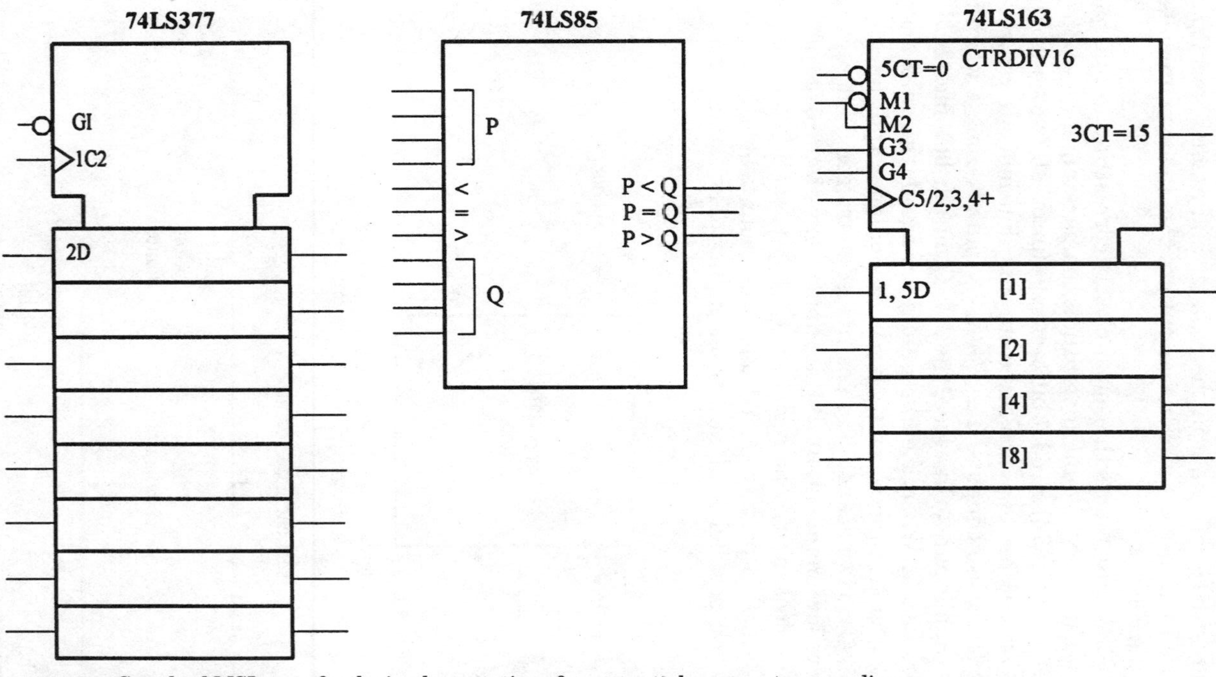

Figure 9.62 Standard MSI parts for the implementation of a sequential comparator according to the partitioning of Fig. 9.61.

when loading is disabled and count-enable inputs *G3* and *G4* are high. When the count reaches 15, the *5CT* = 15 output becomes **1** and the next clock starts the count from 0.

9.5.2 Description of components

Components needed for this design will be described at the behavioral level. Although the sizes of the 74xx components mentioned above are fixed, we will use parameters for their size. Size parameters specify the number of bits of the comparator, the register, and the counter. These parameters have default values according to the 74xx standard packages, but may be overwritten by higher-level modules.

The Verilog description for the 74LS85 4-bit magnitude comparator is shown in Fig. 9.63. Unlike the comparator descriptions in Chap. 5, this description is purely behavioral, and the comparator's functionality, rather than its hardware details, is evident from the model. Following the module header, parameter n is defined, which sets the size of this comparator module. All **reg** and **wire** declarations will depend on this parameter.

```
`timescale 1ns/100ps

module ls85_comparator (a, b, gt, eq, lt, a_gt_b, a_eq_b, a_lt_b);
parameter n = 4;
input [n-1:0] a, b;
input gt, eq, lt;
output a_gt_b, a_eq_b, a_lt_b;
reg a_gt_b, a_eq_b, a_lt_b;
parameter prop_delay = 10;
    always @(gt or eq or lt or a or b)
        if (a > b) begin
            a_gt_b <= #prop_delay 1;
            a_eq_b <= #prop_delay 0;
            a_lt_b <= #prop_delay 0;
        end else if (a < b) begin
            a_gt_b <= #prop_delay 0;
            a_eq_b <= #prop_delay 0;
            a_lt_b <= #prop_delay 1;
        end else if (a == b) begin
            a_gt_b <= #prop_delay gt;
            a_eq_b <= #prop_delay eq;
            a_lt_b <= #prop_delay lt;
        end
endmodule
```

Figure 9.63 Behavioral description of the 74LS85 4-bit magnitude comparator.

Comparing inputs and setting outputs are done in an **always** block. Relational operators (<, >, >=, and <=) and equality operators (== and !=) are used for comparing vectors. These operators are used in the **always** block that compares *a* and *b* values and assigns values to *a_gt_b, a_eq_b,* and *a_lt_b* comparator outputs. Variables *gt, eq, lt, a,* and *b* form the activity expression that controls activity flow into the sequential statement that assigns output values. An event on any of these variables causes a set of values to be scheduled for circuit outputs to occur after *prop_delay* delay time. After these scheduling activities cease, the comparator waits for another event. If the next event occurs within the delay time, the new values will override pending values. Because we have used nonblocking assignments, the **always** block completes its tasks in zero time and allows other input events to be detected. However, if blocking assignments were used, input activities while the **always** block is blocked would not be detected.

The comparator example offers a convenient coding style for combinational circuits. In general, an **always** block contains all logic of the combinational circuit. High-level sequential statements in a sequential block describe the activity. An event control statement controls the flow of activity into the sequential block. The activity expression of this event control statement is formed by an **or** of all combinational circuit inputs.

Figure 9.64 shows the Verilog description for the 74LS377 register. As with the comparator, parameter *n* sets the size of this register. This parameter is used for declaring inputs, outputs, and internal variables if needed. The description shown is at the behavioral level and uses a single **always** block. Enclosed in this block is an **if** statement that assigns 8-bit inputs to register outputs when the register is enabled by its active low *g_bar* input. The positive edge of the clock controls the flow of activities into the if statement. When program flow reaches

```
`timescale 1ns/100ps

module ls377_register (clk, g_bar, d8, q8);
parameter n = 8;
input clk, g_bar;
input [n-1:0] d8;
output [n-1:0] q8;
reg [n-1:0] q8;
parameter prop_delay = 7;
    always @(posedge clk)
        if (g_bar == 0) q8 = #prop_delay d8;
endmodule
```

Figure 9.64 Behavioral description of the 74LS377 8-bit clocked register.

the assignment to *q8* output, the value of *d8* is read, and after *prop_delay* it is assigned to *q8*.

A description of the 4-bit binary counter is presented in Fig. 9.65. This is a parametrized description that uses *n* for the size of the counter and *m* for its maximum count. On the positive edge of the clock in the *counting* sequential block, a 4-bit *internal_count* **reg** is assigned a value according to the active control mode of the counter. An active *clr_bar* clears the count to 0, *ld_bar* loads parallel data into the *internal_count,* and active *enp* and *ent* control inputs cause the internal count to be incremented by 1. When the internal count reaches 1111, the carry output of the counter, *rco,* becomes **1.** The value of the internal count **reg** is put in the 4-bit *q_abcd* output after *prop_delay* delay.

9.5.3 Design implementation

Figure 9.66 shows the graphical representation of the internal structure of the sequential comparator. This implementation is accomplished according to the partitioning shown in Fig. 9.61, which uses two 74LS85 packages for realizing an 8-bit comparator, a 74LS377 for the register, and a 74LS163 for the counter. The figure shows a top-level module to specify the wiring of components and another module on top of it for bindings and generic parameter specifications.

```
`timescale 1ns/100ps

module ls163_counter (clk, clr_bar, ld_bar, enp, ent, abcd, q_abcd, rco);
parameter n = 4, m = 4'b1111;
input clk, clr_bar, ld_bar, enp, ent;
input [n-1:0] abcd;
output [n-1:0] q_abcd;
output rco;
reg [n-1:0] q_abcd;
reg rco;
parameter prop_delay = 12;
reg [n-1:0] internal_count;
    always @(posedge clk) begin : counting
        if (clr_bar == 0) internal_count = 0;
        else if (ld_bar == 0) internal_count = abcd;
        else if (enp == 1 && ent == 1) internal_count = internal_count + 1;
    if (internal_count == m && ent == 1)
            rco = #prop_delay 1;
        else rco = 0;
        q_abcd = #prop_delay internal_count ;
    end
endmodule
```

Figure 9.65 Behavioral description of the 74LS163 4-bit counter.

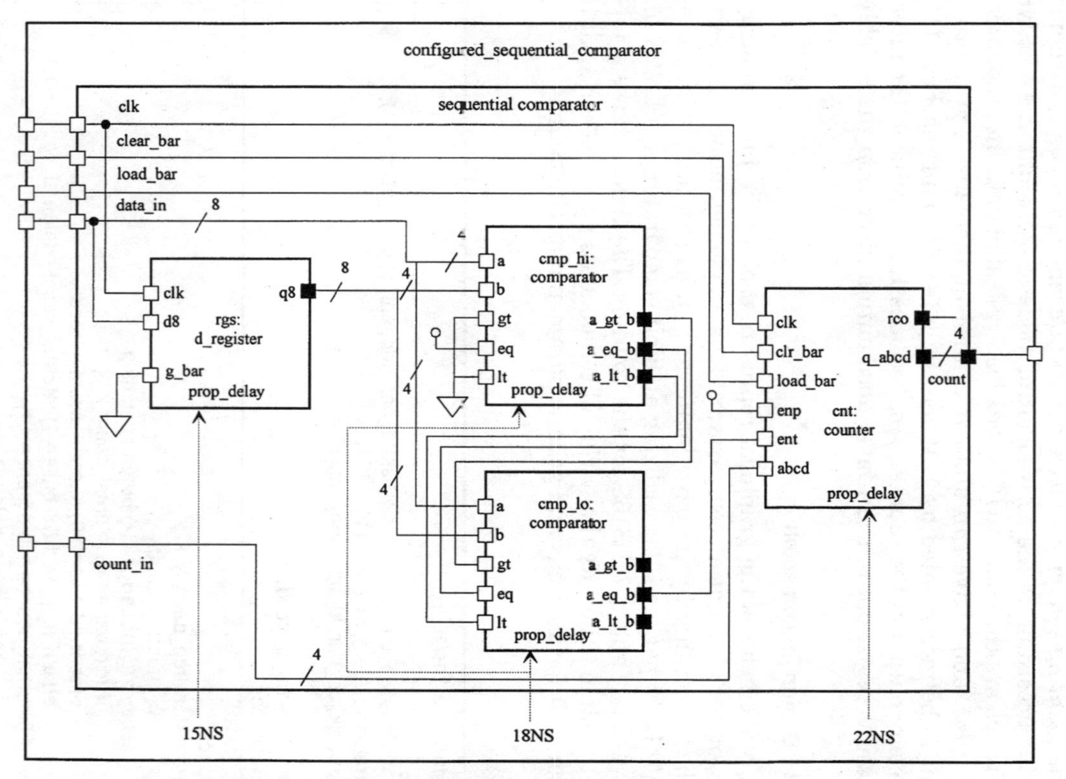

Figure 9.66 Composition aspect of the sequential comparator.

The Verilog description for the sequential comparator circuit is shown in Fig. 9.67. In the *sequential_comparator* module, a **specify** block issues warning messages whenever short glitches (smaller than 1000 ns) are observed on the clock input of the circuit. This circuit uses two instances of the comparator of Fig. 9.63. However, parameter *n* of *LS85_comparator* may be overwritten by 8 to make an 8-bit comparator. This would require only one comparator instead of cascading two 4-bit units. However, for consistency with the 7485 standard package, 4-bit comparators are used.

The *configured_sequential_comparator* module in Fig. 9.68 instantiates the module in Fig. 9.67. Using a **defparam** block, this upper-level module overwrites the *prop_delay* delay parameters of the individual

```
`timescale 1ns/100ps

module sequential_comparator (data_in, clk, clear_bar, load_bar, count_in, count);
input [7:0] data_in;
input clk, clear_bar, load_bar;
input [3:0] count_in;
output [3:0] count;
supply0 gnd; supply1 vdd;
    specify
        $width (posedge clk, 1E3);
    endspecify
    wire [7:0] old_data;
    wire compare_out, gt_i, eq_i, lt_i;
    ls377_register rgs (clk, gnd, data_in, old_data);
    ls85_comparator cmp_lo (data_in [3:0], old_data [3:0], gnd, vdd, gnd, gt_i, eq_i, lt_i);
    ls85_comparator cmp_hi (data_in [7:4], old_data [7:4], gt_i, eq_i, lt_i,, compare_out, );
    ls163_counter cnt (clk, clear_bar, load_bar, vdd, compare_out, count_in, count, );
endmodule
```

Figure 9.67 Structural implementation of the sequential comparator.

```
`timescale 1ns/100ps

module configured_sequential_comparator (data_in, clk, clear_bar, load_bar, count_in, count);
input [7:0] data_in;
input clk, clear_bar, load_bar;
input [3:0] count_in;
output [3:0] count;
   sequential_comparator std (data_in, clk, clear_bar, load_bar, count_in, count);
   defparam
      std.rgs.prop_delay = 15,
      std.cmp_lo.prop_delay = 18,
      std.cmp_hi.prop_delay = 18,
      std.cnt.prop_delay = 22;
endmodule
```

Figure 9.68 Configuring the structural description of the *sequential_comparator*.

7400 series parts. In order to reach the delay parameters of the individual parts, hierarchical naming is used.

Figure 9.69 shows a simple test bench for verifying the basic operations of the *sequential_comparator* module. An **always** block generates a periodic signal on the *ck* clock input, and an **initial** block assigns values to the control and data inputs of the sequential comparator for testing most of the comparator's functionalities.

Using generic-size standard packages, a Verilog model for an expanded version of the design of Fig. 9.66 can easily be formed. In order for the sequential comparator to be able to handle 16-bit input data and to keep a modulo-255 count of matched data, the size parameters in Figs. 9.63, 9.64, and 9.65 can be overwritten by an expanded sequential comparator module that instantiates them. Sizing these modules from an upper-level module eliminates the need for recompiling individual units. Figure 9.70 shows the *expanded_sequential_comparator* module, which uses *i, j,* and *k* parameters to size the register, the comparator, and the counter that it instantiates. This design uses a single comparator instead of cascading two such units as was done in the original version of the sequential comparator.

9.6 Summary

This chapter presented descriptions of hardware at the behavioral level and discussed how **initial** and **always** blocks can be used to describe

```
`timescale 1us/100ns

module test_sequential_comparator ();
reg [7:0] data;
reg ck, cl_bar, ld_bar;
reg [3:0] cnt;
wire [3:0] cnt_out;
    initial    begin
        ck <= 0;
        cl_bar <= 0; cl_bar <= #2.5 1; cl_bar <= #60 0;
        ld_bar <= 1; ld_bar <= #50 0; ld_bar <= #55 1;
        cnt <= 4'b1111; cnt <= #40 4'b1011; cnt <= #55 4'b0111;
        data <= 8'b00000000; data <= #3 8'b01110111;
        data <= #5 8'b10101100; data <= #25 8'b01010100;
        #70 $stop;
    end
    always #2 ck = ~ck;
    configured_sequential_comparator mfi (data, ck, cl_bar, ld_bar, cnt, cnt_out);
endmodule
```

Figure 9.69 Test bench for testing the configured sequential comparator of Fig. 9.68.

```
`timescale 1ns/100ps

module expanded_sequential_comparator (data_in, clk, clear_bar, load_bar, count_in, count);
parameter i = 16, j = 8, k = 255;
input [i-1:0] data_in;
input clk, clear_bar, load_bar;
input [j-1:0] count_in;
output [j-1:0] count;
supply0 gnd; supply1 vdd;
    specify
        $width (posedge clk, 1E3);
    endspecify
    wire [i-1:0] old_data;
    wire compare_out;
    defparam std.rgs.n = i, std.cmp.n = i, std.cnt.n = j, std.cnt.m = k;
    ls377_register rgs (clk, gnd, data_in, old_data);
    ls85_comparator cmp (data_in, old_data, gnd, vdd, gnd,, compare_out, );
    ls163_counter cnt (clk, clear_bar, load_bar, vdd, compare_out, count_in, count, );
endmodule
```

Figure 9.70 Expanded version of the sequential comparator.

the main functionality of a module. In the early part of the chapter, syntax and semantics for various sequential statements were described. We then showed how event control statements are used to describe controlling hardware, handshaking, and data acquisition. Various forms of event control statements were extensively used in these descriptions.

Although Verilog simulation environments provide ways of observing internal and external module variables, the system tasks presented here, along with the behavioral constructs of Verilog, provide a great deal of flexibility in generation of output file formats. This chapter demonstrated the use of behavioral constructs in conjunction with file output system tasks. The last part of the chapter presented a structural design based on generic components that were described behaviorally. This demonstrates how material presented in earlier chapters is used with the behavioral descriptions presented here.

Further Reading

Hill, F. J., and G. R. Peterson, *Digital Systems: Hardware Organization and Design,* 3d ed., John Wiley, New York, 1987.

IEEE Standard Hardware Description Language Based on the Verilog Hardware Description Language, IEEE Std. 1364-1995, Institute of Electrical and Electronic Engineers, New York, 1996.

Navabi, Zainalabedin, *VHDL: Analysis and Modeling of Digital Systems,* McGraw-Hill, New York, 1998.

Palnitkar, S., *Verilog HDL: A Guide to Digital Design and Synthesis,* Prentice-Hall, Upper Saddle River, N.J., 1996.

Smith, Douglas J., *A Practical Guide for Designing, Synthesizing and Simulating ASICs and FPGAs Using VHDL or Verilog,* Doone Publications, Madison, Ala., June 1996.

Thomas, D. E., and P. R. Moorby, *The Verilog Hardware Description Language,* 3d ed., Kluwer Academic Publishers, Norwell, Mass., 1996.

Wakerly, J. F., *Digital Design Principles and Practices,* 2d ed., Prentice-Hall, Englewood Cliffs, N.J., 1993.

Problems

9.1 Write a Verilog description for a D-type flip-flop with a d input, asynchronous set and reset inputs, and q and qb outputs. Use three delay parameters for set-input to q, reset-input to q, and d-input to q.

9.2 In a **specify** block, write a statement to issue a message if a negative pulse shorter than 1 μs appears on the input clock of the flip-flop of Prob. 9.1.

9.3 In a **specify** block, write a statement to issue a warning message if the frequency of the observing clock is lower that 100 kHz. Apply this to the flip-flop of Prob. 9.2.

9.4 Modify the description in Fig. 9.32 to one for a Mealy machine detecting the same sequence. Write a test bench and compare the Mealy and Moore machine outputs.

9.5 Write a behavioral description for a Mealy machine that continuously monitors its x input for the 11010 sequence. When the sequence is found, the output becomes **1,** and it returns to **0** with the clock. The circuit has a synchronous reset input that resets the circuit to its initial state when it becomes **1.**

9.6 Write a description for an asynchronous circuit that generates one positive pulse for every two complete positive pulses that appear on its input. Use event control sequential statements.

9.7 Write a behavioral description for a divide-by-n circuit in which n is passed to it via a parameter. The circuit has an x input and a z output. For every n positive pulses on x, one positive pulse should appear on z.

9.8 Generate two phases of a clock using a single triggering signal as input. The width of the short pulses on the triggering signal determines the time that both phases are zero. Use event control sequential statements.

9.9 Write behavioral Verilog code for modeling an asynchronous circuit. The circuit has inputs x and y and output z. If a **0**-to-**1** transition on x is immediately followed by a **1**-to-**0** transition on y (with no other transitions on either input between these two transitions), the output becomes **1.** The output stays high until either x changes to **0** or y changes to **1.**

9.10 Write a module to output consecutive numbers in hexadecimal format to a file. With every edge of a clock input, the 8-bit number is incremented and written to a file. Generate a test bench for this module.

9.11 A 4-bit shift register has a *mode,* a *serial_input,* and *clock* inputs as well as four *parallel_input* lines. The four lines of outputs are *parallel_output.* When *mode* is high, the shift register is in the right-shift mode, and on the falling edge of the *clock,* the *serial_input* is clocked into the shift register. When *mode* is low, on the falling edge of the *clock,* the *parallel_input* is loaded into the shift register. For this shift register, write a module with a delay parameter. After this delay, the proper output appears on the *parallel_output* on the falling edge of the clock.

9.12 Write a module for a behavioral description of a toggle flip-flop. The flip-flop has a single *t* input and two *q* and *nq* outputs. After the rising edge of the *t* input, the two outputs will be complemented. The *q* output has a low-to-high propagation delay of *q_tplh* and a high-to-low propagation delay of *q_tphl.* The *nq* output has a low-to-high propagation delay of *nq_tplh* and a high-to-low propagation delay of *nq_tphl.* Use parameters for all delay values.

9.13 In this problem, you will configure the toggle flip-flop of the previous problem. Write a description of an *n*-bit *t_register* using *t_ffs* of Prob. 9.12. For the *t_register,* write a top-level module that uses the behavioral *t_ff* with *q_tplh, q_tphl, nq_tplh,* and *nq_tphl* delay values of 2, 4, 3, and 5 ns, respectively. The output of a toggle flip-flop has a frequency half of that of its input. Use a configured *t_register* to build divide-by-16 circuit.

9.14 Develop a behavioral model of an 8-bit sequential multiplier. The 4-bit version of this multiplier was discussed in Sec. 1.3.2. Use the same interface and signaling as the multiplier in Chap. 1, i.e., use *dataready, busy,* and *done.* The circuit receives two 8-bit operands when the input *dataready* becomes **1.** This causes the multiplication process to begin and the *busy* flag to become active. Using the add-shift method, the multiplier takes one or two clock pulses for each bit of the multiplicand. When the process is completed, *done* becomes **1** for one clock period and *busy* returns to zero. The circuit receives two operands from its *inputbus* and produces the result on its 8-bit *result* output. Your behavioral code for this circuit should model it at the clock level. That is, the number of clock pulses that the behavioral model takes for multiplication of two numbers should be the same as that of an actual circuit using the add-shift method.

9.15 Write a description of a T flip-flop with a clock (*clk*) and a *t* input. Toggling is done on the rising edge of *clk* when *t* is **1.** Include parameters for the flip-flop delay, the pulse width on *t,* and the flip-flop identification number. The flip-flop should be able to report an error message (use standard output) if a glitch of less than the specified parameter is detected on its *t* input. The message should include the flip-flop identification number.

9.16 A memory unit has *cs* and *rw* control inputs, 10-bit *addr* address lines, and 8-bit bidirectional *iodata* data lines. Both *cs* of **1** and *rw* of **1** will initiate a read from the memory. The completion of this operation will be indicated with the *ready* signal from the memory, which will stay **1** until *cs* is **0.**

Similarly, *rw* of **0** will cause a write to the memory, the completion of which will be indicated by a **1** on *ready.* As in read, the *ready* signal stays asserted until *cs* is back to **0.** A controller circuit uses this memory to implement a push-pop stack. The stack has three stack operations, *push, pop,* and *tos,* as well as a *reset* operation. The stack I/O lines are called *stackdata.* The *push* operation pushes an 8-bit data word onto the stack. The *pop* operation pops an 8-bit word, and the *tos* operation places the top of the stack on the stack output lines (*stackdata*). The *reset* input resets the stack counter. Write a behavioral description for the stack controller. The controller should provide a *hold* output to indicate that the stack is not ready for a *push, pop,* or *tos* operation.

Chapter 10

CPU Modeling for Discrete Design

Concepts of the Verilog hardware description language, the syntax and semantics of its constructs, and various ways in which a hardware component can be described in Verilog were discussed in the previous chapters. No additional Verilog constructs are presented in this chapter; instead, we will use the constructs of earlier chapters to describe a simple 8-bit processor. A CPU structure represents a large class of digital systems, and its description involves the use of many important Verilog constructs. The next chapter uses this CPU in a board-level design. Board-level components will be emphasized there.

The CPU will be described at a level appropriate for implementation by use of discrete gates and components. Design methods developed in this chapter for a Verilog-based design apply most to a manual design environment, and are not synthesis-oriented. However, a synthesizable description of the complete machine will be described to illustrate Verilog synthesizable styles.

This chapter begins by introducing and describing a CPU example at a high level of abstraction; it then captures this high-level information in a behavioral Verilog description. Following that, the data path of this machine and its structural details are designed, and the information is then used to develop the dataflow description for our example CPU. The last part of this chapter develops a test bench for testing the behavioral or dataflow models. Various descriptions of this CPU are presented in Appendix D. The end-of-chapter problems suggest ways for enhancing the capabilities, and also improving its test bench.

10.1 Defining a Comprehensive Example

The CPU that we use in this chapter is *a r*educed *p*rocessor, which we refer to as PAR-1, (pronounced and written as "Parwan"). Parwan, first developed to teach computer hardware to novice logic design students, employs a reduced hardware requirement and a simple instruction set. The implementation of the machine in terms of medium-scale integrated circuits (MSI) and small-scale integrated circuits (SSI) parts was illustrated for this purpose. Later, a senior design project for students in a VLSI design course capitalized on this processor. Using standard public-domain CAD tools, Parwan was designed as a full-custom VLSI chip. Other implementations of Parwan include MSI- and FPGA-based designs. Figure 10.1 shows a general outline of this CPU. Because of its reduced architecture and simple instruction set, the hardware details of Parwan are easier to explore, and students are able to see the inner workings of a CPU down to its transistors and gates.

We will use Parwan to illustrate the use of Verilog as a language for design and verification of CPU-like architectures. The use of this simple architecture enables us to show various styles and applications of hardware description languages without overshadowing these concepts with the complexity of an architecture. Building a CPU with discrete components is emphasized.

The behavioral description of Parwan, presented later in this chapter, is written according to the functionality of this CPU as it is first described to a user, or in our case to a student. This description follows a synthesizable style that can be used as input to a commercial synthesis tool. The dataflow description considers register transfer level hardware details and utilizes the same partitioning that was previously employed when generating the layout. The dataflow description and the behavioral description have the same functionality and input-output ports.

The two alternatives for design of this system that are presented here are a behavioral synthesis-based description and a register transfer level description using register and logic unit components. While the former alternative offers a simple design and implementation solution, the latter gives the designer better control of the bus and register structure of the system. In addition to its being a complete synthesis input, we will use the behavioral description as a guide in partitioning the CPU into its dataflow level components.

10.2 Parwan CPU

Because of the size of its data registers and busses, Parwan is generally considered to be an 8-bit processor. This machine has an 8-bit

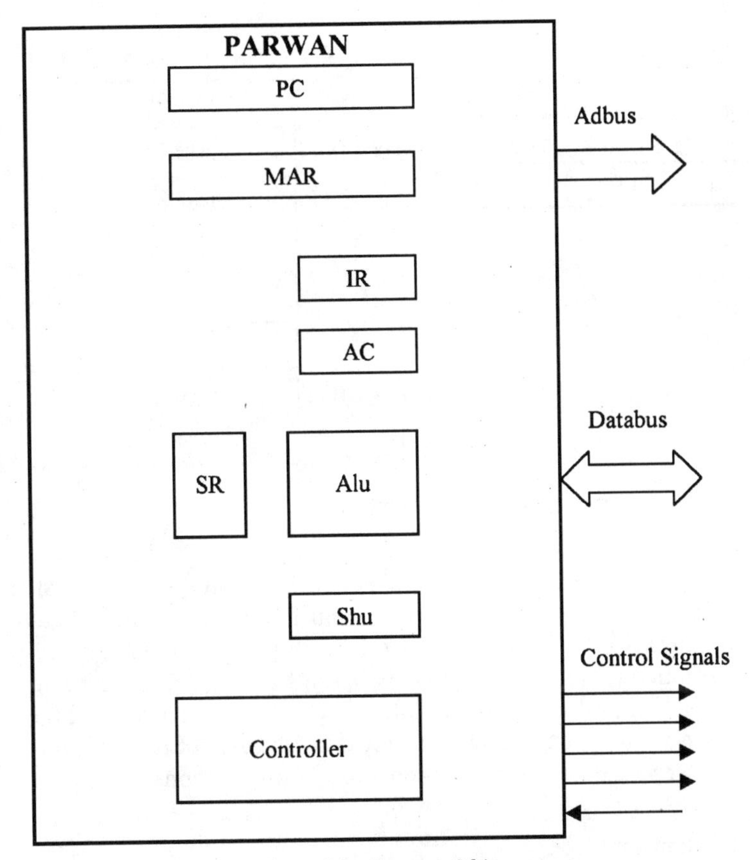

Figure 10.1 A general outline of the Parwan 8-bit processor.

external data bus and a 12-bit address bus. It has the basic arithmetic and logical operations and several jump and branch instructions, along with direct and indirect addressing modes. Parwan also has a simple subroutine call instruction and an input interrupt that resets the machine.

10.2.1 Memory organization of Parwan

Parwan is capable of addressing 4096 bytes of memory through its 12-bit address bus (*adbus*). This memory is partitioned into 16 pages of 256 bytes each. As shown in Fig. 10.2, the 4 most-significant bits of *adbus* constitute the page address and its 8 least-significant bits are the offset.

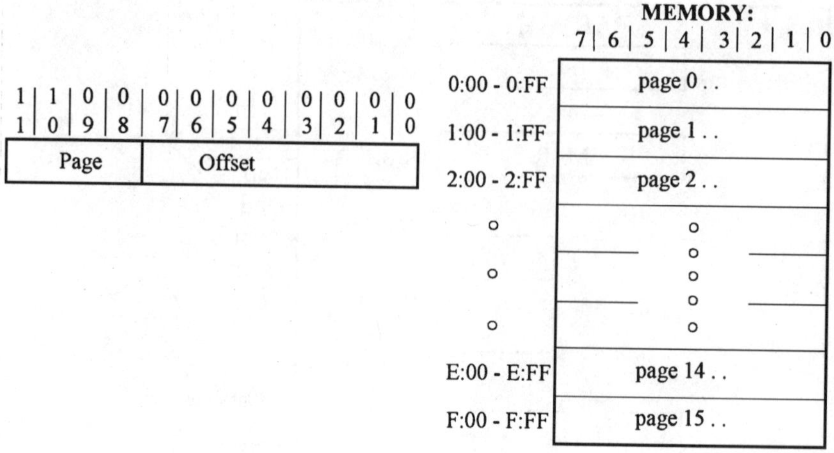

1 1	1 0	0 9	0 8	0 7	0 6	0 5	0 4	0 3	0 2	0 1	0 0
Page				Offset							

MEMORY: 7 | 6 | 5 | 4 | 3 | 2 | 1 | 0

Address	Memory
0:00 - 0:FF	page 0 . .
1:00 - 1:FF	page 1 . .
2:00 - 2:FF	page 2 . .
o o o	o o o o o
E:00 - E:FF	page 14 . .
F:00 - F:FF	page 15 . .

Figure 10.2 Page and offset parts of Parwan addresses.

In this figure, and in the future examples of this chapter, we use hexadecimal numbers for the addresses. We separate the page and offset parts of the address by a colon. In spite of the 16 pages of memory partitioning, Parwan's memory is treated as a contiguous 4K memory in which page crossing is done automatically. This memory is also used for communication with input and output devices. Because of its memory-mapped I/O, Parwan does not have separate I/O instructions.

10.2.2 Instruction set

With two addressing modes, Parwan has a total of 24 instructions, as summarized in Fig. 10.3. The main and only CPU data register is the accumulator, which is used in conjunction with most instructions. This machine has *overflow, carry, zero,* and *negative* flags (*v, c, z,* and *n*). These flags may be modified by specific flag instructions or by the instructions that alter the contents of the accumulator.

The *lda* instruction loads the accumulator with the contents of memory, while the *and, add,* and *sub* instructions access memory for an operand, perform the specified operation (ANDing, adding, and subtracting), and load the results into the accumulator. Flags *z* and *n* are set or reset based on the results of *lda, and, add,* and *sub*. Instructions *add* and *sub* also influence the *v* and *c* flags (overflow and carry), depending on the outcome of the corresponding operations. The *sta* instruction stores the contents of the accumulator into the specified memory location. Execution of the *jmp* instruction causes the next instruction to be received from the address specified by the instruction.

Instructions *lda, and, add, sub, jmp,* and *sta* use 12-bit addresses and can be used with the indirect addressing mode. We refer to these instructions as having a full-addressing scheme.

The addressing scheme of *jsr* and branch instructions is page addressing. These instructions can point only to the page that they appear in. The *jsr* instruction with an 8-bit top of subroutine (*tos*) address causes the next instruction to be received from memory location *tos*+1 of the current page. At the end of a subroutine, a return from the subroutine can be accomplished by an indirect jump to *tos*. Four branch instructions, *bra_v, bra_c, bra_z,* and *bra_n,* cause the next instruction to be received from the specified location of the current page if the respective flag *v, c, z,* or *n* is set.

Instructions *nop, cla, cma, cmc, asl,* and *asr* are nonaddress instructions and perform operations on the internal registers of the CPU flags. The *nop* instruction performs no operation, *cla* clears the accumulator, *cma* complements the accumulator, *cmc* complements the *c* flag, and *asl* and *asr* cause an arithmetic left or right shift of the contents of the accumulator. When shifting left, the most-significant bit of the accumulator is shifted into the carry flag, and the overflow flag is set if the sign of the accumulator changes. The *asr* instruction extends the sign of the accumulator and shifts out its least-significant bit. Both shift instructions affect the zero and negative flags.

Instruction Mnemonic	Brief Description	Address Bits	Address Scheme	Indirect Address	Flags Use	Flags Set
LDA loc	Load AC w/(loc)	12	FULL	YES	----	--zn
AND loc	AND AC w/(loc)	12	FULL	YES	----	--zn
ADD loc	Add (loc) to AC	12	FULL	YES	-c--	vczn
SUB loc	Sub (loc) from AC	12	FULL	YES	-c--	vczn
JMP adr	Jump to adr	12	FULL	YES	----	----
STA loc	Store AC in loc	12	FULL	YES	----	----
JSR tos	Subroutine to tos	8	PAGE	NO	----	----
BRA_V adr	Branch to adr if V	8	PAGE	NO	v---	----
BRA_C adr	Branch to adr if C	8	PAGE	NO	-c--	----
BRA_Z adr	Branch to adr if Z	8	PAGE	NO	--z-	----
BRA_N adr	Branch to adr if N	8	PAGE	NO	---n	----
HLT	Halt operation	-	NONE	NO	----	----
CLA	Clear AC	-	NONE	NO	----	----
CMA	Complement AC	-	NONE	NO	----	--zn
CMC	Complement carry	-	NONE	NO	-c--	-c--
ASL	Arith shift left	-	NONE	NO	----	vczn
ASR	Arith shift right	-	NONE	NO	----	--zn

Figure 10.3 Summary of Parwan instructions.

Instruction Mnemonic	Opcode Bits 7 6 5	D/I Bit 4	Bits 3 2 1 0
LDA loc	0 0 0	0/1	Page adr
AND loc	0 0 1	0/1	Page adr
ADD loc	0 1 0	0/1	Page adr
SUB loc	0 1 1	0/1	Page adr
JMP adr	1 0 0	0/1	Page adr
STA loc	1 0 1	0/1	Page adr
JSR tos	1 1 0	-	- - - -
BRA_V adr	1 1 1	1	1 0 0 0
BRA_C adr	1 1 1	1	0 1 0 0
BRA_Z adr	1 1 1	1	0 0 1 0
BRA_N adr	1 1 1	1	0 0 0 1
HLT	1 1 1	0	0 0 0 0
CLA	1 1 1	0	0 0 0 1
CMA	1 1 1	0	0 0 1 0
CMC	1 1 1	0	0 1 0 0
ASL	1 1 1	0	1 0 0 0
ASR	1 1 1	0	1 0 0 1

Figure 10.4 Parwan instruction opcodes.

10.2.3 Instruction format

As shown in Fig. 10.3, there are three groups of Parwan instructions. Full-address instructions requiring 2 bytes can access all of Parwan's memory and be used with indirect addressing. Page-address instructions requiring 2 bytes can access the current page but cannot use indirect addressing. Instructions in the third group, which are nonaddress instructions, do not use the memory for their operands. Figure 10.4 shows the opcodes and format of these instructions.

Full-address instructions. The opcode specifying the operation of a full-address instruction is formed by the 3 most-significant bits of the first byte. The next bit (bit 4) specifies the direct or indirect addressing mode (0 for direct, 1 for indirect), and the other 4 bits (the least-significant 4) contain the page number of the operand of the instruction. The second byte of a full-address instruction specifies the offset address which together with the page address completes a 12-bit address for the operand. Figure 10.5 shows the formation of a complete 12-bit address for this group of Parwan instructions.

Page-address instructions. Figure 10.6 shows the format for *jsr* and branch instructions. These instructions reference memory within the page in which they appear. The opcode of *jsr* is 110, and the other 5 bits of the first instruction byte are ignored. The opcode field of a branch instruction contains 111. Bit 4 is always 1 and its least-significant bits specify the condition for a branch. The second byte of *jsr* and branch instructions specify the jump address within the current page.

For a branch example, consider the instruction shown in Fig. 10.7. At location 0D of page 5, a *bra_c* causes the next instruction to be received from location 6A of page 5 if the carry flag is set, or from location 0F of page 5 if the carry flag is zero.

Figure 10.8 shows the execution of a *jsr* instruction at location 5:11. The byte at location 5:12 specifies that the subroutine begins at location 33 of the current page (5:33). The first location of a subroutine is reserved for the return address, and the programmer is not allowed to use it for program information, i.e., code or data. The programmer is

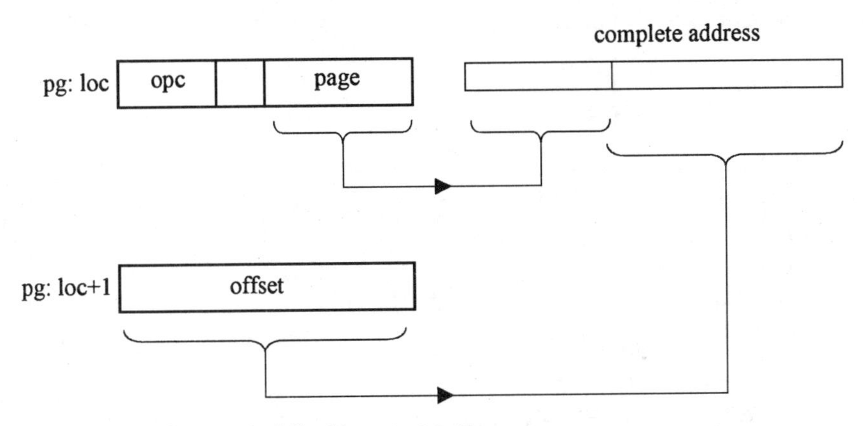

Figure 10.5 Addressing in full-address instructions.

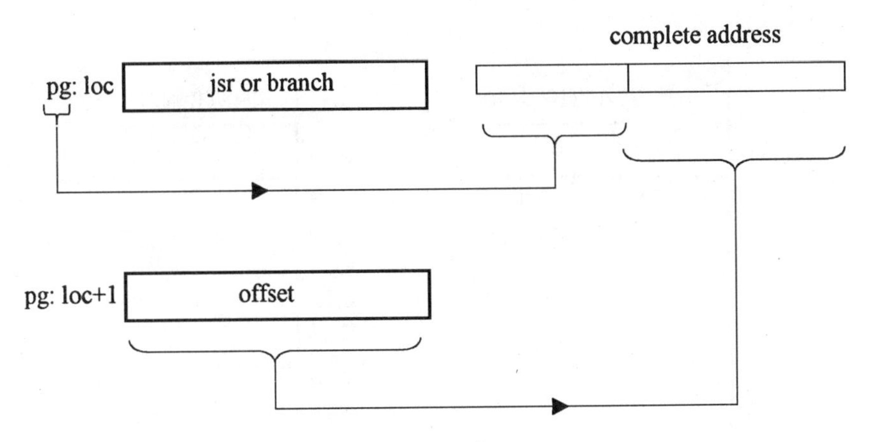

Figure 10.6 Addressing in page-address instructions.

required to use an indirect jump instruction at the end of a subroutine to return from it. Figure 10.8 shows a *jmp-indirect* instruction at locations 5:55 and 5:56. After the execution of *jsr,* the return address (the address of the instruction that follows *jsr* in memory, which in this example is location 5:13) is placed in the first location of the subroutine (location 5:33). The indirect jump at location 5:55 causes the program flow to return to location 5:13 after the subroutine completes.

Nonaddress instructions. Nonaddress instructions are the last group of Parwan instructions. These instructions occupy 1 byte whose most-significant 4 bits are 1110. The other 4 bits specify *nop, cla, cma, cmc, asl, or asr.*

Indirect addressing in Parwan. If bit 4 of the first byte of a full-address instruction is **1**, the address specified by this instruction is the indirect

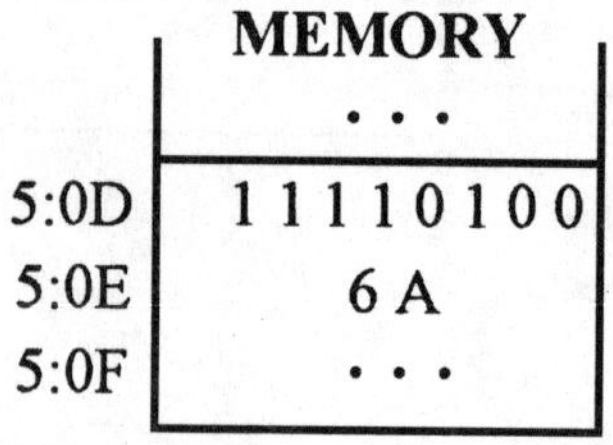

	MEMORY
	. . .
5:0D	1 1 1 1 0 1 0 0
5:0E	6 A
5:0F	. . .

Figure 10.7 A branch instruction.

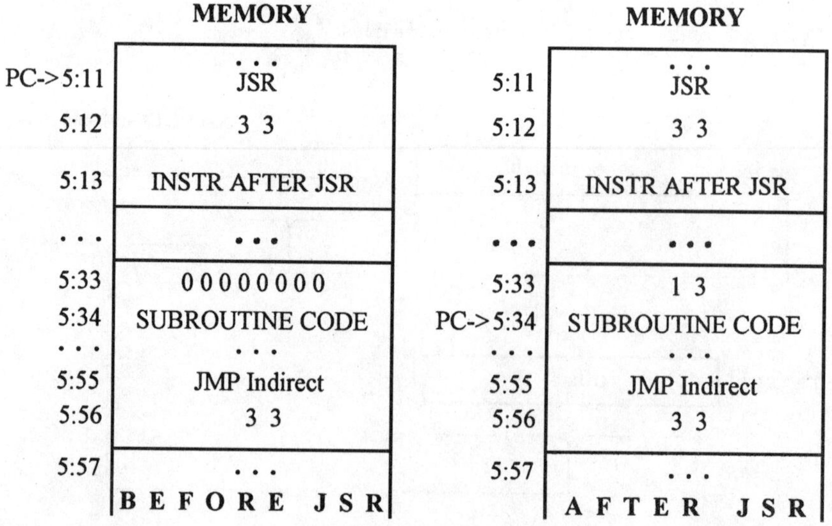

	MEMORY
	. . .
PC->5:11	JSR
5:12	3 3
5:13	INSTR AFTER JSR
. . .	. . .
5:33	0 0 0 0 0 0 0 0
5:34	SUBROUTINE CODE
. . .	. . .
5:55	JMP Indirect
5:56	3 3
5:57	. . .
	BEFORE JSR

	MEMORY
	. . .
5:11	JSR
5:12	3 3
5:13	INSTR AFTER JSR
. . .	. . .
5:33	1 3
PC->5:34	SUBROUTINE CODE
. . .	. . .
5:55	JMP Indirect
5:56	3 3
5:57	. . .
	AFTER JSR

Figure 10.8 An example for the execution of *jsr.* Memory and *pc,* before and after *jsr.*

address of the operand. Indirect addressing uses a 12-bit address to receive an 8-bit offset from the memory. This offset, together with the page number of the indirect address, makes a complete address for the actual operand of the instruction.

Figure 10.9 shows an example of indirect addressing in Parwan. It is assumed that a full-address instruction with indirect addressing is in locations 0:25 and 0:26. The 12-bit address of this instruction (6:35) points to 1F in the memory, which is used with page number 6 to form 6:1F as the actual address of the operand.

10.2.4 Programming in Parwan assembly

For a better understanding of Parwan instructions, consider the program shown in Fig. 10.10. The assembly code of this figure adds 10 data bytes, which are stored in memory starting from location 4:25, and stores the result at 4:03. The code begins at location 0:15 and assumes that constants 25, 10, and 1 are stored in 4:00, 4:01, and 4:02, respectively. Although Parwan does not have an immediate addressing mode, a series of shifts and adds can generate any necessary constant. Modifying this processor to handle the immediate addressing mode is dealt with in a series of problems at the end of the chapter.

10.3 Behavioral Description of Parwan

A more compact and far less ambiguous description of the behavior of Parwan than the "word" description of the previous section can be

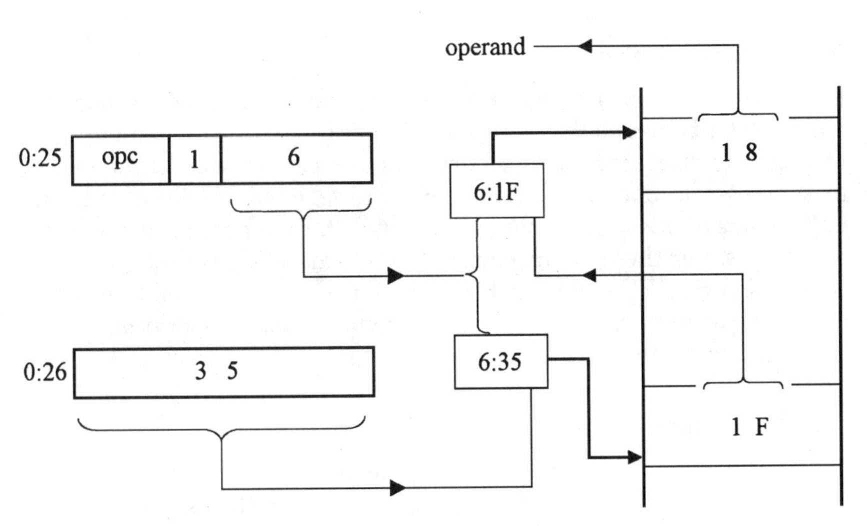

Figure 10.9 An example of indirect addressing in Parwan.

```
                        -- load 25 in 4:00
                        -- load 10 in 4:01
                        -- load 01 in 4:02
0:15  cla               -- clear accumulator
0:16  asl               -- clears carry
0:17  add, i  4:00      -- add bytes
0:19  sta     4:03      -- store partial sum
0:1B  lda     4:00      -- load pointer
0:1D  add     4:02      -- increment pointer
0:1F  sta     4:00      -- store pointer back
0:21  lda     4:01      -- load count
0:23  sub     4:02      -- decrement count
0:25  bra_z   :2D       -- end if zero count
0:27  sta     4:01      -- store count back
0:29  lda     4:03      -- get partial sum
0:2B  jmp     0:17      -- go for next byte
0:2D  nop               -- adding completed
```

Figure 10.10 An example program for the Parwan CPU.

developed using Verilog. This section presents such a behavioral description for our 8-bit machine. The interface description of this machine is kept at the hardware level, using bits for external control signals and memory and data busses. We will use a general synthesizable style of Verilog so that our behavioral description can be synthesized with most available commercial synthesis tools.

10.3.1 Memory timing

In order to simplify the description and concentrate on Verilog code for handling CPU instructions, we are assuming an external memory that responds to read and write requests in one clock cycle. At the end of this chapter, a more general memory interface will be discussed. The CPU has *read_mem* and *write_mem* signals for memory read and write operations. For these operations, on the negative edge of the clock, the corresponding CPU signal will be issued, and it is assumed that memory responds immediately, preparing a byte for a read operation or performing a write for a write operation.

10.3.2 Description style

Chapter 9 introduced several styles for describing state machines. Figure 9.37 showed a state machine description that separates register clocking from the combinational part that handles the determination of the next states of the machine. We will use a generalized format of this style for describing Parwan at the behavioral level. In an **always** block, CPU registers, including control flip-flops, will be set, and in another **always** block, the instruction register will be decoded and appropriate operations that correspond to decoded instructions will be performed.

The sections that follow begin from the top of the Parwan behavioral description and describe each part of the code from top to bottom. The figures that follow show sections of code that form the complete description. For code readability, a define file has been generated. This file, *par_para.v,* shown in Fig. 10.11, defines mnemonics for CPU

```
//------------------------alu_operation_parameters
`define a_and_b 6'b100000
`define b_compl 6'b010000
`define a_input 6'b001000
`define a_add_b 6'b000100
`define b_input 6'b000010
`define a_sub_b 6'b000001
//------------------------par_control_parameters(state)
`define do_initials 4'd1
`define instr_fetch 4'd2
`define do_one_bytes 4'd3
`define opnd_fetch 4'd4
`define do_indirect 4'd5
`define do_two_bytes 4'd6
`define do_jsr 4'd7
`define continue_jsr 4'd8
`define do_branch 4'd9
//----------------------par_control_parameters(code)
`define single_byte_instructions 4'b1110
`define hlt 4'b0000
`define cla 4'b0001
`define cma 4'b0010
`define cmc 4'b0100
`define asl 4'b1000
`define asr 4'b1001
`define jsr 3'b110
`define bra 4'b1111
`define indirect 1'b1
`define jmp 3'b100
`define sta 3'b101
`define lda 3'b000
`define ann 3'b001
`define add 3'b010
`define sbb 3'b011
`define jsr_or_bra 2'b11
```

Figure 10.11 Defining Parwan parameters mnemonics, contents of par_para.v.

opcodes, defines names for groups of operators, and assigns names to CPU control states. This file will be included in the behavioral description of Parwan.

10.3.3 Parwan input/output ports

The main Parwan busses are the bidirectional 8-bit *databus* and the 12-bit *adbus.* In addition to these busses, other input/output pins are used for CPU external control signals. As shown in the partial code of Fig. 10.12, seven control signals appear on the Parwan module input output port list. The *clk* and *interrupt* signals are inputs to the CPU. The *read_mem* and *write_mem* are outputs for memory read and write requests. The output *halted* becomes **1** when the CPU executes a halt instruction. It is required that the CPU be reset by the *interrupt* input to come out of the halt state. The *grant* input becomes **1** when the CPU is allowed to use its external data and address busses, and the *ready* input becomes **1** to inform the CPU that a requested memory operation is complete. As discussed before, this CPU description assumes a one-clock memory and does not utilize *ready* and *grant* inputs.

10.3.4 Parwan registers

The partial code in Fig. 10.12 begins the behavioral description of Parwan. Following this part, the declarations in Fig. 10.13 declare state flip-flops and main CPU registers. As discussed previously, the style used here consists of an **always** block for register clocking and another one for instruction execution, as shown in Fig. 10.14. The registers declared in Fig. 10.13 are used in the feedback path of the **always** block that handles instruction decoding and execution.

Parwan registers in the behavioral description include a 4-bit state register that can define up to 16 unique states of the machine. Other registers are *ac* (accumulator), *ir* (instruction register), *sr* (status register), *pc* (program counter), and *mar* (memory address register).

```
module Parwan_behavioral
    (clk, interrupt, read_mem, write_mem, databus, adbus, halted, ready, grant);
output read_mem, write_mem, halted;
reg read_mem, write_mem, halted;
output [11:0] adbus;
reg [11:0] adbus;
inout [7:0] databus;
input clk, ready, grant, interrupt;
```

Figure 10.12 Parwan input/output ports.

```
reg [3:0] present_state, next_state;
reg [7:0] ac, next_ac, ir, next_ir;
reg [4:0] sr, next_sr;
reg [11:0] pc, next_pc, mar, next_mar;
reg [9:0] ten_bit;
```

Figure 10.13 Parwan behavioral register declarations.

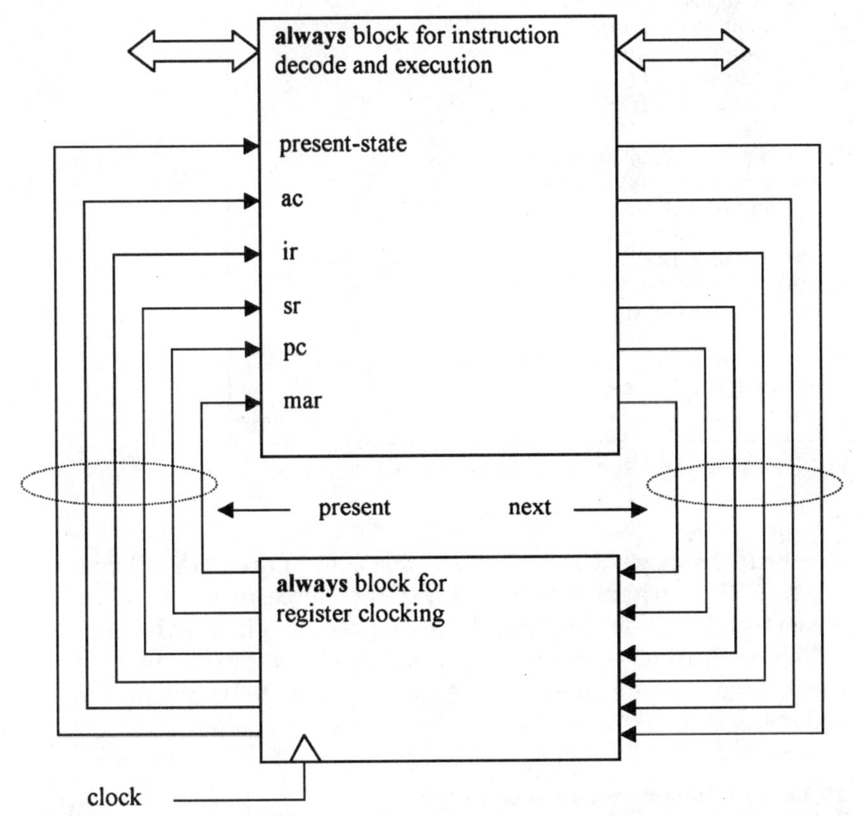

Figure 10.14 Two **always** blocks in the behavioral code for instruction execution and clocking.

10.3.5 Register clocking

Following declaration of registers, the Parwan behavioral description continues with initialization of control flip-flops (Fig. 10.15) and an **always** block for register clocking (Fig. 10.16).

Figure 10.15 shows that the present and next values of control flip-flops are set to 0001 (`*do_initials*), which is a stable state of the CPU.

```
initial begin
    present_state <= `do_initials;
    next_state <= `do_initials;
end
```

Figure 10.15 Initializing CPU control flip-flops.

```
always @(negedge clk) begin
    if (interrupt) begin
        present_state <= `do_initials;
        ac <= 8'b00000000;
        ir <= 8'b00000000;
        sr <= 4'b0000;
        pc <= 12'b000000000000;
        mar <= 12'b000000000000;
    end else begin
            ac <= next_ac;
            ir <= next_ir;
            sr <= next_sr;
            pc <= next_pc;
            mar <= next_mar;
            present_state <= next_state;
    end
end
```

Figure 10.16 Register clocking for feedback path in Fig. 10.14.

Figure 10.16 shows that with the falling edge of the clock, all next values of CPU registers are clocked into their present values. This performs the clocked feedback in the style presented in Fig. 10.14.

Figure 10.16 also shows that an active-high *interrupt* input resets the internal registers and sets the machine state to the *do_initials* initial state.

10.3.6 Instruction execution block

Following the partial codes shown in Figs. 10.11, 10.12, 10.13, 10.15, and 10.16, the Verilog behavioral description of Parwan continues with an **always** block that corresponds to the upper box in Fig. 10.14. This Verilog code, shown in Fig. 10.17, performs instruction decoding and executions and represents a large combinational circuit.

In this block, a **case** statement checks the present state of the machine and, based on this state and the contents of the instruction register, makes assignments to CPU registers and control flip-flops. According to Fig. 10.14, in the combinational part, reading a register value or looking up the present state of the machine is done from the inputs that are present values, i.e., *ac, ir, sr, pc, mar,* and *present_state*; while writing new values to CPU registers is done to combinational block outputs that are next register values, i.e., *next_ac, next_ir, next_sr, next_pc, next_mar,* and *next_state.*

The code shown in Fig. 10.17 is only an outline for the combinational block of Fig. 10.14. In this code, references are made to partial codes for performing individual CPU tasks, which will be discussed next.

```
always @(present_state or interrupt or databus or ac or ir or sr or pc or mar)
begin
    // ... Register and bus initializations; Fig. 10.18 //

    case(present_state)

    `do_initials: begin
        // ... Interrupt handling; Fig. 10.19 //

    `instr_fetch: begin
        // ... Fetch instruction byte into IR, increment PC; Fig. 10.20 //

    `do_one_bytes: begin
        // ... Perform single-byte instructions; Fig. 10.21 //

    `opnd_fetch: begin
        // ... Read operand into MAR, increment PC; Fig. 10.22 //
        // ... Branch out to indirect, jsr, bra, and two-bytes; Fig. 10.22 //

    `do_indirect: begin
        // ... Handle indirect addressing; Fig. 10.23 //

    `do_two_bytes: begin
        // ... Perform two-byte instructions ... //
        if (ir[7:5] == `jmp) begin
            // ... Handle the jmp instruction; Fig. 10.24 //
        end else begin
            if (ir[7:5] == `sta) begin
                // ... Handle the sta instruction; Fig. 10.24 //
            end else begin
                if (~ir[7]) begin
                    // ... Read memory into databus; Fig. 10.25, top ... //
                    if (~ir[6]) begin
                        // ... Load temporary for lda or and; Fig. 10.25, middle ... //
                    end else begin
                        // ... Load temporary for add or sub; Fig. 10.25, bottom ... //
                    end
    end

    `do_jsr: begin
        // ... Start and complete the jsr instruction; Fig. 10.26 //

    `do_branch: begin
        // ... Perform bra instructions; Fig. 10.27 //
end
```

Figure 10.17 Parwan behavioral description outline.

The complete **always** block that corresponds to the combinational part of Fig. 10.14 consists of the outline of Fig. 10.17 with referenced partial codes inserted where indicated.

Figure 10.18 shows initializations performed in the **always** block of Fig. 10.17. All combinational block outputs that connect to register inputs on the feedback path are initialized with their corresponding inputs. This way, if a register value does not receive a value in the **case** statement in this **always** block, the input value will propagate to the output, causing register contents to remain unchanged. Outputs corresponding to register inputs, e.g., *next_ac,* will be different from inputs, e.g., *ac,* only when an explicit assignment is made to them within the *case_statement.* Figure 10.18 also shows inactive values (high impedance) assigned to data and address busses. This guarantees that busses float unless they are driven during a clock cycle.

With each clock, every time a register value input to the **always** block changes, activity flow begins in the **always** block. After the initializations of Fig. 10.18, the present state of the machine is checked. As shown in Fig. 10.19, before a fetch begins, the status of the interrupt input is checked, and if it is active, the machine remains in its initial state. If not interrupted, *pc* is transferred to *mar,* and the state of the machine is set to *'instr_fetch.*

```
next_pc <= pc;
dbus <= 8'bZZZZZZZZ;
adbus <= 12'bZZZZZZZZZZZZ;
read_mem <= 1'b0;
write_mem <= 1'b0;
halted <= 1'b0;
next_ir <= ir;
next_ac <= ac;
next_mar <= mar;
next_sr <= sr;
ten_bit = 10'b0000000000;
```

Figure 10.18 Initialization part of Fig. 10.17.

```
if (interrupt) begin
    next_pc <= 12'b000000000000;
    next_state <= `do_initials;
end else begin
    next_mar <= pc;
    next_state <= `instr_fetch;
end end
```

Figure 10.19 Code inserts for interrupt handling and starting instruction fetch in Fig. 10.17.

```
read_mem <= 1'b1;
adbus <= mar;
next_ir <= databus;
next_pc <= pc + 1;
next_state <= `do_one_bytes;
```

Figure 10.20 Reading the first instruction byte; code is part of Fig. 10.17.

In the instruction fetch state, shown in Fig. 10.20, *read_mem* is issued; assuming that memory responds immediately, *databus* is read into the instruction register. This will read the first byte of the instruction and increment the program counter for the next byte.

After the first byte of an instruction is read, the *present_state* is set to *'do_one_bytes* and activity flow in the **always** block of Fig. 10.17 reaches the section of code shown in Fig. 10.21. This code segment checks bits 3 to 0 of the instruction byte for *nop, cla, cma, cmc, asl,* or *asr* and performs appropriate operations. When this is done, the next state of CPU is set to *'instr_fetch* for fetching the next instruction. As seen in the upper part of Fig. 10.21, for non-single-byte instructions, the state of the CPU is set to *'opnd_fetch* to fetching the second byte of a 2-byte instruction.

Figure 10.22 shows operations performed in the operand-fetch state. A memory read is issued, and while *mar* is on the address bus, *databus* is read into the *next_mar* so that it will contain the operand in the next clock cycle. While this is being done, the code in Fig. 10.22 assigns next states to *next_state* according to bits 4 and 5 of the instruction register. With the next clock, at the same time that *mar* is loaded with the instruction operand, activity flow in the **always** block of Fig. 10.17 reaches one of the states branched to in Fig. 10.22.

If the *ir* bit 4 is **1,** after performing the above code, the state of the CPU becomes *'do_indirect.* In this case, the code shown in Fig. 10.23 will be executed. As shown, a memory read is performed, and the operand in the least-significant 8 bits of *mar* is overwritten by the newly read data. As discussed in Sec. 10.2.3, indirect addressing is only within the current instruction page, and as a result, the page part of *mar* is kept intact. After completion of this address, the actual address of the operand remains in *mar.*

Figure 10.24 shows partial code for the part of Fig. 10.17 that handles *jmp* and *sta* instructions. Activity flow reaches this code when *mar* has the direct operand address. This address is placed on the *pc* inputs for the *jmp* instruction. For the *sta* instruction, the address in *mar* is used for writing *ac* to the memory.

```
next_mar <= pc; // prepare for next memory read
if (ir[7:4] != `single_byte_instructions)
      next_state <= `opnd_fetch;
else begin
      case(ir[3:0])
      `cla: next_ac <= 8'b00000000;
      `cma: begin
            next_sr[0] <= ac[7];
            next_sr[1] <= ~|( ~ac );
            next_ac <= ~ ac;
      end
      `cmc: next_sr[2] <= ~ sr[2];
      `asl: begin
            next_sr[0] <= ac[6];
            next_sr[1] <= ~|ac[6:0];
            next_sr[2] <= ac[7];
            next_sr[3] <= ac[6] ^ ac[7];
            next_ac <=  {ac[6:0],1'b0};
      end
      `asr: begin
            next_sr[0] <= ac[6];
            next_sr[1] <= ~|ac[7:1];
            next_sr[2] <= ac[7];
            next_sr[3] <= ac[6] ^ ac[7];
            next_ac <=  {1'b0, ac[7:1]};
      end
      `hlt: halted <= 1'b1;
      default: halted <= 1'b1;
      endcase
      next_state <= `instr_fetch;
end
```

Figure 10.21 Executing single-byte instructions; code is inserted in Fig. 10.17.

```
read_mem <= 1'b1;
adbus <= mar;
next_mar[7:0] <= databus;
if (ir[7:6] != `jsr_or_bra) begin
      next_mar[11:8] <= ir[3:0];
      if (ir[4] == `indirect)  next_state <= `do_indirect;
      else  next_state <= `do_two_bytes;
end else begin
      if ( ~ir[5]) next_state <= `do_jsr;
      else  next_state <= `do_branch;
end
next_pc <= pc + 1;
```

Figure 10.22 Fetching the second byte and setting the state for four groups of 2-byte instructions. Code is part of Fig. 10.17.

```
read_mem <= 1'b1;
adbus <= mar;
next_mar[7:0] <= databus;
next_state <= `do_two_bytes;
```

Figure 10.23 Handling indirect addressing in Fig. 10.17.

```
if (ir[7:5] == `jmp) begin
      next_pc <= mar;
      next_state <= `instr_fetch;
end else begin
      if (ir[7:5] == `sta) begin
            write_mem <= 1'b1;
            adbus <= mar;
            dbus <= ac;
            next_state <= `do_initials;
      end
end
```

Figure 10.24 Handling *jmp* and *sta* in *do_two_bytes* state, part of Fig. 10.17.

Execution of 2-byte full-address instructions causes the activity flow in Fig. 10.17 to reach the partial code shown in Fig. 10.25. The first two lines of this partial code read the instruction operand using the address in *mar.* The data read become available on the *databus.* Following this memory read, depending on bits 6 and 5 of the instruction register, *lda, and, add,* and *sub* instructions are performed. The code shown in Fig. 10.25 places a 10-bit result in the temporary variable *ten_bit.* The 2 most-significant bits of this variable are used for overflow and carry results of the operations performed. These bits are assigned to appropriate bits of the *next_sr* input to the *sr* register. The 8 least-significant bits of *ten_bit* are assigned to the accumulator. This is shown in the last part of Fig. 10.25.

Figure 10.26 shows operations performed for the *jsr* instruction in two consecutive clock cycles. In the first cycle, the *'do_jsr* state writes *pc* to the memory and loads *pc* with *mar.* In the next cycle, *pc* is incremented in order to point to the first instruction in the subroutine. The operations performed here implement *jsr* as shown in Fig. 10.8.

For performing branch instructions, the flow of activity reaches *'do_branch* in Fig. 10.17. Operations performed for this are shown in Fig. 10.27. As shown, if a bit of *sr* matches a corresponding bit of *ir[3:0], pc* is loaded with *mar* for the branch to be taken. Otherwise, *next_pc* retains its value, causing the program flow to continue from the instruction after the branch.

```
read_mem <= 1'b1;
adbus <= mar;

if (~ir[6]) begin
      if (~ir[5])
            ten_bit = {sr [3:2], databus};
      else
            ten_bit = {sr [3:2], (ac & databus)};
end else begin
      if (~ir[5]) begin
            ten_bit = ac + databus + sr[2];
            if (ac[7] == databus[7] && ten_bit[7] != ac[7])
                  ten_bit[9]=1'b1;
            else ten_bit[9]=1'b0;
      end else begin
            ten_bit = ac - databus - sr[2];
            if (ac[7] == ~databus[7] && ten_bit[7] != ac[7])
                  ten_bit[9] = 1'b1;
            else ten_bit[9]=1'b0;
      end
end
next_sr <= {ten_bit [9], ten_bit [8], ~|ten_bit[7:0], ten_bit [7]};
next_ac <= ten_bit [7:0];
next_state <= 'do_initials;
```

Figure 10.25 Performing full-address instructions. Code is inserted in Fig. 10.17.

```
write_mem <= 1'b1;
adbus <= mar;
dbus <= pc [7:0];
next_pc <= mar;
next_state <= `continue_jsr;
// ... after one clock do the following ... //
next_pc <= pc + 1;
next_state <= `do_initials;
```

Figure 10.26 Executing *jsr.* Code is part of Fig. 10.17.

```
if (|{sr & ir[3:0]}) next_pc <= mar;
next_state <= `do_initials;
```

Figure 10.27 Branch instructions. Code inserts in Fig. 10.17.

The above paragraphs discussed code fragments in the sequential block of Fig. 10.17. The code for the behavioral description of Parwan consists of Figs. 10.11, 10.12, 10.13, 10.15, 10.16, and 10.17 with the code fragments of Figs. 10.18 to 10.27 inserted. The complete code for the Parwan processor is shown in Appendix D. This description is synthesizable and may be synthesized to available FPGAs and ASIC netlists.

10.4 Parwan Bussing Structure

The bussing structure of a CPU describes the way its registers and logic units are connected. The first step in the hardware design process of a CPU is the design of this structure. Figure 10.28 shows the bussing structure of Parwan. This diagram is useful when performing a detailed study of timing for the individual machine instructions and will be used in the development of a dataflow description of Parwan in the next section. In this diagram, the names of major busses and registers appear in capital letters, and all other signal names are in lowercase letters. Only signal names to be referenced in this section are shown.

10.4.1 Interconnection of components

The major components of Parwan are *ac, ir, pc, mar, sr, alu,* and *shu.* Data flow between these components through busses and hard-wired interconnections. Figure 10.28 uses a hollow triangle to show controlled interconnection of a register or logic unit output to a bus and uses an arrow for permanent wired connections. For example, the output of *shu,* labeled *obus,* connects to *dbus* only when a signal named *obus_on_dbus* is active. On the other hand, connection of the accumulator output (*ac_out*) to the input of *alu* is hard-wired. In general, connections to busses with multiple sources must be controlled.

10.4.2 Global view of Parwan components

Of the seven Parwan components, *ac, ir, pc, mar,* and *sr* are registers, and *alu* and *shu* are combinatorial logic units. Each component has a set of inputs and outputs and several control lines. In addition, register structures have clock inputs that are all connected to the main system clock.

The accumulator, *ac,* is an 8-bit register that provides one operand of *alu.* The instruction register, *ir,* receives data from *dbus* through *alu* and provides instruction bits to the controller and a page address to the address bus (*adbus*). The 12-bit program counter register, *pc,* is a binary up-counter that provides instruction addresses to *adbus* through the memory address register (*mar*). This register is an address buffer. The *mar* and *pc* registers have page and offset parts that are identified by *mar_page, pc_page, mar_offset,* and *pc_offset.* Page numbers are stored in their 4 most-significant bits.

The arithmetic logic unit, *alu,* is a combinatorial logic unit with two sets of 8-bit inputs, four flag inputs, and six control inputs. The outputs of this unit are connected to the inputs of the *shu* unit. The shifter (*shu*) is also a combinatorial logic unit, and it performs right and left shiftings of its 8-bit operand. The status register, *sr,* has four inputs

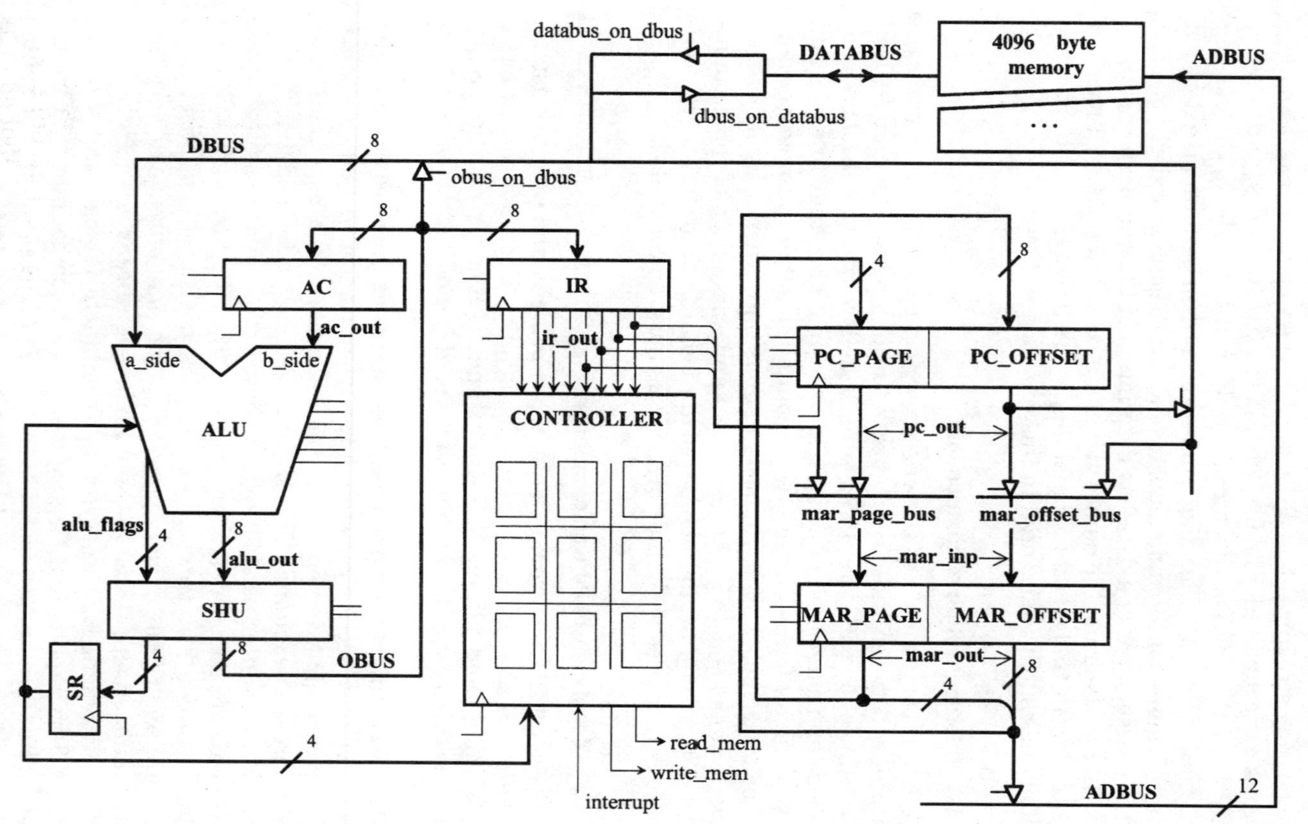

Figure 10.28 Parwan bussing structure.

and four outputs. Outputs of this register pass through *alu* and *shu* and circle back to the register's own inputs. This allows flags to be modified by either one of these logical units.

Figure 10.28 also shows a controller; a unit that generates control signals for the data components and busses. These signals cause the movement of data through system busses and storage of these data into registers. The controller makes its decisions based on its state, external interrupt, and bits of *ir* and *sr.*

10.4.3 Instruction execution

The bussing structure in Fig. 10.28 provides the necessary registers and the data path for executing the Parwan instructions. Based on an instruction in *ir,* the controller generates control signals in an appropriate order for the proper execution of the instruction. For an illustration of this mechanism, we describe the sequence of events for execution of an *lda* instruction.

Initially, the program counter contains the address of the instruction to be fetched. Fetching begins by moving the address from *pc* to *mar* and incrementing *pc.* For this move to take place, the address in *pc* must be placed on *mar_bus,* and *mar* must be enabled and clocked. When this is completed, the controller activates the signal that places *mar* on *adbus,* and at the same time it asserts *read_mem.* This causes the byte from the memory to appear on *databus,* which must now pass through *dbus, alu, shu,* and *obus* to reach *ir.* For this purpose, the controller activates the *databus_on_dbus* control signal, instructs *alu* to place its *a_side* on its output, and instructs the shifter unit to place its data input on its output without shifting it. The data on the output of *shu* become available for loading into *ir.*

Following an instruction fetch, the controller makes appropriate decisions based on the bits of *ir.* In our example, the most-significant bits of *ir* are 0000, indicating a direct-address *lda* instruction. To complete the address for the full-address *lda* instruction, the controller causes the current contents of *pc* to be clocked into *mar* and then be placed on *adbus* while it asserts *read_mem.* The byte read from the memory will pass through *dbus* and *mar_bus* to reach the input of the *mar_offset* register. At the same time, the controller activates a signal which will place *ir* on the *mar_page* register. Clocking *mar* while its load input is enabled causes the full address of the operand of *lda* to be clocked into this register.

The next read from the memory places the operand of the *lda* instruction on *databus.* This 8-bit data item passes through *dbus* and becomes available on the *a_side* of *alu.* The controller instructs *alu* to place its *a_side* input on its output, and it causes *shu* to pass its input to its output without shifting. The *lda* operand now appears on *obus,* the controller enables the loading of *ac,* and, on the edge of the clock, the operand of *lda* is clocked into the accumulator.

Execution of other instructions is done in a manner similar to the procedure described for *lda.* For the *and, add,* and *sub* instructions, when the operand of instruction becomes available on the *a_side* of *alu,* the controller signals *alu* to perform *and, add,* or *sub* operations instead of passing the operand through to the output of *alu.* For indirect addressing, the controller causes an extra read from the memory before performing the operation.

10.5 Dataflow Description of Parwan

The behavioral description in Sec. 10.3, the outline of which was shown in Fig. 10.17, presented an unambiguous description of the correct operation of Parwan. Hardware implementation of this machine, or even its bussing structure, is not apparent from the description in Fig. 10.17. This section presents a description of Parwan which is closer to its actual hardware. This description consists of the structural interconnection of the data registers and logic units, and it uses a dataflow description for the controller. Since the overall description deals with controlling the flow of data through registers and buses, we refer to it as the dataflow description of Parwan. The description presented here provides a better hardware correspondence than that of Sec. 10.3.

10.5.1 Data and control partitioning

Figure 10.29 shows the data and control partitioning that we use for the dataflow description of Parwan. The data section contains the

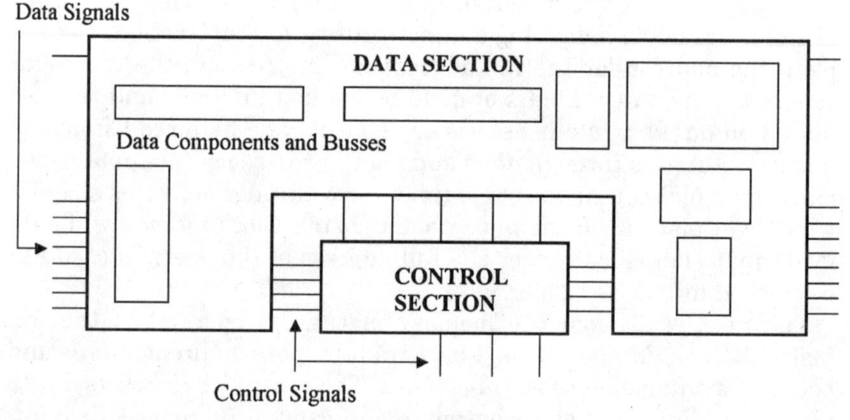

Figure 10.29 Data and control sections of Parwan CPU.

Applies To	Category	Signal Name	Functionality
AC	Register Control	load_ac,	Loads *ac*
		zero_ac	Resets *ac*
IR	Register Control	load_ir	Loads *ir*
PC	Register Control	increment_pc,	Increments *pc*
		load_page_pc,	Loads page part of *pc*
		load_offset_pc,	Loads offset part of *pc*
		reset_pc	Resets *pc*
MAR	Register Control	load_page_mar,	Loads page part of *mar*
		load_offset_mar	Loads offset part of *mar*
SR	Register Control	load_sr,	Loads *sr*
		cm_carry_sr	Complements carry flag of *sr*
MAR_BUS	Bus Control	pc_on_mar_page_bus,	Puts page part of *pc* on *mar* page bus
		ir_on_mar_page_bus,	Puts 4 bits of *ir* on *mar* page bus
		pc_on_mar_offset_bus,	Puts offset part of *pc* on *mar* offset bus
		dbus_on_mar_offset_bus	Puts dbus on *mar* offset bus
DBUS	Bus Control	pc_offset_on_dbus,	Puts offset part of *pc* on *dbus*
		obus_on_dbus,	Puts *obus* on *dbus*
		databus_on_dbus	Puts external *databus* on internal *dbus*
ADBUS	Bus Control	mar_on_adbus	Puts all of *mar* on *adbus*
DATABUS	Bus Control	dbus_on_databus	Puts internal *dbus* on external *databus*
SHU	Logic Units	arith_shift_left,	Shifter shifts its input one place to the left
		arith_shift_right	Shifter shifts its input one place to the right
ALU	Logic Units	alu_and,	Output of *alu* becomes and of its two inputs
		alu_not,	Output of *alu* becomes complement of its *b* input
		alu_a,	Output of *alu* becomes the same as its *a* input
		alu_add,	*alu* performs add operation on its two inputs
		alu_b,	Output of *alu* becomes the same as its b input
		alu_sub	*alu* performs subtraction of its two inputs
Others	I/O	read_mem,	Starts a memory read operation
		write_mem,	Starts a memory write operation
		interrupt	Interrupts CPU

Figure 10.30 Inputs and outputs of Parwan control section.

interconnection specification of CPU components. This includes instantiation of individual components and conditional placement of their outputs on appropriate busses. The control section uses external control signals and signals from the data section and generates signals to control conditional assignments of data into registers or busses of the data section.

Figure 10.30 shows a list of control signals generated by the control section to control data movement in the data section. The names are chosen according to the operation that is controlled by the signal. For example, placing the least-significant 4 bits of *ir* into the *mar_page* bus is controlled by the *ir_on_mar_page_bus* signal.

10.5.2 Timing of data and control events

The data and control sections are driven by the same clock signal. On the falling edge of this clock, the control section makes its state transition and the registers of the data section accept their new values. Figure 10.31 shows the timing of control signals relative to the circuit clock.

A control signal becomes active on the falling edge of a clock pulse and remains active until the next negative edge. While a control signal is active, logic units of the data section perform their specified operations, and their results become available at the inputs of their target registers. When the falling edge of a clock arrives, a register, whose load input is enabled, accepts its input. The width of control signals allow for all logic unit and bus propagation delays.

The control section consists of master-slave D-type flip-flops that accept their inputs when the clock becomes **1** and change their outputs when the clock returns to **0**. Control flip-flops and data registers are synchronized with the falling edge of the clock.

10.5.3 General description methodology

The Parwan description is based on the partitioning in Fig. 10.29. The individual components of the data section, shown in Fig. 10.28, are independently described at the behavioral or dataflow level. We describe the data section by wiring its components and bussing com-

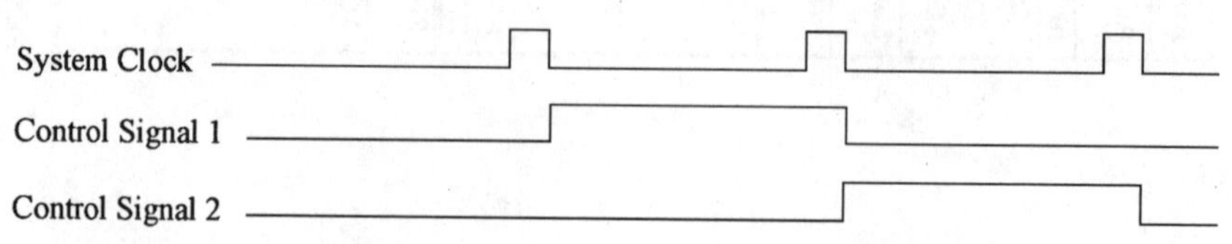

Figure 10.31 Timing of control signals.

ponent outputs according to the bus structure in Fig. 10.28. After the completion of the data section, a state machine description style is used for the description of the control section of Parwan. An overall description wires the data and the control sections.

10.5.4 Description of components

Components of the data section are *alu, shu, sr, ac, ir, pc,* and *mar* and are described in this order. For each component, a functional description, a symbolic notation, a possible gate-level implementation, and Verilog code will be given.

Arithmetic logic unit. The *alu* has two 8-bit operands, six opcode select lines, four flag inputs, eight data outputs, and four flag outputs. The six opcode lines select the operation of *alu* according to the table in Fig. 10.32. This figure also shows the flags that may be affected by *alu* operations.

Figure 10.33*a* shows the logic symbol for the Parwan *alu*. This symbol follows the IEEE standard notation. Figure 10.33*b* shows a possible hardware implementation for 1 bit of the *alu*. Opcode lines of the *alu* select one of the six functions of *alu* to appear on the output. For Verilog code for this *alu*, the *par_para.v* define file in Fig. 10.11 is used. This file defines a name equivalent for the operation codes of *alu* and is merely used for readability of *alu* code.

Figure 10.34 shows the Verilog code for the Parwan *alu*. Following the declarations of inputs, outputs, and temporary **reg** type variables, an **always** block appears in this figure. Activity flow into this block begins when an *alu* input changes. When this happens, status of *alu* control signals is read and checked against all valid *alu* operations. When a valid operation is detected, the corresponding *case_item* executes appropriate statements for the *alu* operation.

Shifter unit. Figure 10.35 shows the logic symbol and hardware of a bit of the shifter unit. This unit has two mode lines that select right- or

Id	Opcode line	Operation	Flags
0	alu_and	a AND b	zn
1	alu_not	NOT b	zn
2	alu_a	a	zn
3	alu_add	b PLUS a	vczn
4	alu_b	b	zn
5	alu_sub	b MINUS a	vczn

Figure 10.32 Operations and flags of *alu*.

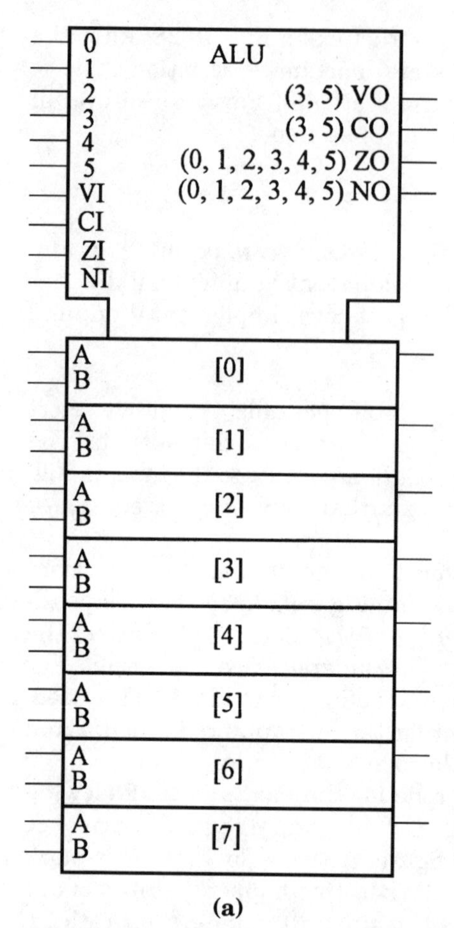

Figure 10.33 Parwan *alu*. (*a*) Logic symbol; (*b*) 1-bit gate-level hardware.

left-shift operations, four flag inputs, eight data inputs, four flag outputs, and eight data outputs. A left-shift operation moves a **0** into the least-significant position of the output, places the shifted input on the output, moves the most-significant bit to the carry output, and sets the overflow based on a sign change. A right-shift operation extends the sign bit of the input, shifts it one place to the right, and makes it available on the output. Both shift operations can affect the negative and zero flags.

The description of *shu* is shown in Fig. 10.36. As in *alu*, a procedural block constitutes the main body of the *shu* module. For shifting, the concatenation operator is used to form the shifted pattern, which is assigned to the temporary variable *tl*. When no shift operation is specified, the input data and flags are assigned to the outputs of the shifter. The concatenation operator is also used for flag assignments.

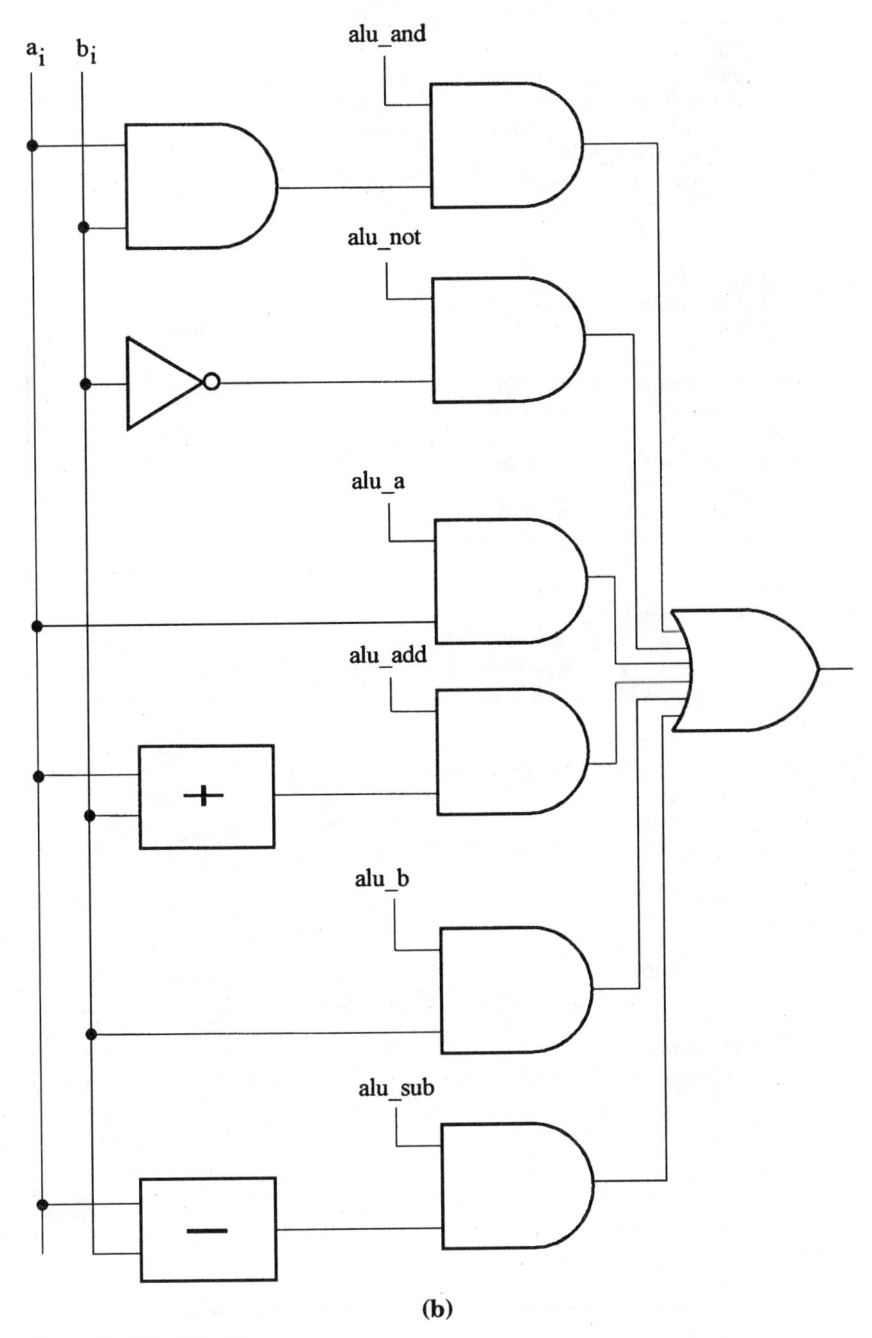

Figure 10.33 *(Continued)*

```
`include "par_para.v"
module arithmetic_logic_unit(a_side, b_side,
          alu_and, alu_not, alu_a, alu_add, alu_b, alu_sub,
          in_flags, z_out, out_flags);
  input  [7:0] a_side, b_side;
  input  alu_and, alu_not, alu_a, alu_add, alu_b, alu_sub;
  input  [3:0] in_flags;
  output  [3:0] out_flags;
  output [7:0] z_out;
  reg [7:0] z_out;
  reg [3:0] out_flags;
  reg [7:0] tl;
  reg v, c, z, n;
always @(a_side or b_side or alu_and or alu_not or alu_a or alu_add
            or alu_b or alu_sub or in_flags)
 begin
 case ({alu_and, alu_not, alu_a, alu_add, alu_b, alu_sub})
      `a_add_b : begin
            {c,tl} = a_side+b_side+in_flags[2];
            if (a_side[7] == b_side[7] && tl [7] != a_side[7])  v = 1'b1;
            else  v = 1'b0;
      end
      `a_sub_b : begin
            {c,tl} = a_side-b_side-in_flags[2];
            if (a_side[7] == ~b_side[7] &&  tl [7] != a_side[7]) v = 1'b1;
            else v = 1'b0;
      end
      `a_and_b : begin
            tl = a_side & b_side; c = in_flags[2]; v = in_flags[3];
      end
      `a_input : begin
            tl = a_side ; c = in_flags[2]; v = in_flags[3];
      end
      `b_input : begin
            tl = b_side ; c = in_flags[2]; v = in_flags[3];
      end
      `b_compl : begin
            tl = ~ b_side; c = in_flags[2]; v = in_flags[3];
      end
      default: begin
            tl = 8'b00000000; c = 1'b0; v = 1'b0;
      end
 endcase
 n = tl[7];
 z = ~|tl;
 z_out <= tl;
 out_flags <= {v, c, z, n};
 end
endmodule
```

Figure 10.34 Verilog code for Parwan *alu*.

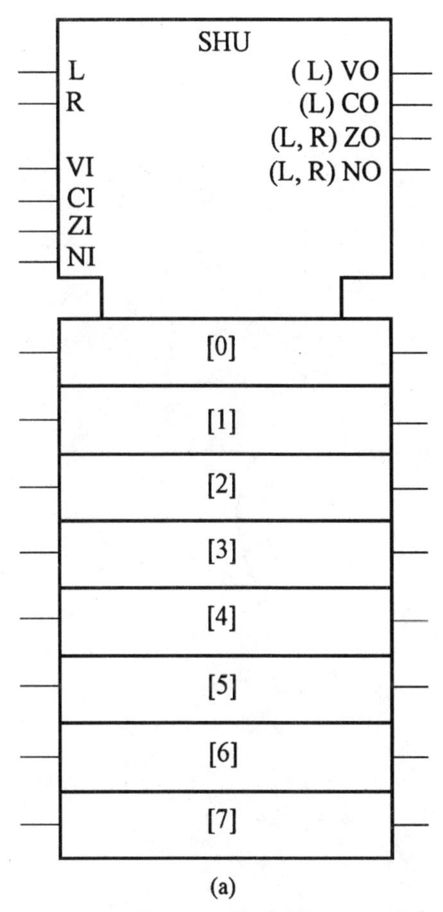

Figure 10.35 Parwan *shu*. (*a*) Logic symbol; (*b*) 1-bit hardware.

Status register unit. The status register is a synchronous, negative-edge-trigger 4-bit register. As shown in the logic symbol in Fig. 10.37*a*, the data loaded into the flags are synchronously controlled by *load* and *cm_carry* inputs. When *load* is active, all four input flags are loaded into the flag flip-flops, and when *cm_carry* is active, the *c* flag is loaded with the complement of its present value. The hardware of the *c* bit of this register is shown in Fig. 10.37*b*.

The Verilog code for the status register, shown in Fig. 10.38, uses a procedural block, the activity flow into which is controlled by the negative edge of the clock. On the negative edge of the clock, if *load* is active, all four flags will be loaded; otherwise, if *load* is not active and *cm_carry* input is active, the carry flag will be complemented.

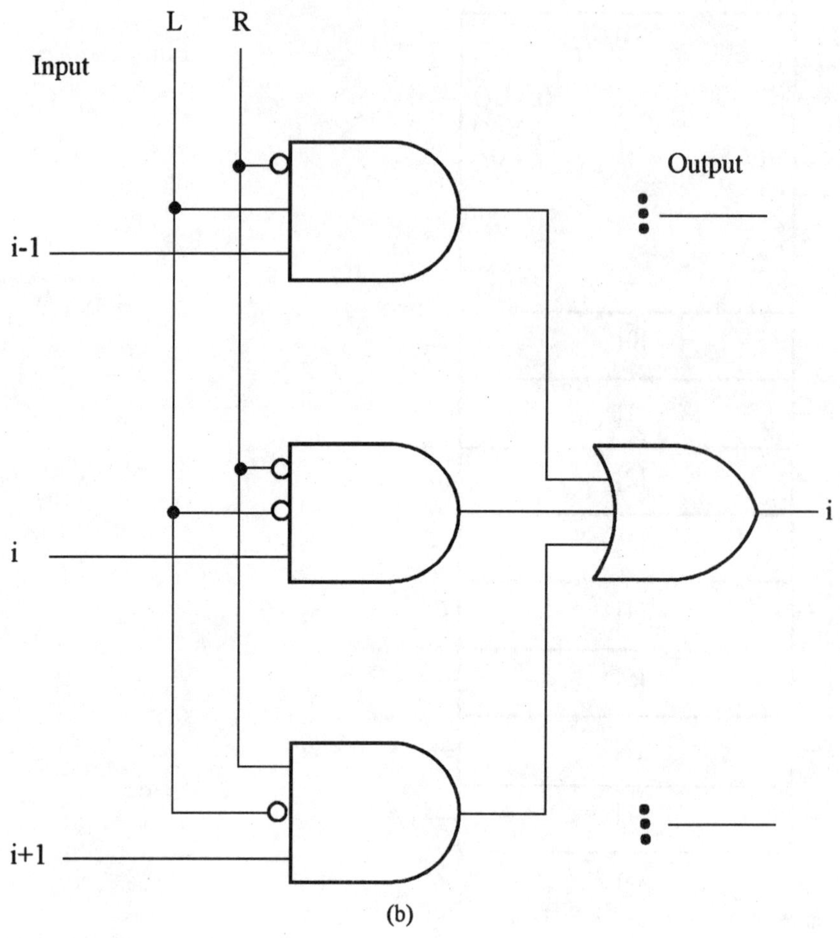

Figure 10.35 ***(Continued)***

Accumulator. The Parwan accumulator is an 8-bit register with synchronous loading and zeroing inputs. As shown in Fig. 10.39, loading of external data into the register is done on the falling edge of the clock when the *load* input is active and the *zero* input is disabled. Simultaneous activation of the *load* and *zero* inputs causes synchronous resetting of the register.

Figure 10.40 shows the Verilog code for the accumulator. In the *accumulator_unit* module, when *load* is **1** on the falling edge of *ck,* zero or the *i8* input is assigned to output *o8.*

Instruction register. The instruction register (*ir*) is an 8-bit synchronous register with an active high load input. The load input enables the clock and causes the register to be loaded on the falling edge of the clock input. The logic symbol of *ir* and the hardware of 1 bit of this register are shown in Fig. 10.41.

Figure 10.42 shows the Verilog description of the instruction register that corresponds to its logic symbol. In the *instruction_register_unit* module, a sequential block controlled by the negative edge of the clock assigns 8-bit inputs to the register outputs.

Program counter. The program counter is a 12-bit synchronous up-counter with one reset and two load inputs. The *load_page* input

```
module shifter_unit
   (alu_side, arith_shift_left, arith_shift_right, in_flags, obus_side, out_flags);
 input [7:0] alu_side;
 input [3:0] in_flags;
 input arith_shift_left, arith_shift_right;
 output [7:0]obus_side;
 output [3:0]out_flags;
 reg [7:0] obus_side;
 reg [3:0] out_flags;
 reg [7:0] tl;
 reg c, v, z, n;
always @ (alu_side or arith_shift_left or arith_shift_right or in_flags)
begin
      if (arith_shift_left) begin
            tl = {alu_side[6:0], 1'b0};
            n = tl[7];
            z = ~|tl;
            c = alu_side[7];
            v = ^{alu_side[6], alu_side[7]};
      end else begin
            if (arith_shift_right) begin
                  tl = {alu_side[7], alu_side[7:1]};
                  n = tl[7];
                  z = ~|tl;
                  c = in_flags[2];
                  v = in_flags[3];
            end else begin
                  tl = alu_side[7:0];
                  n = in_flags[0];
                  z = in_flags[1];
                  c = in_flags[2];
                  v = in_flags[3];
            end
      end
      obus_side <= tl;
      out_flags <= {v, c, z, n};
end
endmodule
```

Figure 10.36 Behavioral description of the Parwan shifter unit.

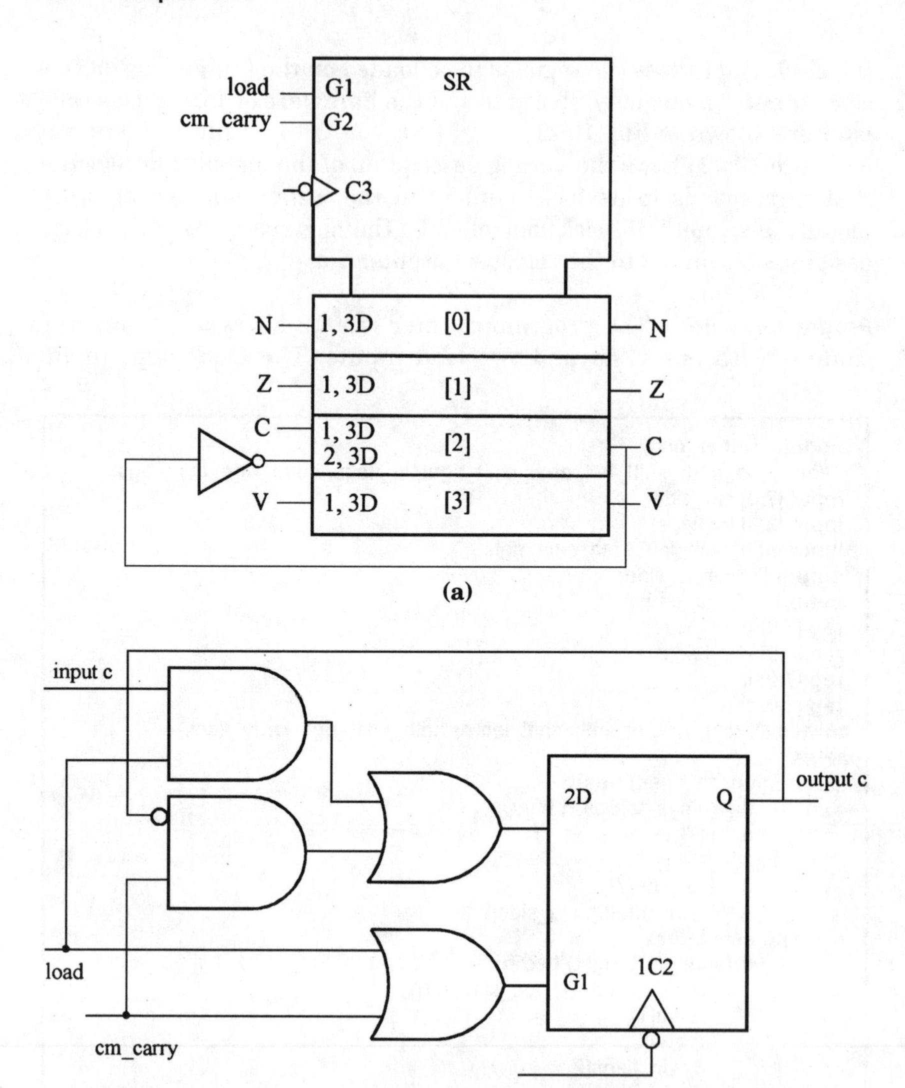

Figure 10.37 The status register. (*a*) Logic symbol; (*b*) *carry_bit* hardware.

synchronously loads input data into the most-significant 4 bits of the register, and the *load_offset* loads input data into the least-significant 8 bits. The synchronous reset input resets the entire register. The logic symbol for this unit appears in Fig. 10.43*a*. A possible hardware implementation of the program counter uses cascaded toggle flip-flops. Such a counter, a bit of which is shown in Fig. 10.43*b*, is slow but easy to build and lay out.

The Verilog description of the program counter in Fig. 10.44 consists of an **initial** statement, an *always* statement and a continuous assign statement. The **initial** block initializes an internal count and sets the *pc* output to **0.** The first **always** block uses an event control statement to control activity flow into a sequential statement that sets the internal *count* variable. When *increment* is **1,** *count* is incremented. The continuous assign statement block assigns *count* to *o12* outputs any time *count* changes.

Memory address register. The memory address register (*mar*) is a 12-bit register with two synchronous load inputs. The *load_page* input loads the parallel data into the most-significant nibble of the register, and the *load_offset* loads data into its least-significant byte. The logic symbol and hardware of 1 bit of this unit are presented in Fig. 10.45.

Figure 10.46 shows the Verilog code for *mar.* Like other register descriptions, the module in Fig. 10.46 uses an **always** statement that encloses an event control statement that checks for the negative edge of the clock. Control signals *load_page* and *load_offset* control clocking of the page and offset parts of *mar.*

10.5.5 Data section of Parwan

We completed our discussion of the Parwan data unit components in the previous section. The data unit specifies the interconnection of these components and defines the bussing structure of Parwan. The inputs of this unit are the data bus, the clock signal, and the signals from the control section. Control signals specify the operation of the components in the data section, control their clocking, and enable

```
module status_register_unit (in_flags, out_status, load, cm_carry, ck);
  input [3:0] in_flags;
  input load, cm_carry,ck;
  output [3:0] out_status;
  reg [3:0] status;
initial  status = 4'b0000;
always @( negedge ck )
   if (load)
      status <= in_flags;
   else begin
      if (cm_carry)
         status[2] <= ~status[2];
   end
assign out_status = status;
endmodule
```

Figure 10.38 Behavioral description of the Parwan status register.

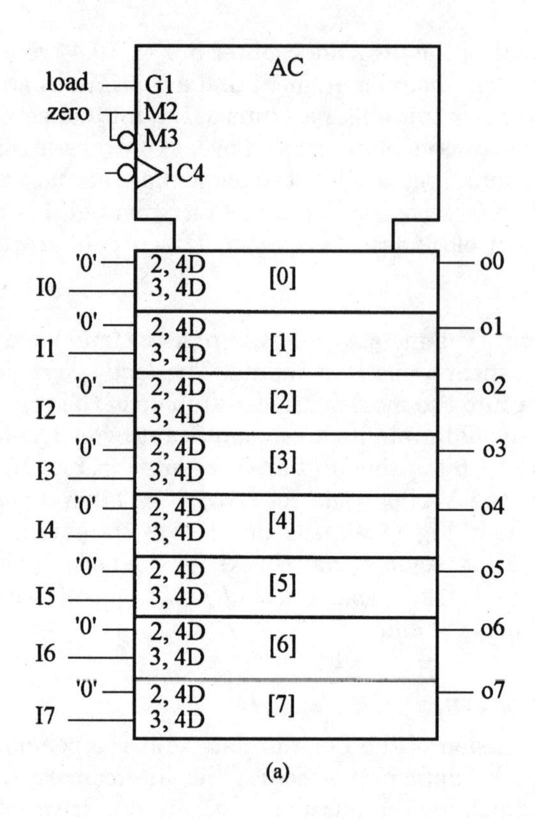

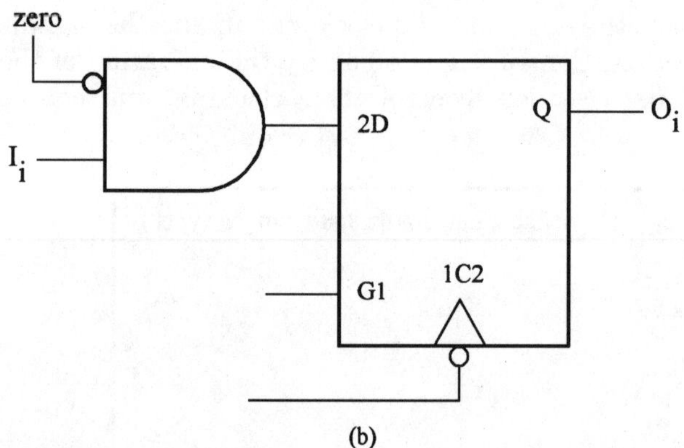

Figure 10.39 Parwan accumulator. (*a*) Logic symbol; (*b*) 1-bit hardware.

```
module accumulatur_unit (i8, o8, load, zero, ck);
  input [7:0] i8;
  input load, zero, ck;
  output [7:0] o8;
  reg [7:0] o8;
always @ ( negedge ck )
   if (load) begin
      if (zero)
         o8 <= 8'b00000000;
      else
         o8 <= i8;
   end
endmodule
```

Figure 10.40 Description of the Parwan accumulator.

bussing of data to their inputs. The outputs of the data section are the data and address busses, bits of *ir*, and four status flags. An outline of the module describing the data section (*par_data_path*) of Parwan is shown in Fig. 10.47.

As shown in Fig. 10.47, the first part of *par_data_path* specifies inputs, outputs, and bidirectional inout lines. This is followed by declaration of **net**s for component interconnections and internal CPU busses. Components described in the previous section are instantiated, and, using declared **nets**, they are interconnected according to the structure shown in Fig. 10.28. In the last part of Fig. 10.47, three-state bus assignments, also shown in Fig. 10.28, are done. Details of code fragments that are referred to in Fig. 10.47 will be described in the following paragraphs.

Following the **module** keyword and module name in the description of the data path, a list of *par_data_path* inputs and outputs appears, as shown in Fig. 10.48. This list consists of two system busses, *databus* and *adbus*; input control lines from the controller; and outputs to the control unit. The *databus* is an 8-bit bidirectional bus and is declared as **inout,** while *adbus* is unidirectional and is driven only from within the data unit. Control input lines coming from the controller are declared as **input** and connect to structures within the data path. Data path outputs *ir_lines* and *status,* which are declared as **output,** are inputs to the controller circuit.

Figure 10.49 shows internal bus declarations as well as wires for interconnection of data path components. All **net**s in this figure are declared as three-state wires, **wire.** However, only *dbus* and *mar_bus* are used as three-state busses; all other **net**s are merely used for connecting components.

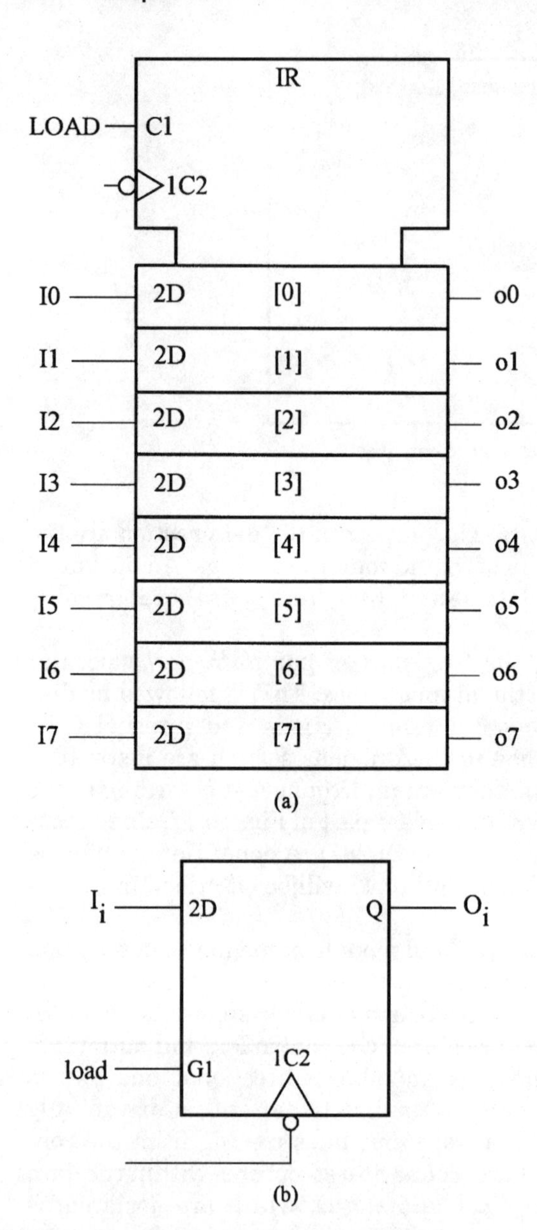

Figure 10.41 The Parwan instruction register. (*a*) Logic symbol; (*b*) 1-bit hardware.

```
module instruction_register_unit (i8, o8, load, ck);
  input [7:0] i8;
  input load,ck;
  output [7:0] o8;
  reg [7:0] o8;
always @ ( negedge ck )
   if (load)  o8 <= i8;
endmodule
```

Figure 10.42 Verilog code for Parwan instruction register.

Following the bus declarations in Fig. 10.47, the code fragment for component instantiations appears. Figure 10.50 shows instantiation of the data path components. These components were discussed in Sec. 10.5.4. Wires declared in Fig. 10.49, such as *ac_out,* are used for component interconnections. For example, *ac_out* connects the accumulator, *r1,* output and to the arithmetic logic unit (component *l1*) input. The other input of *l1* instantiation is directly driven by *dbus.* Control inputs declared in Fig. 10.48 connect to control inputs of instantiated components. For example, the *load_offset_pc* input of the data path connects to the fifth port of *program_counter_unit,* which, according to the Verilog code in Fig. 10.44, is an input of this unit.

The last part of Fig. 10.47 for data path description is bus assignments. This code fragment is shown in Fig. 10.51. Continuous assignments are used for internal and external bus connections. On the right-hand side of an assignment to a bus, a condition operator is used that selects either the source of the bus or **Z** float values. Assigning **Z**s to the busses is possible because **wire** type is used for their declarations. Selection of the source of the bus is conditioned with control signals that are outputs of the controller circuit.

Selecting one of several sources for a bus is implemented with multiple continuous assignments to the bus. For example, the *mar_bus[11:8]* bus segment in Fig. 10.51 may be driven by *ir_out[3:0]* or *pc_out[11:8]* (fifth and sixth lines in this figure). These sources are selected by the *ir_on_mar_page_bus* and *pc_on_mar_page_bus* control signals, respectively. If none of these signals are active, *4'bZZZZ* values will be selected by both condition operations on the right-hand sides of the two continuous assignments to *mar_bus[11:8].* If only one control signal is active, the selected source values override the high-impedance values selected by the inactive control signal. For a bus with multiple sources, it is not expected that more than one source drives the bus at any one time.

The style of coding used for the Parwan data path is synthesizable and is accepted by most commercial synthesis tools.

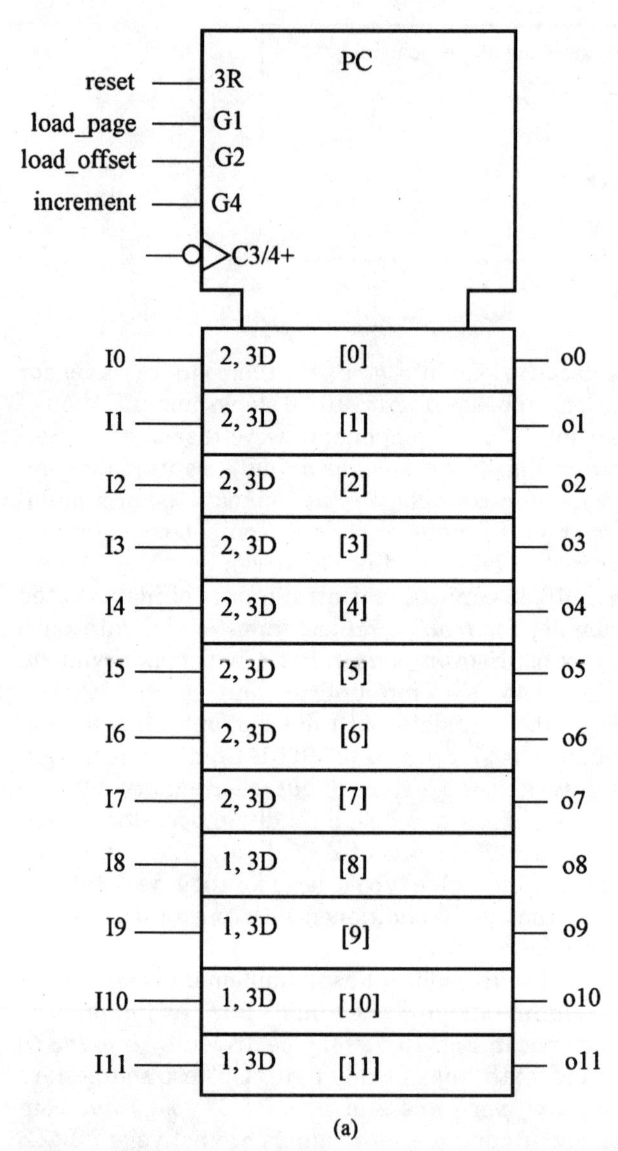

Figure 10.43 Parwan program counter. (*a*) Logic symbol; (*b*) 1-bit hardware.

10.5.6 Control section of Parwan

In the previous sections, hardware implementations for data components and the data path of Parwan were explicitly specified or were implied by Verilog constructs. For the data components, we showed example hardware implementations, and for the data path, continuous

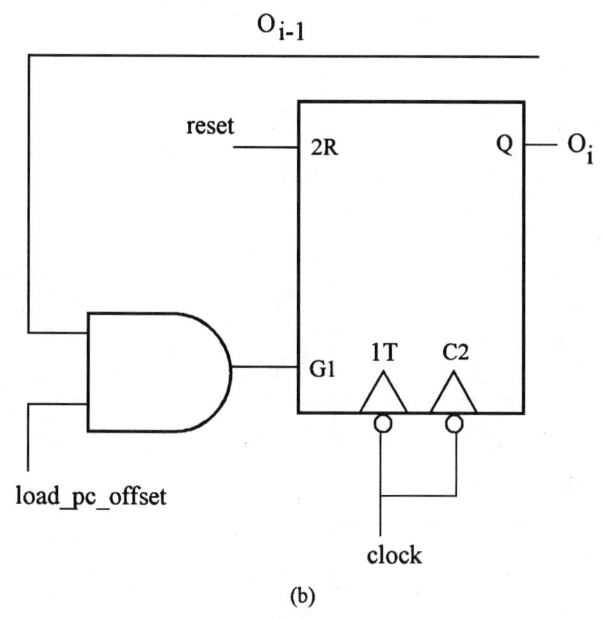

Figure 10.43 ***(Continued)***

```
module program_counter_unit
    (i12, o12, increment, load_page, load_offset, reset, ck);
  input [11:0] i12;
  input increment, load_page, load_offset, reset, ck;
  output [11:0] o12;
  reg [11:0] count ;
initial  count = 12'b000000000000;
always @( negedge ck )
    if(reset)
      count = 12'b000000000000;
    else begin
      if (increment)
          count <= count+1;
      else begin
        if(load_page)
            count[11:8] <= i12[11:8];
        if(load_offset)
            count[7:0] <= i12[7:0];
      end
    end
assign o12 = count;
endmodule
```

Figure 10.44 Behavioral description of Parwan program counter.

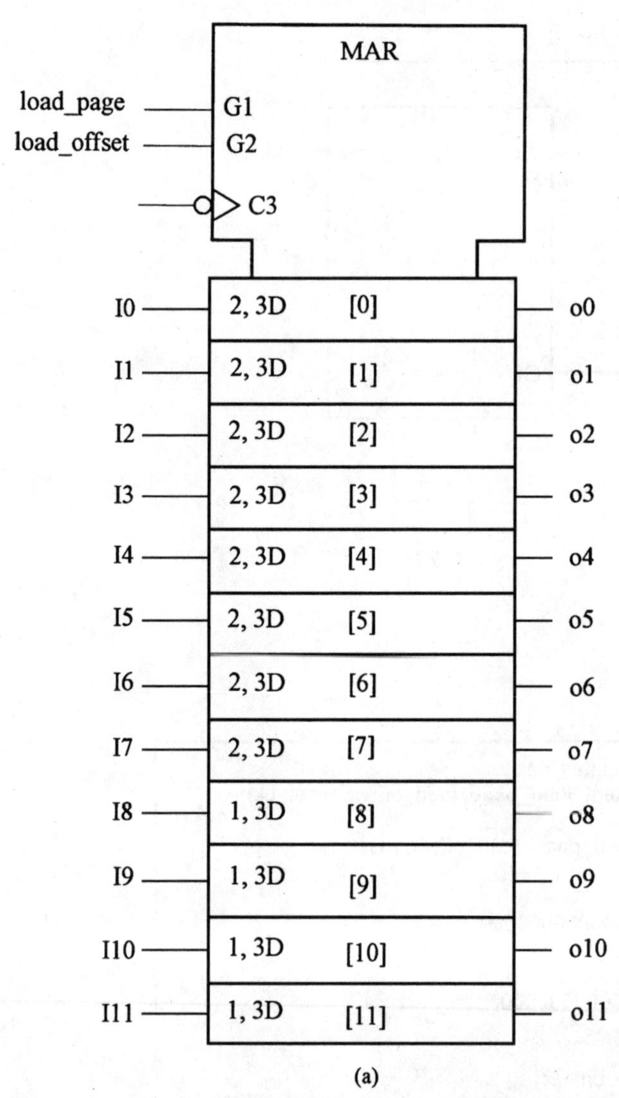

Figure 10.45 Memory address register of Parwan. (*a*) Logic symbol; (*b*) 1 bit of hardware.

assignments with clear hardware correspondence were used. The intent of this section is to present Parwan controller Verilog code and at the same time show a corresponding hardware for this unit.

Hardware style. The controller consists of a series of control flip-flops that drive control signals and other control flip-flops. In order to be able to treat each state of the controller independently and to be able

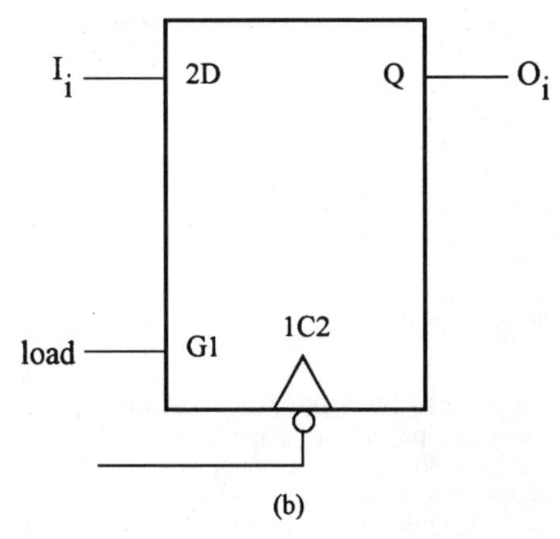

Figure 10.45 ***(Continued)***

```
module memory_address_register_unit (i12, o12, load_page, load_offset, ck);
  input [11:0] i12;
  input load_page, load_offset, ck;
  output [11:0] o12;
  reg [11:0] o12;
always @ ( negedge ck ) begin
    if( load_page )
      o12[11:8] <= i12[11:8];
    if( load_offset )
      o12[7:0] <= i12[7:0];
end
endmodule
```

Figure 10.46 Behavioral description of the Parwan memory address register.

```
module par_data_path (
// Specification and declaration of inputs and outputs; Fig. 10.48 ...
    . . .

// Declaration of busses and wires; Fig. 10.49 ...
    . . .

// Component instantiations; Fig. 10.50 ...
    . . .

// Bus assignments and interconnections; Fig. 10.51 ...
    . . .

endmodule
```

Figure 10.47 Parwan data path outline.

```
        databus, adbus, clk, load_ac, zero_ac, load_ir,
        increment_pc, load_page_pc, load_offset_pc, reset_pc,
        load_page_mar, load_offset_mar, load_sr, cm_carry_sr,
        pc_on_mar_page_bus, ir_on_mar_page_bus,
        pc_on_mar_offset_bus, dbus_on_mar_offset_bus,
        pc_offset_on_dbus, obus_on_dbus, databus_on_dbus,
        mar_on_adbus,
        dbus_on_databus,
        arith_shift_left, arith_shift_right,
        alu_and, alu_not, alu_a, alu_add, alu_b, alu_sub,
        ir_lines, status);
inout [7:0] databus;
input clk, load_ac, zero_ac, load_ir, increment_pc, load_page_pc, load_offset_mar;
input load_offset_pc, reset_pc, load_sr, cm_carry_sr, pc_on_mar_page_bus;
input ir_on_mar_page_bus, pc_on_mar_offset_bus, dbus_on_mar_offset_bus;
input pc_offset_on_dbus, obus_on_dbus, databus_on_dbus, mar_on_adbus;
input dbus_on_databus, arith_shift_left, arith_shift_right, load_page_mar;
input alu_and, alu_not, alu_a, alu_add, alu_b, alu_sub;
output [7:0] ir_lines;
output [3:0] status;
output [11:0] adbus;
```

Figure 10.48 Specification and declaration of inputs and outputs.

```
wire [7:0] ac_out, ir_out, alu_out, obus;
wire [11:0] pc_out, mar_out;
wire [7:0] dbus;
wire [3:0] alu_flags, shu_flags, sr_out;
wire [11:0] mar_bus;
```

Figure 10.49 Declaration of busses and wires.

```
accumulatur_unit r1 (obus, ac_out, load_ac, zero_ac, clk);
instruction_register_unit r2 (obus, ir_out, load_ir, clk);
program_counter_unit r3 (mar_out, pc_out, increment_pc,
          load_page_pc, load_offset_pc, reset_pc, clk);
memory_address_register_unit r4 (mar_bus, mar_out,
          load_page_mar, load_offset_mar, clk);
status_register_unit r5 (shu_flags, sr_out, load_sr, cm_carry_sr, clk);
arithmetic_logic_unit l1 (dbus, ac_out,
          alu_and, alu_not, alu_a, alu_add, alu_b, alu_sub,
          sr_out, alu_out, alu_flags);
shifter_unit l2 (alu_out, arith_shift_left, arith_shift_right,
          alu_flags, obus, shu_flags);
```

Figure 10.50 Component instantiations.

```
assign mar_bus[7:0] = dbus_on_mar_offset_bus ? (dbus) : (8'bZZZZZZZZ);
assign databus = dbus_on_databus ? (dbus) : (8'bZZZZZZZZ);
assign dbus = obus_on_dbus ? (obus) : (8'bZZZZZZZZ);
assign dbus = databus_on_dbus ?(databus) : (8'bZZZZZZZZ);
assign mar_bus[11:8] = ir_on_mar_page_bus ? (ir_out[3:0]) : (4'bZZZZ);
assign mar_bus[11:8] = pc_on_mar_page_bus ? (pc_out[11:8]) : (4'bZZZZ);
assign mar_bus[7:0] = pc_on_mar_offset_bus ? (pc_out[7:0]) : (8'bZZZZZZZZ);
assign dbus = pc_offset_on_dbus ? (pc_out[7:0]) : (8'bZZZZZZZZ);
assign adbus = mar_on_adbus ? (mar_out) : (12'bZZZZZZZZZZZZ);
assign status = sr_out;
assign ir_lines = ir_out;
```

Figure 10.51 Bus assignments and output connections.

to show the Verilog code of each state independent from the code of other states, we have chosen a one-hot implementation for the Parwan controller.

Like any state machine implementation, a one-hot state machine consists of a logic block and a register block that stands on a feedback path from the output of the logic block back to its input, as shown in Fig. 10.52. However, unlike in binary encoding or other state assignment formats, each flip-flop of a one-hot state machine corresponds to exactly one state of the machine. Therefore, under normal operating conditions, of all n flip-flops of a one-hot state machine, one and only one is active at any one time.

Each state flip-flop of a one-hot implementation enables a combinational logic block. These logic blocks provide the necessary conditions for activating other state flip-flops and for issuing control signals for the data section or the memory. Figure 10.53 zooms on the hardware surrounding a typical control flip-flop. The logic block in this figure is designated by a bubble.

The inputs of logic blocks come from other state flip-flops and from external inputs that influence state transitions in the control section. The outputs of logic blocks are the control signals, some of which become inputs to the data section. State flip-flops issue various control signals by providing active sources for them. If several control flip-flops issue a control signal, the actual control signal is formed by ORing its various sources together. The *par_data_path* unit uses the result of this OR function as input. Signals used to activate control flip-flops also may have multiple sources; therefore, OR functions also are needed at the inputs of control flip-flops. Although this ORing is required in one-hot implementation, it is not expected that control signals will be driven to **1** by more than one control flip-flop at any one time.

Figure 10.54 shows an example of the control section structure of the Parwan CPU. As shown in this example, all states are synchronized

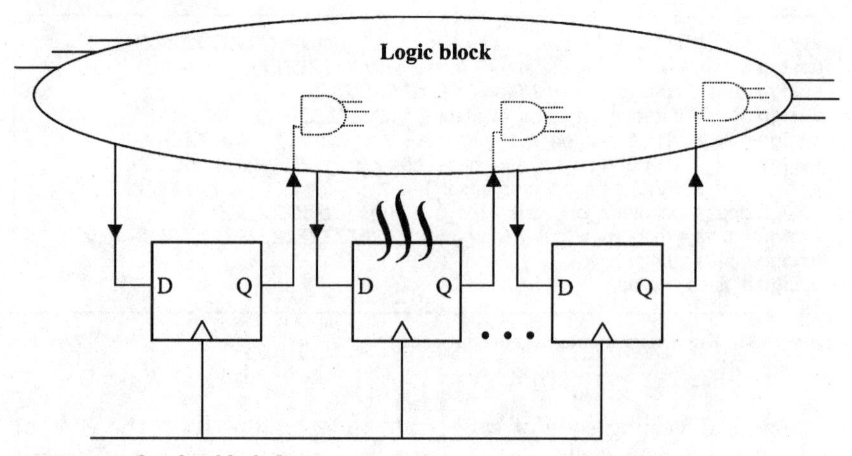

Figure 10.52 One-hot block diagram.

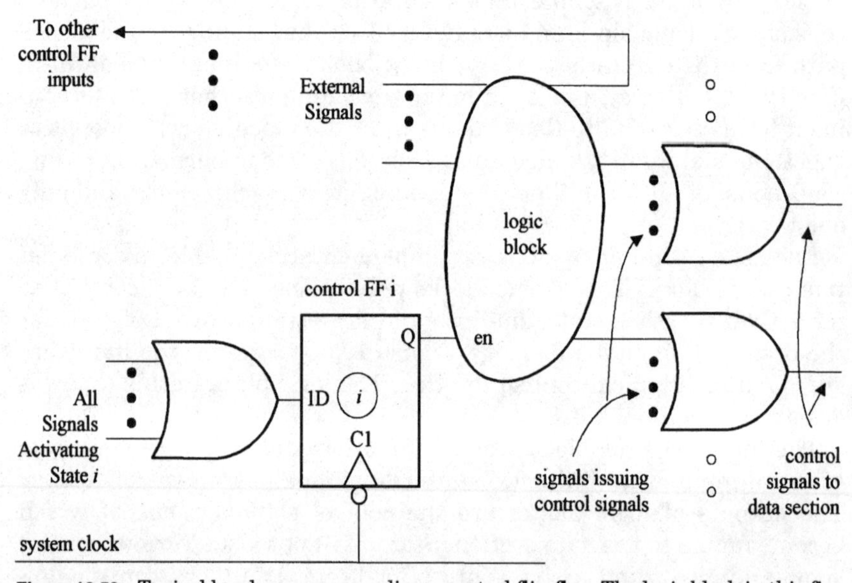

Figure 10.53 Typical hardware surrounding a control flip-flop. The logic block in this figure is designated by a bubble.

with the same clock, and their outputs contribute to the logic for issuing control signals or for activating other control flip-flops. In this figure, state *i* is conditionally activated by itself or by state *k*. State *i* conditionally activates state *j*, and state *k* always becomes active after the clock period during which state *j* is active. Control signal *csx* is always issued when state *k* is active or when state *i* is active and certain conditions are held on the *a*, *b*, and *c* inputs. Control signal *csy* becomes active when control is in state *j* and certain conditions are held on the *d* and *e* inputs.

Coding style. A Verilog coding style that corresponds to the block diagram of Fig. 10.52 uses an **always** block for state flip-flop clocking and another for sequencing between the states that correspond to the logic block in Fig. 10.52. In addition, an **initial** block sets the present state such that the first flip-flop is set to **1** and all others to **0.** A **case** statement in the **always** statement corresponding to the logic block has *case_item* for the logic blocks that are enabled by the state flip-flops. Statements in each *case_item* issue control signals and correspond to logic blocks such as that of Fig. 10.53.

As illustrated in Figs. 10.53 and 10.54, control signals issued by individual states are ORed together to generate actual control signal outputs of the controller. To implement this ORing effect in our style of coding, all control signals are set to **0** at the beginning of the **always** block that implements the controller logic. In each *case_item,* appropriate control signals are set to **1** conditioned upon the active control state. Therefore, with activation of a state, a control signal remains **0** unless set by a state flip-flop.

Controller coding. The Verilog module describing the Parwan controller begins with a listing of inputs and outputs, as shown in Fig.

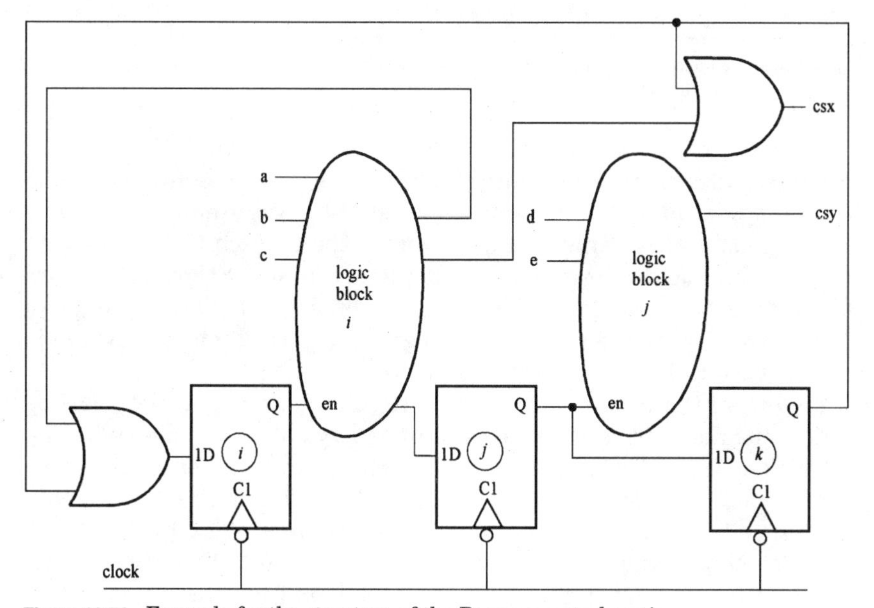

Figure 10.54 Example for the structure of the Parwan control section.

```
`include "par_para_1hot.v"
module par_control_unit (clk, load_ac, zero_ac, load_ir,
        increment_pc, load_page_pc, load_offset_pc, reset_pc,
        load_page_mar, load_offset_mar, load_sr, cm_carry_sr,
        pc_on_mar_page_bus, ir_on_mar_page_bus,
        pc_on_mar_offset_bus, dbus_on_mar_offset_bus,
        pc_offset_on_dbus, obus_on_dbus, databus_on_dbus,
        mar_on_adbus,
        dbus_on_databus,
        arith_shift_left, arith_shift_right,
        alu_and, alu_not, alu_a, alu_add, alu_b, alu_sub,
        ir_lines, status,
        read_mem, write_mem, interrupt, halted, ready, grant);
output load_ac, zero_ac, load_ir, increment_pc, load_page_pc, load_offset_mar;
reg load_ac, zero_ac, load_ir, increment_pc, load_page_pc, load_offset_mar;
output load_offset_pc, reset_pc, load_sr, cm_carry_sr, pc_on_mar_page_bus;
reg load_offset_pc, reset_pc, load_sr, cm_carry_sr, pc_on_mar_page_bus;
output ir_on_mar_page_bus, pc_on_mar_offset_bus, dbus_on_mar_offset_bus;
reg ir_on_mar_page_bus, pc_on_mar_offset_bus, dbus_on_mar_offset_bus;
output pc_offset_on_dbus, obus_on_dbus, databus_on_dbus, mar_on_adbus;
reg pc_offset_on_dbus, obus_on_dbus, databus_on_dbus, mar_on_adbus;
output dbus_on_databus, arith_shift_left, arith_shift_right, load_page_mar;
reg dbus_on_databus, arith_shift_left, arith_shift_right, load_page_mar;
output read_mem, write_mem, halted;
reg read_mem, write_mem, halted;
output alu_and, alu_not, alu_a, alu_add, alu_b, alu_sub;
reg alu_and, alu_not, alu_a, alu_add, alu_b, alu_sub;
input [7:0] ir_lines;
input [3:0] status;
input clk, ready, grant, interrupt;
reg [9:1] present_state, next_state;
```

Figure 10.55 Controller inputs, outputs, and state declarations.

10.55. Controller outputs are declared as **reg** so that they can be assigned values in a procedural block. The last line in Fig. 10.55 declares *present_state* and *next_state* as 9-bit registers. These variables model control flip-flop inputs and outputs. Each bit corresponds to one flip-flop for our one-hot implementation. State values are defined in the *"par_para_1hot.v"* include file, which is similar to *"par_para.v"* except for its state assignment part. Each state value is a 9-bit vector of only one **1** and eight **0**s.

Following the declarations of Fig. 10.55, Parwan controller Verilog code continues with an **initial** and an **always** block, as shown in Fig. 10.56. The **initial** statement sets the present and next state variables to 000000001, which sets the first control flip-flop of the one-hot controller to **1** and sets all others to **0.** The **always** block in Fig. 10.56 models the register shown in Fig. 10.52. On the falling edge of *clk,* control flip-flops are updated by data provided from the logic block.

The logic block in Fig. 10.52 is modeled by an **always** statement, the header of which is shown in Fig. 10.57. As shown in Fig. 10.52, control flip-flop outputs (*present_state*) and controller inputs from the data unit control activity into this block. Following the **begin** keyword, all control signals that are mostly outputs of the controller to the Parwan data unit are initialized to **0.** As discussed previously, setting these lines to **0** and then selectively driving them to **1** in a control state during a clock period implements the ORing function required for control signals, as shown in Fig. 10.53. All instances of procedural assignment of a **1** to the same control signal appear in hardware as individual inputs of the OR gate driving that control signal. This correspondence is represented in Fig. 10.58.

The Verilog code for the Parwan controller continues with nine *case_items*, which are included in the **always** block of Fig. 10.57. The nine *case_items* implement the control flip-flops of Fig. 10.52.

```
initial begin
    present_state <= `do_initials;
    next_state <= `do_initials;
end

always @(negedge clk) begin
    if (interrupt)
        present_state <= `do_initials;
    else present_state <= next_state;
end
```

Figure 10.56 State initialization and clocking.

```
always @(present_state or ir_lines or status or interrupt or ready or grant)
begin
    {load_ac, zero_ac} <= 2'b00; load_ir <= 0;
    {increment_pc, load_page_pc, load_offset_pc, reset_pc} <= 4'b0000;
    {load_page_mar, load_offset_mar} <= 2'b00;
    {load_sr, cm_carry_sr} <= 2'b00;
    {pc_on_mar_page_bus, ir_on_mar_page_bus} <= 2'b00;
    {pc_on_mar_offset_bus, dbus_on_mar_offset_bus} <= 2'b00;
    {pc_offset_on_dbus, obus_on_dbus, databus_on_dbus} <= 3'b000;
    {mar_on_adbus, dbus_on_databus} <= 2'b00;
    {arith_shift_left, arith_shift_right} <= 2'b00;
    {alu_and, alu_not, alu_a, alu_add, alu_b, alu_sub} <= 6'b000000;
    {read_mem, write_mem} <= 2'b00; halted <= 0;
    case (present_state)
```

Figure 10.57 Partial code corresponding to the logic block of the Parwan controller.

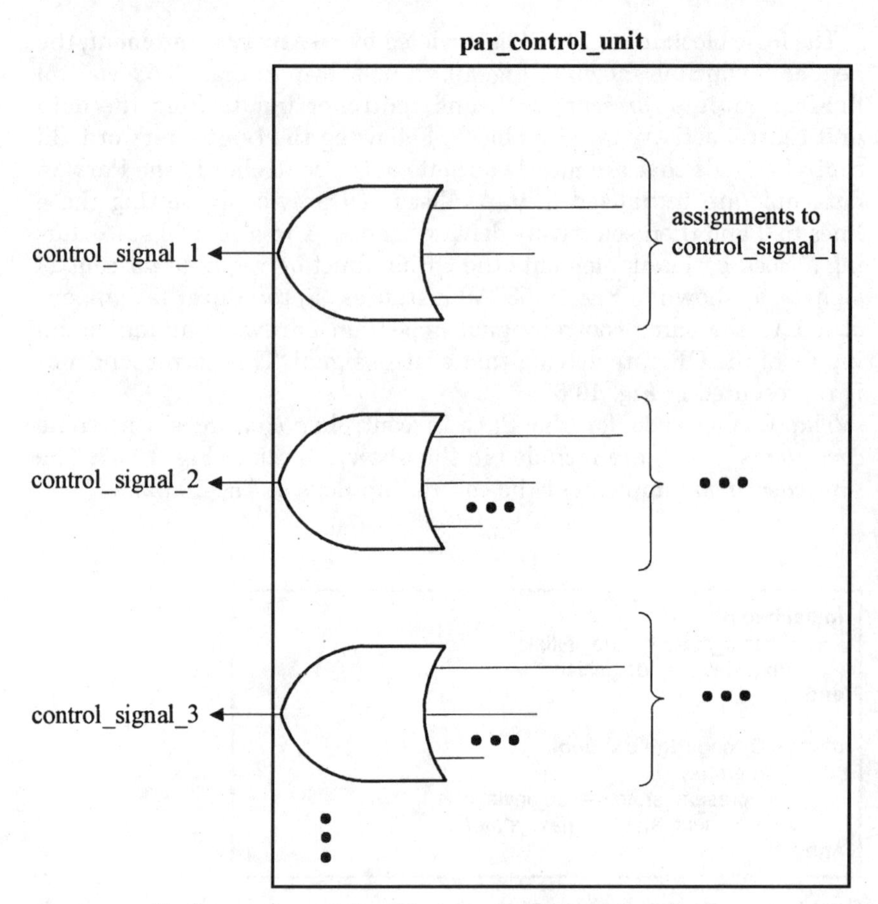

Figure 10.58 Hardware correspondence for assigning 1s to control signals.

Control state 1. Figure 10.59*a* shows the *s1* procedural block for state 1. Figure 10.59*b* shows hardware corresponding to this block. State 2 (shown by the number 2, circled) is activated when state 1 is active and *interrupt* is zero. This output of the AND gate generating this signal becomes an input to the OR gate at the input of the state 2 flip-flop.

In state 1, a fetch begins by placing *pc* on the *mar* bus and then initiating the transfer of these data into *mar.* If the interrupt input is active, the *pc* reset input is issued and control returns to state 1. If the CPU is not interrupted, state 2 becomes active on the falling edge of the clock. Also on this edge of the clock, *mar* receives its new value. In coding the data registers and control states, we have made certain that they are all synchronized with the falling edge of the clock.

```
`do_initials: begin : s1
    if (interrupt) begin
        reset_pc <= 1;
        next_state <= `do_initials;
    end else begin
        {pc_on_mar_page_bus, pc_on_mar_offset_bus} <= 2'b11;
        {load_page_mar, load_offset_mar} <= 2'b11;
        next_state <= `instr_fetch;
    end
end
```

(a)

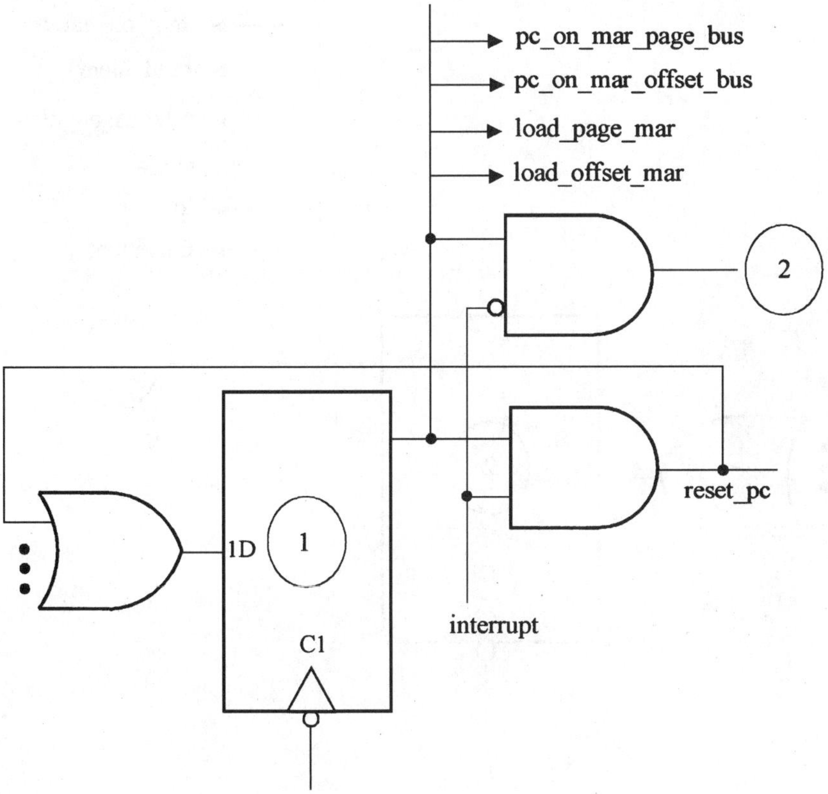

(b)

Figure 10.59 State 1: Starting a fetch. (*a*) Verilog code; (*b*) hardware.

```
`instr_fetch: begin : s2
      read_mem <= 1;
      mar_on_adbus <= 1;
      databus_on_dbus <= 1;
      alu_a <= 1;
      load_ir <= 1;
      increment_pc <= 1;
      next_state <= `do_one_bytes;
end
```

(a)

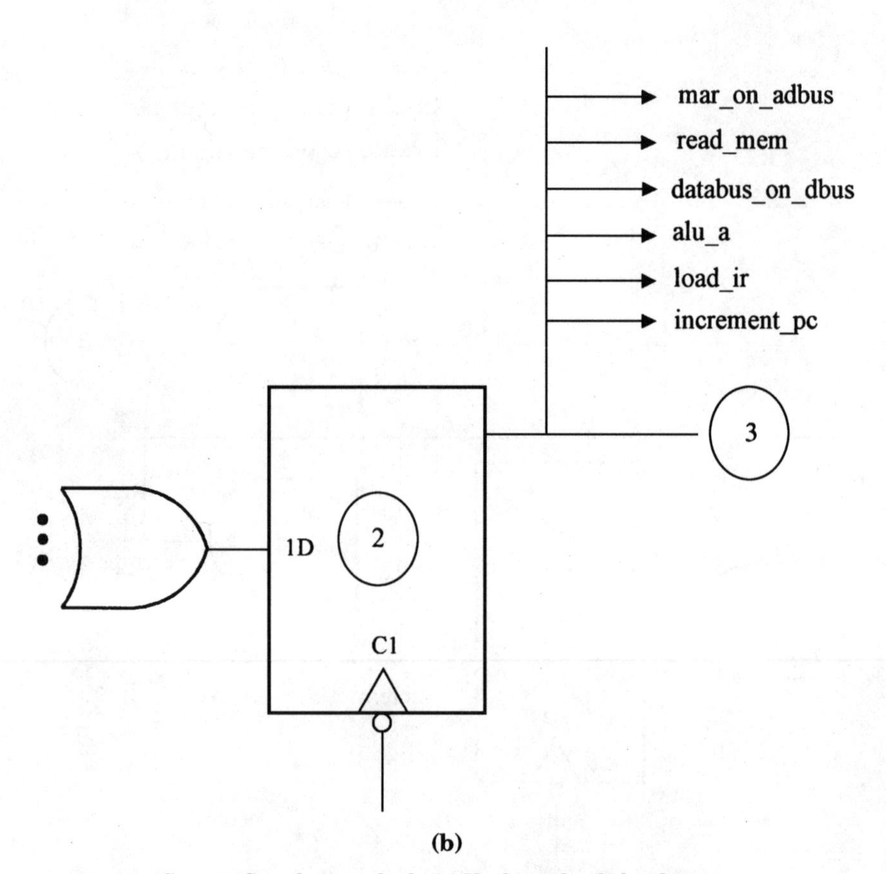

(b)

Figure 10.60 State 2: Completing a fetch. (*a*) Verilog code; (*b*) hardware.

Control state 2. When in state 2, the *mar* bus has received the new value that was scheduled for it in state 1. State 2 of Parwan, as shown in Fig. 10.60*a*, completes the fetch operation by placing *mar* on *adbus* and issuing a *read_mem*. The contents of memory that appear on *databus* must be placed in *ir*. For this purpose, the *databus_on_dbus* control signal is issued, the *a_input* function of *alu* is selected, and the load input of *ir* (*load_ir*) is enabled. On the edge of the clock, control state 3 becomes active and *ir* will have its new value. Also in state 2, the increment function of *pc* is selected so that its value gets incremented on the next falling edge of the clock. The control flip-flop shown in Fig. 10.60*b* generates inputs to six control output OR gates and one OR gate at the input of control flip-flop 3.

Control state 3. When state 3 becomes active, as shown in Fig. 10.61, the newly read instruction is in *ir*. State 3 starts the process of reading the next byte from the memory. At the same time, it checks for the number of bytes in the current instruction. If the instruction is a 2-byte instruction, the next byte becomes its address, and control state 4 is activated to continue the execution of 2-byte instructions. On the other hand, if the current instruction is a nonaddress instruction and does not require a second byte, state 3 performs its execution and activates state 2 for fetching the next instruction.

Nonaddress (single-byte) instructions perform operations on the accumulator and flags. For their execution, appropriate *alu* and *shu* functions are selected, and the target register or flag is enabled. For example, for *asr* (arithmetic shift right), the following steps are taken:

1. The *b_input* function of *alu* is selected so that the *alu* output becomes the contents of *ac*.
2. The *arith_shift_right* function of *shu* is selected so that it shifts its inputs (the contents of *ac*) one place to the right.
3. The *load_sr* signal (load input of status register) is enabled so that new values of flags, generated by *shu,* are loaded into *sr*.
4. The load input of *ac* (*load_ac*) is enabled so that this register gets loaded with the output of *shu* (shifted *ac*).

Discrete gate hardware implementation of state 3 is shown in Fig. 10.61*b*. The control signal outputs shown here drive the corresponding OR gates on the outputs of the control unit.

Control state 4. State 4 becomes active when a full-address or page-address instruction (instructions requiring 2 bytes) is being executed.

```
`do_one_bytes: begin : s3
    {pc_on_mar_page_bus, pc_on_mar_offset_bus} <= 2'b11;
    {load_page_mar, load_offset_mar} <= 2'b11;
    if (ir_lines[7:4] != `single_byte_instructions)
        next_state <= `opnd_fetch;
    else begin
        if (ir_lines[3:0] == `asl) arith_shift_left <= 1;
        if (ir_lines[3:0] == `asr) arith_shift_right <= 1;
        if (ir_lines[1] == 1) alu_not <= 1;
        else alu_b <= 1;
        if ((ir_lines[3] | ir_lines[1]) == 1) load_sr <= 1;
        if ((ir_lines[3] | ir_lines[1] | ir_lines[0]) == 1 load_ac <= 1;
        if (ir_lines[2] == 1) cm_carry_sr <= 1;
        if ((ir_lines[3] == 0) & (ir_lines[0] == 1)) zero_ac <= 1;
        if (ir_lines[3:0] == `hlt) halted <= 1;
        next_state <= `instr_fetch;
    end
end
```

(a)

Figure 10.61 State 3: Preparing for address fetch, execution of single-byte instructions. (*a*) Verilog code; (*b*) hardware.

The preparations for reading the address byte (the second byte of the instruction) were done in state 3. State 4, shown in Fig. 10.62, completes the read operation and makes the newly read byte available at the input of the offset part of the *mar.* Because *load_offset_mar* has become active, this byte will be clocked into *mar* on the next falling edge of the clock. If the instruction being executed is a full-address instruction (*lda, and, add, sub, jmp,* or *sta*), the page number from *ir* becomes available at the input of the page part of *mar* to be clocked into this register with the next clock. If the instruction being executed is *jsr* or *bra* (page-address instructions), the *mar_page* register retains its current value. This is because these instructions address only within the current page.

State 4 activates state 5 or 6 for handling the indirect or direct addressing mode of full-address instructions, and it activates state 7 or 9 for *jsr* or *bra* instructions, respectively. Figure 10.62*b* shows control signals and signals for activation of control states.

Control state 5. The falling edge of the clock that activates state 5 also loads a full 12-bit address into *mar.* State 5, shown in Fig. 10.63, handles the indirect addressing mode. In this state, the memory location pointed to by *mar* is read on the *databus* and is made available on the

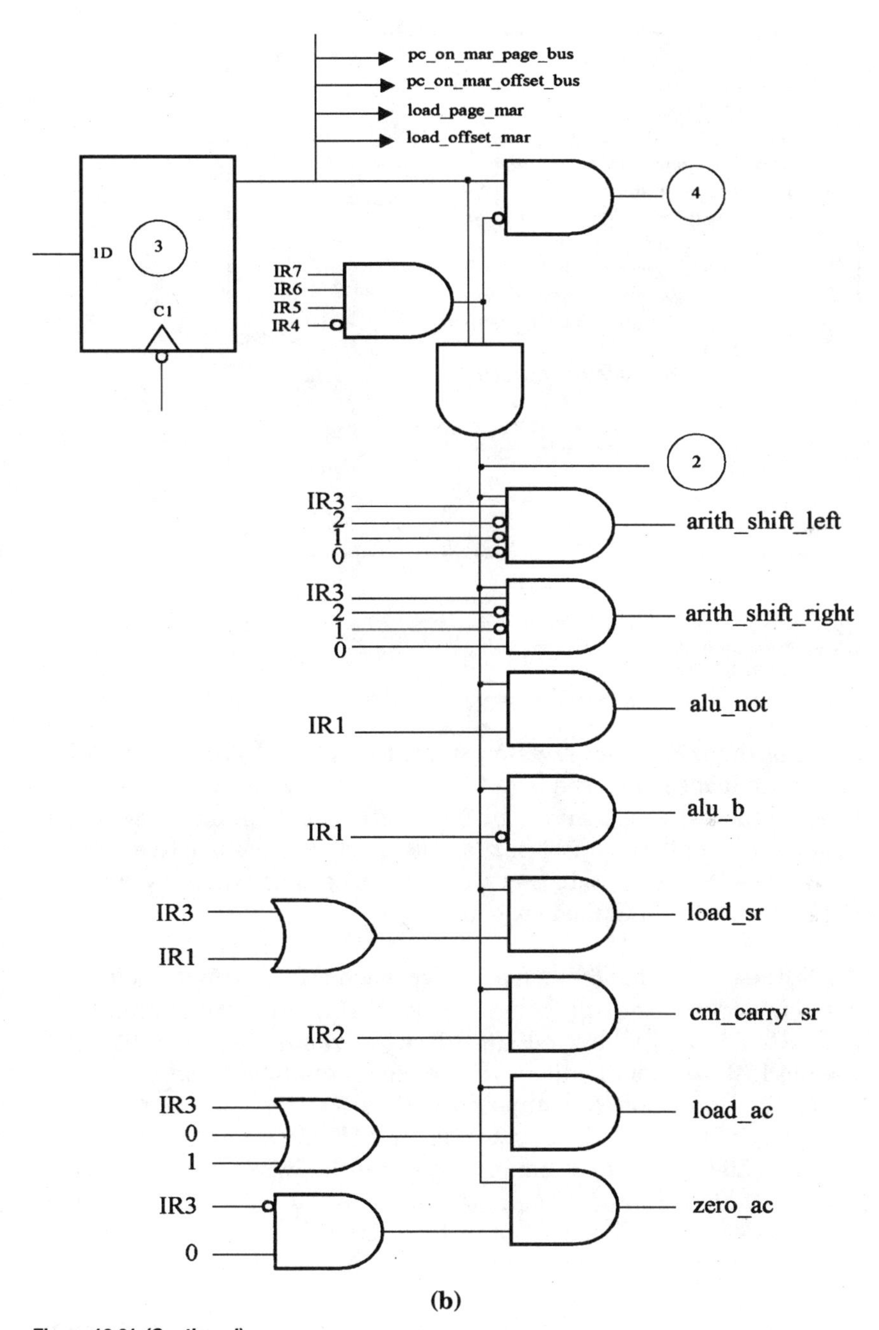

(b)

Figure 10.61 ***(Continued)***

```
`opnd_fetch: begin : s4
    read_mem <= 1;
    mar_on_adbus <= 1;
    databus_on_dbus <= 1;
    dbus_on_mar_offset_bus <= 1;
    load_offset_mar <= 1;
    if (ir_lines[7:6] != `jsr_or_bra) begin
        ir_on_mar_page_bus <= 1;
        load_page_mar <= 1;
        if (ir_lines[4] == `indirect)
            next_state <= `do_indirect;
        else
            next_state <= `do_two_bytes;
    end else begin
        if (ir_lines[5]) next_state <= `do_jsr;
        else next_state <= `do_branch;
    end
    increment_pc <= 1;
end
```

(a)

Figure 10.62 State 4: Completing address of full-address instructions; branching for indirect, direct, *jsr,* and *branch.* (*a*) Verilog code; (*b*) hardware.

input of the *mar_offset* register. Activation of *load_offset_mar* causes *mar* to be loaded with the byte from the memory on the next negative edge of the clock. As shown in Figure 10.9, the indirect addressing mode affects only the offset part of the address. State 5 activates state 6, which is the same state that would have been activated by state 4 if the direct addressing mode was used.

Control state 6. State 6 becomes active when the instruction being executed is *jmp, sta, lda, and, add,* or *sub.* In this state, *mar* contains the complete operand address. Figure 10.64 shows three nested blocks in the block statement of state 6. These blocks are labeled *jm, st,* and *rd.* As in the other states, a flip-flop with discrete gates generates the hardware of this state, shown in Fig. 10.64*b*.

In Fig. 10.64*a,* an **if** statement checks for the *jmp* instruction. In this case, the *pc* load input, *load_pc,* is enabled to cause the contents of *mar* to be loaded in *pc.* This is followed by activation of state 2 for fetching a new instruction from the memory location pointed to by the new contents of *pc.*

If the opcode of the *sta* instruction is detected on the most-significant bits of *ir_lines,* a write to memory is initiated and the contents of *ac* are routed through *alu* to reach the *databus* so that they can be written into the memory.

The third possible option taken in the procedural code of Fig. 10.64*a* is *rd* for *lda, and, add,* and *sub*. This part of the code is entered when bit 7 of *ir* is 0. For these instructions, the actual operand is read from the memory location pointed to by *mar.* When the *databus_on_dbus* control signal is issued, this operand becomes available on the *a_side* of *alu*. (Recall that *dbus* is directly connected to the *a_side* of *alu* in Fig. 10.28.) Based on the instruction being executed, an if–then–else statement selects an appropriate function of *alu*. For example, for the *add* instruction, the *alu_add* function is selected. This causes the result of adding *ac* and data on the *a_side* of *alu* to become available on the output of *alu*. This result is loaded in *ac* by issuing the *load_ac* control signal. Activation of *load_sr* in the *rd* block of state 6 updates flags with values that resulted from the *alu* operation. Upon completion of full-address instructions, control branches to state 1 for the next instruction fetch.

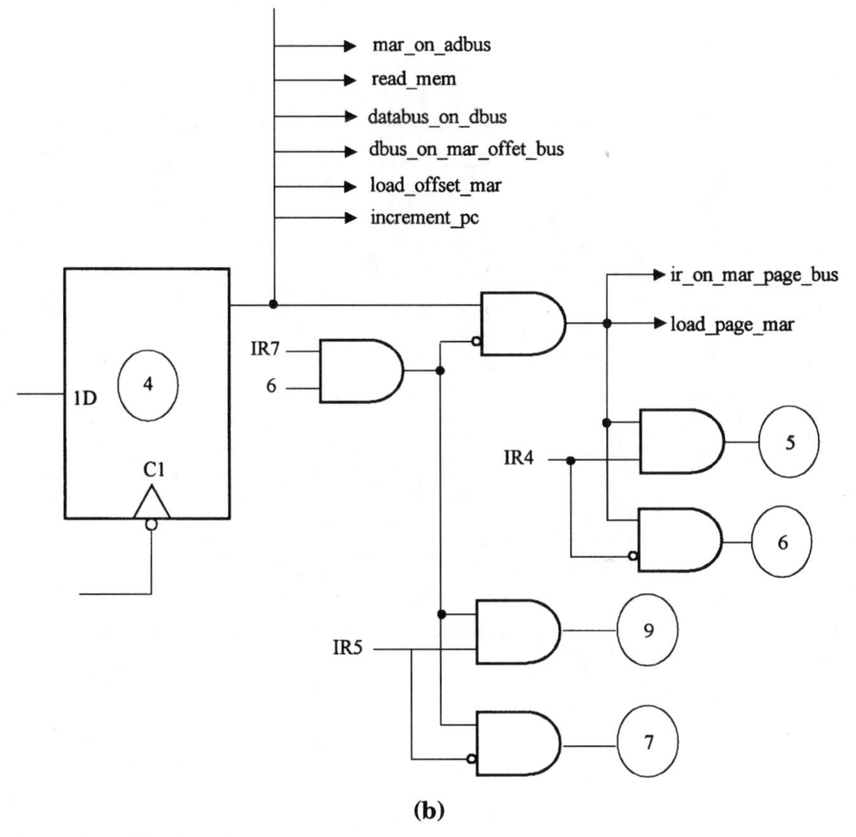

Figure 10.62 *(Continued)*

```
`do_indirect: begin : s5
    read_mem <= 1;
    mar_on_adbus <= 1;
    databus_on_dbus <= 1;
    dbus_on_mar_offset_bus <= 1;
    load_offset_mar <= 1;
    next_state <= `do_two_bytes;
end
```

(a)

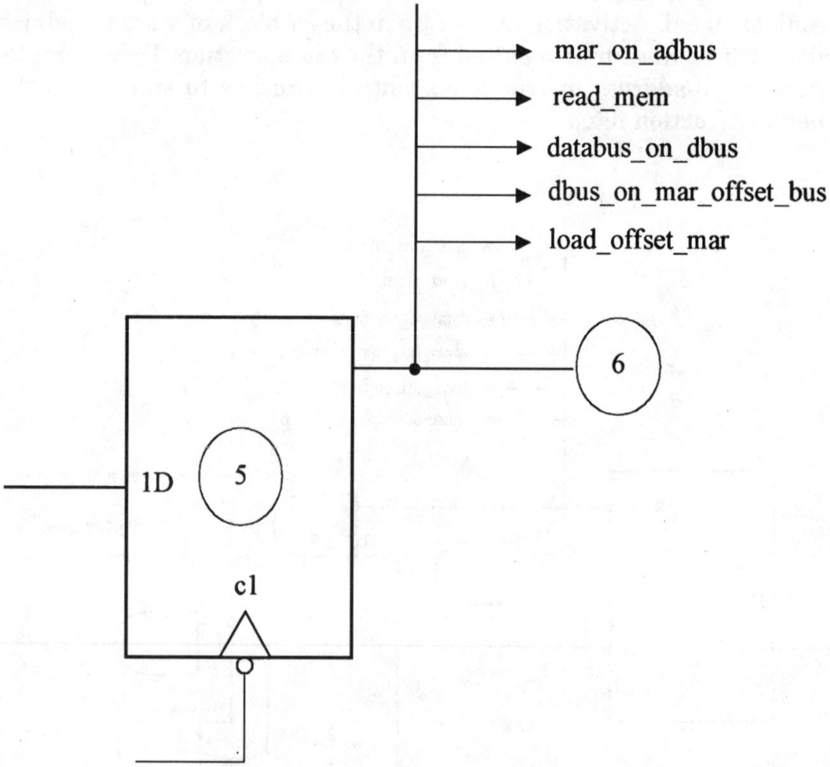

(b)

Figure 10.63 State 5: Taking care of indirect addressing. (*a*) Verilog code; (*b*) hardware.

Control state 7. State 4 of the Parwan controller caused a memory read operation and targeted the information read from the memory into the *mar_offset* register. If the *jsr* instruction is being executed, the address in the *mar_offset* register becomes the address of the top of the subroutine. State 7, shown in Fig. 10.65, continues the execution of *jsr*. This state writes the contents of *pc* to the top of the subroutine (pointed by *mar*) and at the same time targets the address of the top of the subroutine (12 bits of *mar*) for the *pc* register.

Control state 8. Following state 7 on the falling edge of the clock, state 8 of Fig. 10.66 becomes active to complete the execution of *jsr*. In this state, *pc* now contents the first location of the subroutine. Since the

```
`do_two_bytes: begin : s6
    if (ir_lines[7:5] == `jmp) begin : jm
        {load_page_pc, load_offset_pc} <= 2'b11;
        next_state <= `instr_fetch;
    end else begin
        if (ir_lines[7:5] == `sta) begin      : st
            write_mem <= 1;
            mar_on_adbus <= 1;
            alu_b <= 1;
            obus_on_dbus <= 1;
            dbus_on_databus <= 1;
            next_state <= `do_initials;
        end else begin
            if (~ir_lines[7]) begin : rd
                read_mem <= 1;
                mar_on_adbus <= 1;
                databus_on_dbus <= 1;
                load_ac <= 1;
                load_sr <= 1;
                if (~ir_lines[6]) begin
                    if (~ir_lines[5]) alu_a <= 1;
                    else alu_and <= 1;
                end else begin
                    if (~ir_lines[5]) alu_add <= 1;
                    else alu_sub <= 1;
                end
                next_state <= `do_initials;
            end else next_state <= `do_initials;
        end
    end
end
```

(a)

Figure 10.64 State 6: Reading the actual operand and executing *jmp, sta, lda, and, add,* and *sub* instructions. (*a*) Verilog code; (*b*) hardware.

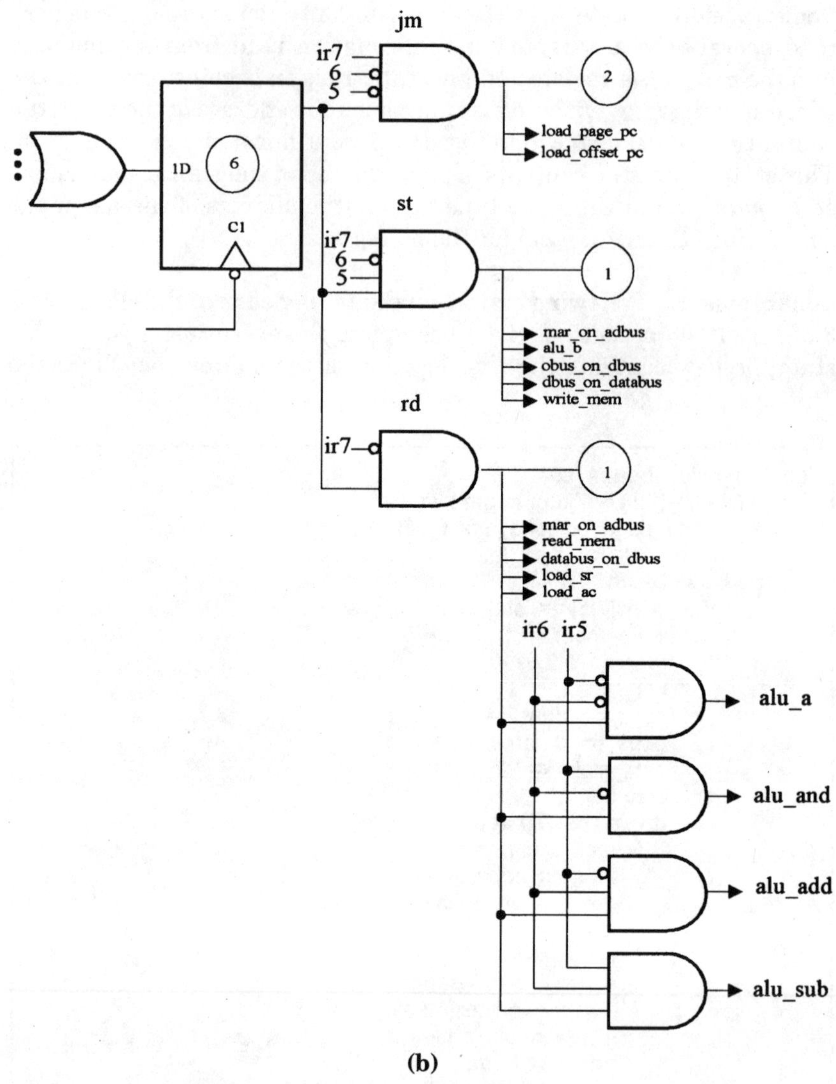

Figure 10.64 ***(Continued)***

actual subroutine code begins in the location after the top of the subroutine, state 8 issues the *increment_pc* signal and activates state 1 for fetching the first instruction of the subroutine.

Control state 9. Control reaches state 9, shown in Fig. 10.67, when state 4 is active and a branch instruction is being executed. When state 9 becomes active, the branch address is in the *mar* register. State 9 loads *mar* into *pc* if a match is found between the branch directive (bits 3 to 0) and the status register bits (*v, c, z,* and *n* flags). If the branch condition is not satisfied, *pc* retains its value—this value points to the memory location that follows the branch instruction. In either case, control returns to state 1 for fetching the next instruction. The discrete logic shown in Fig. 10.67*b* uses AND-OR logic to activate loading of *pc.*

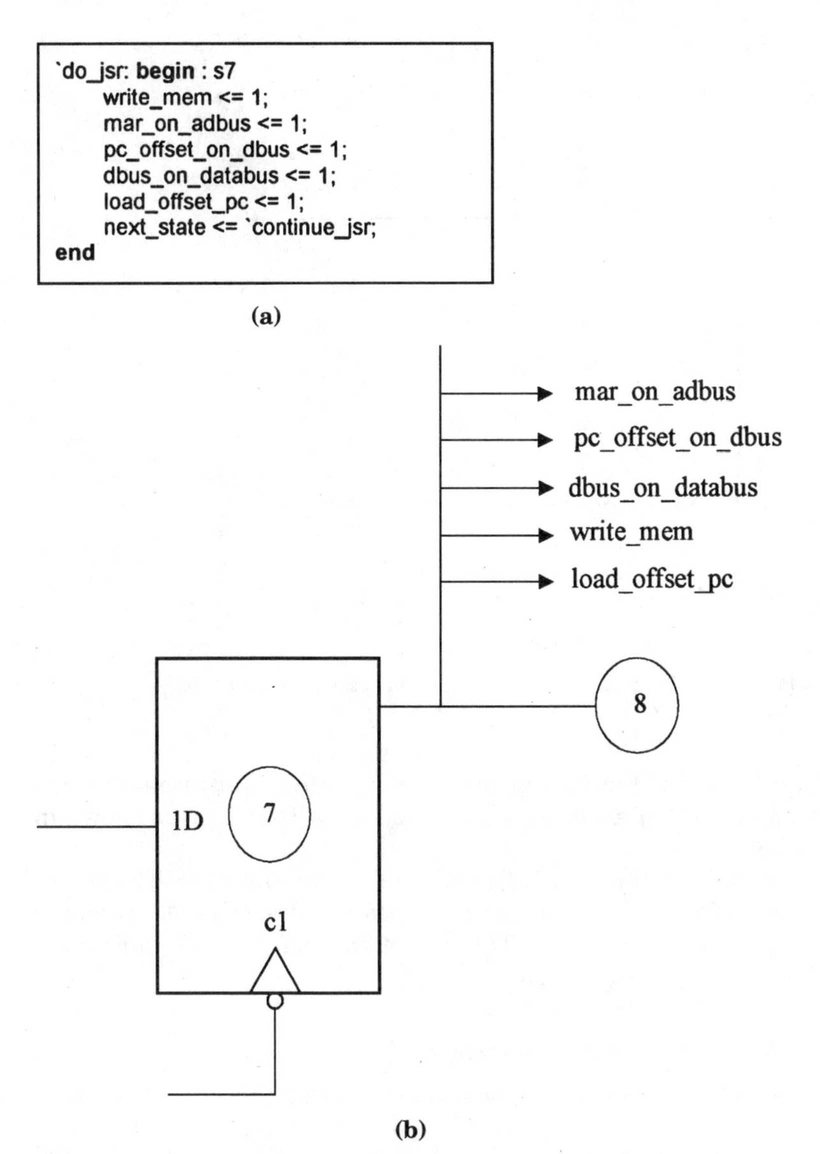

Figure 10.65 State 7: Writing the return address of the subroutine; making *pc* point to the top of the subroutine. (*a*) Verilog code; (*b*) hardware.

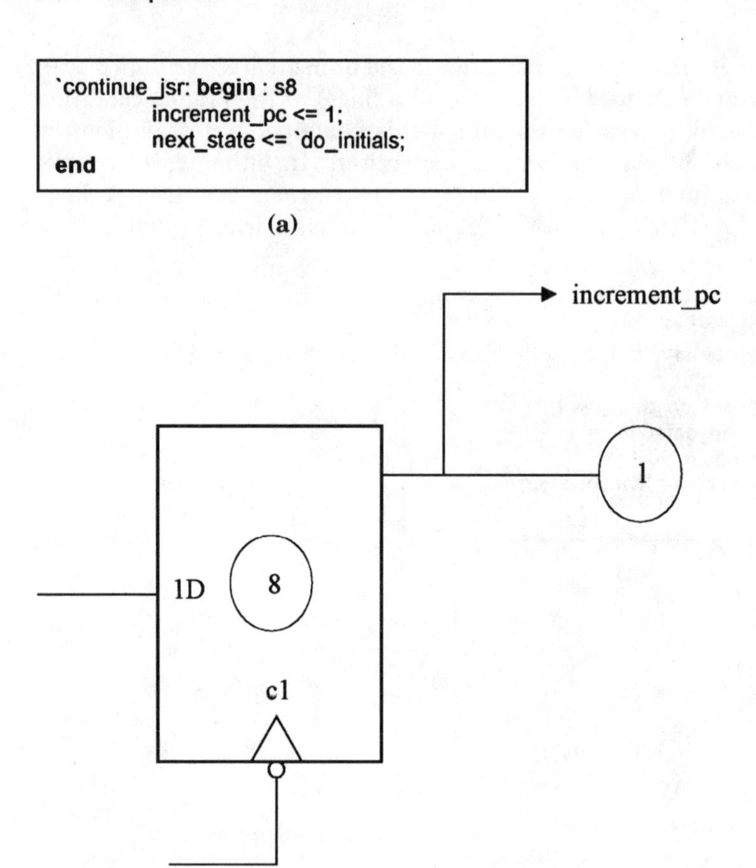

Figure 10.66 State 8: Incrementing *pc* to skip the location reserved for the return address. (*a*) Verilog code; (*b*) hardware.

The last part of the Verilog code of *par_control_unit* resets the controller to state 1 if an invalid code is detected in *ir.* This is shown in Fig. 10.68.

This completes the description of the Parwan controller. The general outline of the circuit corresponding to the descriptions presented above is shown in Fig. 10.69. This circuit diagram uses the same notations we employed in Fig. 10.53.

10.5.7 Wiring data and control sections

The complete dataflow description of the Parwan processor consists of the *par_data_path* of Sec. 10.5.5 and the *par_control_unit* of Sec. 10.5.6. The header part of the Verilog module for describing this CPU is shown in Fig. 10.70.

Figure 10.71 shows the general outline of the *par_central_processing_unit* module. This description specifies interconnections between the *par_data_path* and *par_control_unit* modules. The signal names omitted from this figure are the same as those used in Figs. 10.48 and 10.55 for the data and control sections, respectively.

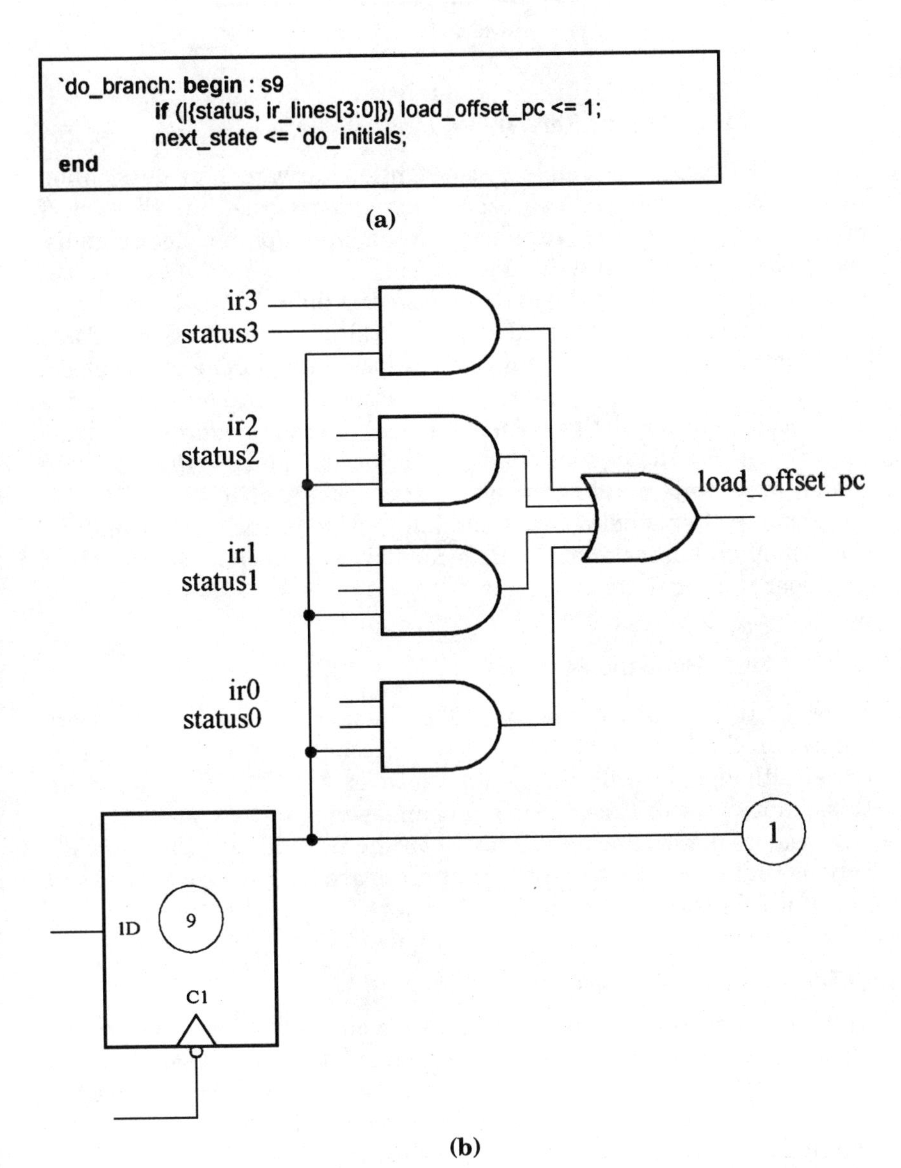

Figure 10.67 State 9: Conditional loading of *pc* for branch instructions. (*a*) Verilog code; (*b*) hardware.

```
        default: begin : invalid_state
                next_state <= `do_initials;
        end
    endcase
end
endmodule
```

Figure 10.68 Code fragment for an invalid state and ending of the controller description.

10.6 A Test Bench for Parwan

Figure 10.72 shows a simple test bench for Parwan. This description instantiates the *par_central_processing_unit* module, initializes part of its memory, generates waveforms on the interrupt and clock signals, and performs read and write operations.

In an **initial** block in the *parwan_tester* module, a 15-byte portion of Parwan memory is initialized with several of its instructions. For a read operation, after a short delay after the falling edge of the clock, *read_mem* is monitored, and if it is asserted, the contents of the addressed location of the memory will be placed on *databus*. These data remain on the bus for only one clock cycle, and *databus* is set to a high-impedance state at the end of the clock. A write operation also takes place after a delay after the falling edge of the clock. Dataflow and behavioral descriptions of Parwan drive the bus for only one clock and float it after that.

10.7 A More Realistic Parwan

In order to emphasize the main CPU issues, several details were ignored in the Parwan design that was presented in the previous sections. Although we will not be able to cover a real CPU structure in this chapter, we will make certain changes to it so that we can use it in a near-realistic board-level design in the next chapter. The modifications that are necessary for this purpose are for memory timing and halt control signals.

10.7.1 CPU control signals

In the previous sections, to focus attention on instruction-level timing, we assumed that the timing of the memory was such that it would respond to CPU read and write requests in one clock cycle. However, in most actual circuits, this is not the case, and some kind of handshaking between the memory and the CPU is necessary. Therefore, the CPU input *ready* is used. This input is issued by the memory and

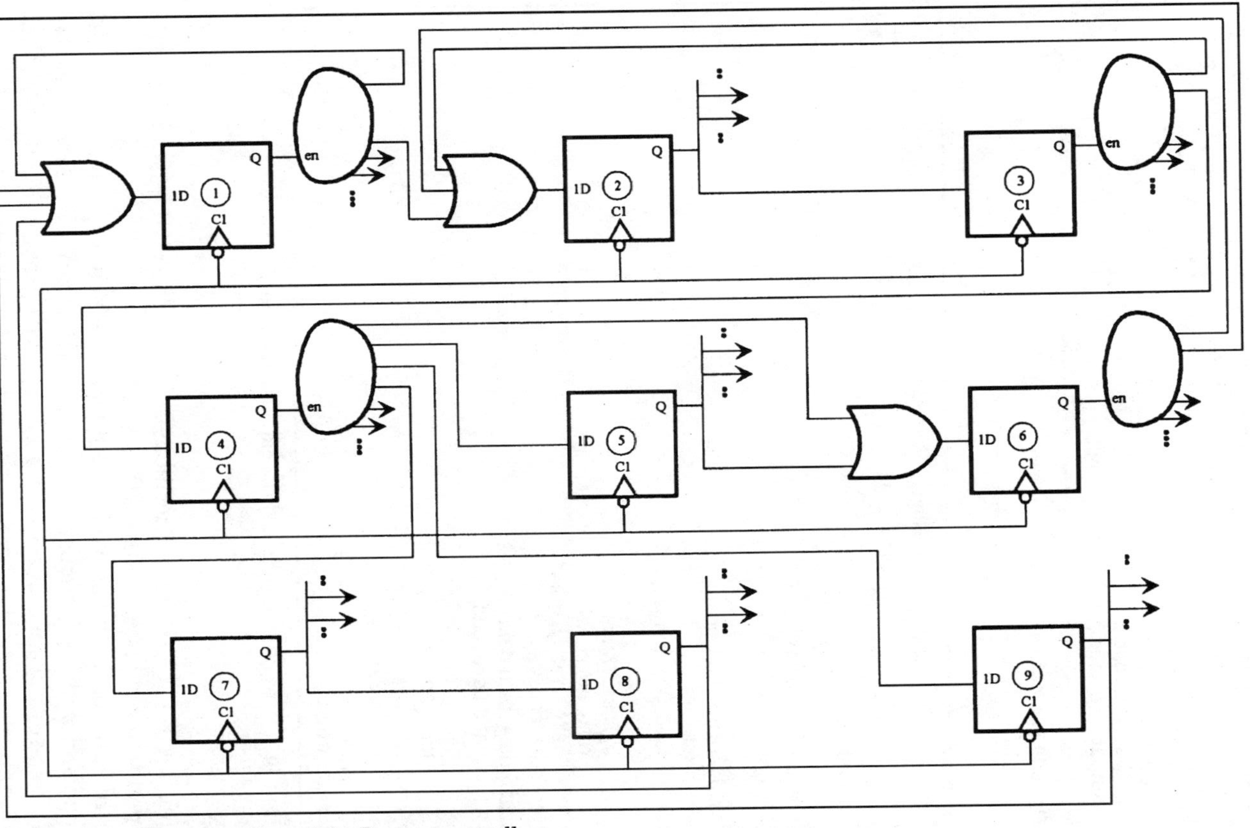

Figure 10.69 General outline of the Parwan controller.

```
module par_central_processing_unit (clk, interrupt, read_mem, write_mem,
                                     databus, adbus, halted, ready, grant);
input clk, interrupt, ready, grant;
inout [7:0]databus;
output [11:0] adbus;
output read_mem, write_mem, halted;
```

Figure 10.70 Module header for Parwan.

```
module par_central_processing_unit (clk, interrupt, read_mem, write_mem,
. . .
wire load_ac, zero_ac;
wire load_ir;
. . .
wire [7:0] ir_lines;
wire [3:0] status;

par_control_unit ctrl_u
      (clk,
       load_ac, zero_ac,
       . . .
       ir_lines, status,
       read_mem, write_mem, interrupt,
       halted, ready, grant);

par_data_path data_u
      (databus, adbus,
       clk,
       load_ac, zero_ac,
       . . .
       ir_lines, status);

endmodule
```

Figure 10.71 The general outline of the dataflow description of Parwan.

informs the CPU when the memory has completed a read or write operation requested by the CPU.

In addition to the speed of the memory, which may slow down CPU read and write operations, another issue that the CPU needs to resolve is the use of system busses. In a system with several I/O devices that may be accessing the memory through system busses, the CPU needs to know if other devices are using system busses before it can transfer its request to the memory. For this purpose, the *grant* input is used to tell the CPU if it has been granted the permission to use the bus. The

```
`timescale 1ns/1ns

module parwan_tester;
reg clk, interrupt, ready, grant;
reg [7:0] mem_databus;
wire [7:0] databus = mem_databus;
wire [11:0] adbus;
wire read_mem, write_mem, halted;
reg r, w;
reg [7:0] memory [0:15];
    par_central_processing_unit m (clk, interrupt, read_mem, write_mem,
       databus, adbus, halted, ready, grant);
    initial begin
        memory[0]  = 8'b00000000; // lda
        memory[1]  = 8'b00001011; // 11
        memory[2]  = 8'b10100000; // sta
        memory[3]  = 8'b00001100; // 12
        memory[4]  = 8'b00100000; // and
        memory[5]  = 8'b00001101; // 13
        memory[6]  = 8'b01000000; // add
        memory[7]  = 8'b00001110; // 14
        memory[8]  = 8'b10100000; // sta
        memory[9]  = 8'b00001111; // 15
        memory[10] = 8'b11100000; // nop
        memory[11] = 8'b01001111; // 11
        memory[12] = 8'b00000000; // 12
        memory[13] = 8'b00000001; // 13
        memory[14] = 8'b00010001; // 14
        memory[15] = 8'b00000000; // 15
        //
        ready = 1; grant = 1; interrupt = 1; clk = 0;
        # 3530 interrupt = 0;
        # 30000 $finish;
    end
    always begin
        # 2 r <= clk;
        # 1 w <= write_mem;
        # 500 clk = ~ clk ;
    end
    always @ (negedge r )
        if (read_mem) begin
            if (adbus >= 12'b000000010000) mem_databus = 8'bzzzzzzzz;
            else mem_databus = memory[adbus];
        end else mem_databus = 8'bzzzzzzzz;
    always @(posedge w)
        if (adbus <= 12'b000000010000) memory[adbus] = databus;
endmodule
```

Figure 10.72 A small Parwan test bench.

sequence of events for a read or write operation with *ready* and *grant* signals is as follows:

1. CPU requests memory operation by issuing *mem_read* or *mem_write*.
2. CPU waits for permission to use system busses (waits for *grant* to become **1**).
3. CPU places its address on *adbus* and, for write, also places data on *databus*.
4. Memory is informed of the validity of its address bus (and databus for a write operation).
5. CPU holds its read or write requests asserted and keeps its busses valid until the *ready* signal is issued.
6. CPU removes its request, and both the memory and the CPU release system busses.

Another control signal of Parwan is the *halted* output. The *hlt* instruction causes the CPU to raise its *halted* output. The system clock is disabled when this output is asserted, and resetting the system is necessary if the CPU is to resume its operations.

10.7.2 Hardware modifications

For the signaling discussed above, changes must be made to the CPU controller. For the read operation, *read_mem* must be issued, *grant* must be waited for, and, after the signal for placing *mar* on *adbus* is issued, it is necessary to wait for *ready* to become **1.** At this time, a memory read will take place. Figure 10.73 shows this signaling for the fetch operation.

A similar handshaking with memory signals should be done for all memory write operations. Figure 10.74 shows partial code of state 6 for writing data into the memory. As shown in this figure, after the CPU is granted the bus access, data on memory busses stay valid until the memory issues a *ready* signal, indicating that a write operation has been completed.

10.8 Summary

This chapter showed how Verilog can be used to describe a system at the behavioral level before the system is even designed and at the dataflow level after major design decisions have been made. The behavioral description aids designers as they verify their understanding of the problem, while the dataflow description can be used to verify the

```
`instr_fetch: begin : s2
     read_mem <= 1;
     if (grant) begin
          mar_on_adbus <= 1;
          if (ready) begin
               databus_on_dbus <= 1;
               alu_a <= 1;
               load_ir <= 1;
               increment_pc <= 1;
               next_state <= `do_one_bytes;
          end else
               next_state <= `instr_fetch;
     end else
               next_state <= `instr_fetch;
end
```

Figure 10.73 Memory and bus signaling for the fetch operation of the controller.

```
`do_two_bytes: begin : s6
     . . .
          if (ir_lines[7:5] == `sta) begin       : st
               write_mem <= 1;
               if (grant) begin
                    mar_on_adbus <= 1;
                    alu_b <= 1;
                    obus_on_dbus <= 1;
                    dbus_on_databus <= 1;
                    if (ready)
                         next_state <= `do_initials;
                    else next_state <= `do_two_bytes;
               end else next_state <= `do_two_bytes;
          end else begin
     . . .
end
```

Figure 10.74 Memory and bus signaling for a memory write operation.

bussing and register structure of the design. A design carried to the stage where a dataflow model can be generated is only a few simple steps away from complete hardware realization. For completing the design of Parwan, flip-flop and gate interconnections should replace the component descriptions in the Parwan dataflow model.

We consider the design presented here a manual design. We used one-to-one hardware correspondence so that no intelligent tools would be required for the generation of hardware. The use of Verilog as a top-down partitioning and verification tool has helped us form such a methodology for manual design. The methodology presented here can be applied to designs of much larger magnitude.

Descriptions in this chapter cover the major language issues discussed in the earlier chapters. A complete understanding of these descriptions requires good comprehension of the Verilog syntax and semantics. Readers who understand all the descriptions in this chapter to the point where they can develop similar models can consider themselves proficient in the Verilog hardware description language.

Further Reading

Armstrong, J. R., *Chip-Level Modeling with VHDL,* Prentice-Hall, Englewood Cliffs, N.J., 1988.

Hill, F. J., and G. R. Peterson, *Digital Systems: Hardware Organization and Design,* 3d ed., John Wiley, New York, 1987.

Problems

10.1 Make a list of eight simple instructions you think would be useful and possible to add to Parwan.

10.2 In Parwan assembly code, write a program to move a block of data that is stored in the memory. The data begin at location 4:00 and end at 4:63, and are to be moved to page 5 starting at 5:64.

10.3 A block of data in the Parwan memory begins at location 1:00 and ends at 1:63. Write a program in Parwan assembly language to find the largest positive number in these locations.

10.4 Show the Verilog description of the Mark-1 machine whose ISPS description appeared in Fig. 1.3.

10.5 Modify the behavioral description of the Parwan controller such that the *jsr* instruction can use indirect addressing.

10.6 Suggest a set of instructions for more complete interrupt handling than is presently available in Parwan.

10.7 Parwan can be modified to use only 3 bits for distinguishing between various nonaddress instructions. We can, therefore, reserve bit 3 of the instruction register as an opcode bit for extending Parwan instructions. In addition, two more nonaddress instructions can be added to the Parwan instructions. Modify the behavioral description of Parwan such that nonaddress instructions use 11100*xxx* opcode. For *xxx,* use 000, 001, 010, 011, 100, and 101 for *nop, cla, cma, cmc, asl,* and *asr,* respectively.

10.8 Use the method suggested in Prob. 10.7 to add an instruction for immediate loading of the accumulator. This instruction can use one of the extra opcodes that become available by modifying opcodes of nonaddress instructions. For *ldi* (load accumulator immediate), you may use the 11101000 opcode. The second byte of an *ldi* contains the byte that is to be loaded into the memory. Modify the behavioral description of Parwan for the execution of this instruction.

10.9 Add a stack pointer to the register and bussing structure of Parwan for the implementation of a software stack. Restrict the stack to the last page of the memory. Use the method of opcode expansion suggested in Prob. 10.7 to make room for a new instruction, and use 11101001 for an *lds* instruction that loads the stack pointer with the data in the next instruction byte. Show all bus connections, registers, and necessary control signals.

10.10 Use the method suggested in Prob. 10.7 and the stack pointer of Prob. 10.9 to add two new nonaddress instructions, *push* and *pop*. Use 11100110 for *push* and 11100111 for *pop*. Modify the behavioral description of Parwan for the execution of these instructions.

10.11 Use the stack implementation in Prob. 10.9 to modify the *jsr* instruction such that it pushes the return address onto the top of the stack. Add a new *rts* instruction that causes a return from the subroutine. For this instruction, use the method of opcode expansion suggested in Prob. 10.7, and use 11101010 for its opcode. Modify the behavioral description of Parwan for the execution of these instructions.

10.12 Modify the dataflow description of Parwan for implementing *jsr* as specified in Prob. 10.5.

10.13 Show the dataflow implementation of the opcode extension scheme suggested in Prob. 10.7. Show all required bus connections, and modify the Parwan controller.

10.14 Modify the dataflow description of Parwan for the implementation of the *ldi* instruction as specified in Prob. 10.8. Show all required bus connections and modify the Parwan controller.

10.15 Modify the Parwan dataflow description for the implementation of *lds, push,* and *pop* instructions (see Probs. 10.9 and 10.10). Show all required bus connections, write a description for the stack pointer (*sp*), and insert this unit in the data path description of Parwan. Also, modify the Parwan controller so that it properly executes these instructions.

10.16 Modify the Parwan dataflow description to implement a version of *jsr* that uses the stack in Prob. 10.9 (see Prob. 10.11). Show all required bus connections and modify the Parwan controller to properly execute this instruction.

10.17 Use the stack in Prob. 10.9 to implement a better interrupt handling for Parwan. The new system should have an *int* input which becomes **1** when an interrupt is requested. The CPU identifies an interrupting device by reading the address of its interrupt service routine from location 0:00 of the memory. To service an interrupt, the CPU jumps to predefined memory locations for each of the interrupt sources. Assume that an external priority logic determines the device with the highest priority and generates its service routine address. Your solution to this problem should also include implementation of a *return_from_interrupt* instruction.

Chapter

11

Interface Design and Modeling

Like the previous chapter, this chapter does not include any new Verilog language concepts. Instead, utilization of Verilog for typical applications of this language is emphasized. Chapter 10 covered application of Verilog for low-level logic and register level design. This chapter emphasizes the use of component models and designing at the board level.

This chapter presents an example board-level design. We will develop a board model consisting of a CPU, memory, cache, and DMA. We will show mechanisms for interfacing various components with different handshaking and timing configurations. Most descriptions in this chapter are at the behavioral level and utilize the Verilog constructs discussed in Chap. 9.

11.1 System Overview

The system we are presenting here consists of an 8-bit CPU with 12 address lines, a 4K memory, a set-associative cache, and an asynchronous serial input device. The serial device is interfaced through a DMA controller.

Figure 11.1 shows system components and the bussing arrangement between these components. A bus arbiter is used to share data and address busses for memory accessing between the CPU and the DMA controller.

The DMA device receives serial data via its asynchronous interface, converts them to 8-bit data, and places the data into the memory. The CPU initiates the transfer of data from the DMA device to the memory.

CPU access to the memory is through a two-way set-associative cache system. The cache intercepts all CPU busses and memory access signals.

In the sections that follow, the components shown in Fig. 11.1 will be described. The CPU and the memory model were described in Chap. 10, and only their I/O signals and busses will be presented here. Behavioral descriptions of all other components of Fig. 11.1 will be given in this chapter.

11.2 CPU Timing

In order to be able to present a complete system with a CPU model and its interfaces, we will use the Parwan model of Chap. 10. It is assumed that a CPU model of Parwan with modifications presented in Sec. 10.7 is available for use in a board-level design in this chapter. Timing signals and bussing of this CPU are presented here independent of the internal hardware of Parwan discussed in Chap. 10.

Figure 11.2 shows the bussing and interface of our example CPU. The 12-bit address bus addresses 4096 bytes of data. Signals shown in the upper right-hand side of this figure are used for handshaking and accessing the memory.

As shown in Fig. 11.3*a*, a memory read request is made by the CPU by asserting the *read_mem* signal. This is done synchronously with the

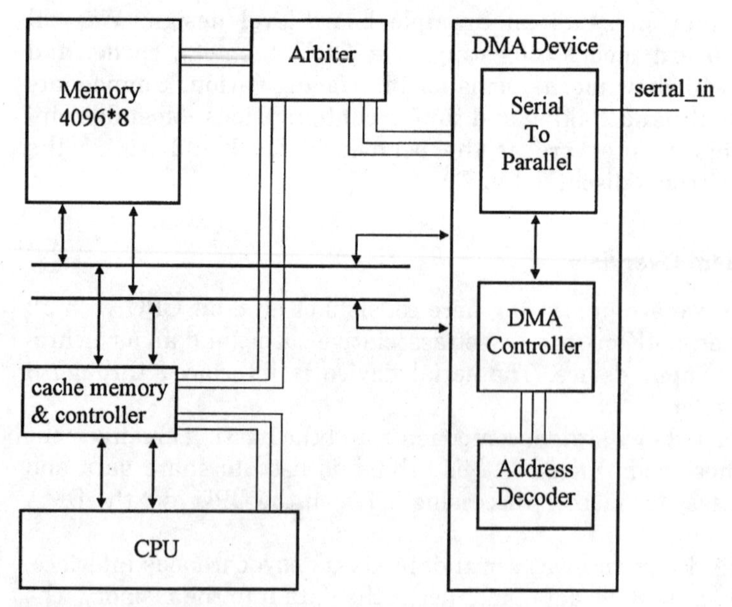

Figure 11.1 Bussing arrangement and system components.

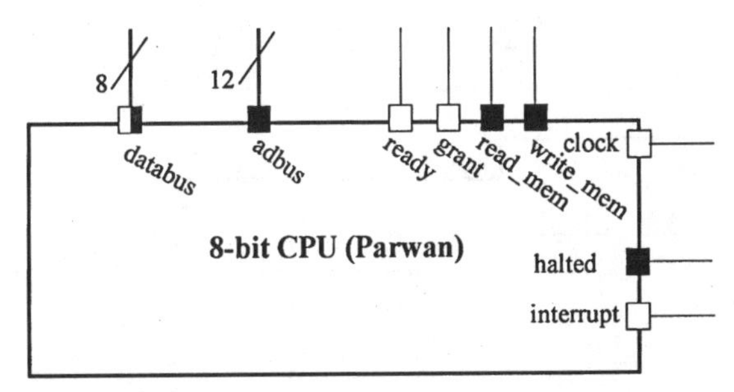

Figure 11.2 CPU interface.

edge of the CPU clock. After asserting this signal, the CPU waits for bus access to be granted to it, which is indicated by the *grant* signal. When this signal is asserted, the CPU places the address of the requested data on its address bus and holds the bus until the memory completes the read operation, which is indicated to the CPU by *ready* being asserted. At this time, the CPU removes its request and frees the address bus. As shown in Fig. 11.3*a,* all CPU handshakings are done synchronously with the falling edge of the clock. Signals driven by the CPU are shown in bold.

The CPU write cycle, shown in Fig. 11.3*b,* is similar to the read cycle except that the CPU drives the *databus* as well as the *adbus.*

11.3 Memory Signals

Ports of the memory used for this interface design are shown in Fig. 11.4. To select the memory, the *cs* signal must be asserted. Memory read or write is indicated by placing a **1** or a **0** on the *rwbar* when the memory is selected. After a delay that depends on the speed of the memory, a read or write operation will be completed. During a read operation, the *databus* is driven by the memory while *cs* is asserted. When writing into the memory, address and data busses must remain valid as long as the *cs* signal is asserted. Figure 11.5 shows the timing of a memory read operation.

11.4 Sharing System Busses

The CPU and the DMA device both require access to the memory through system busses. For the control of simultaneous bus requests, an arbiter is used. All requests for use of data and address busses must go through this component. Bus requests from various devices are granted based on their priorities.

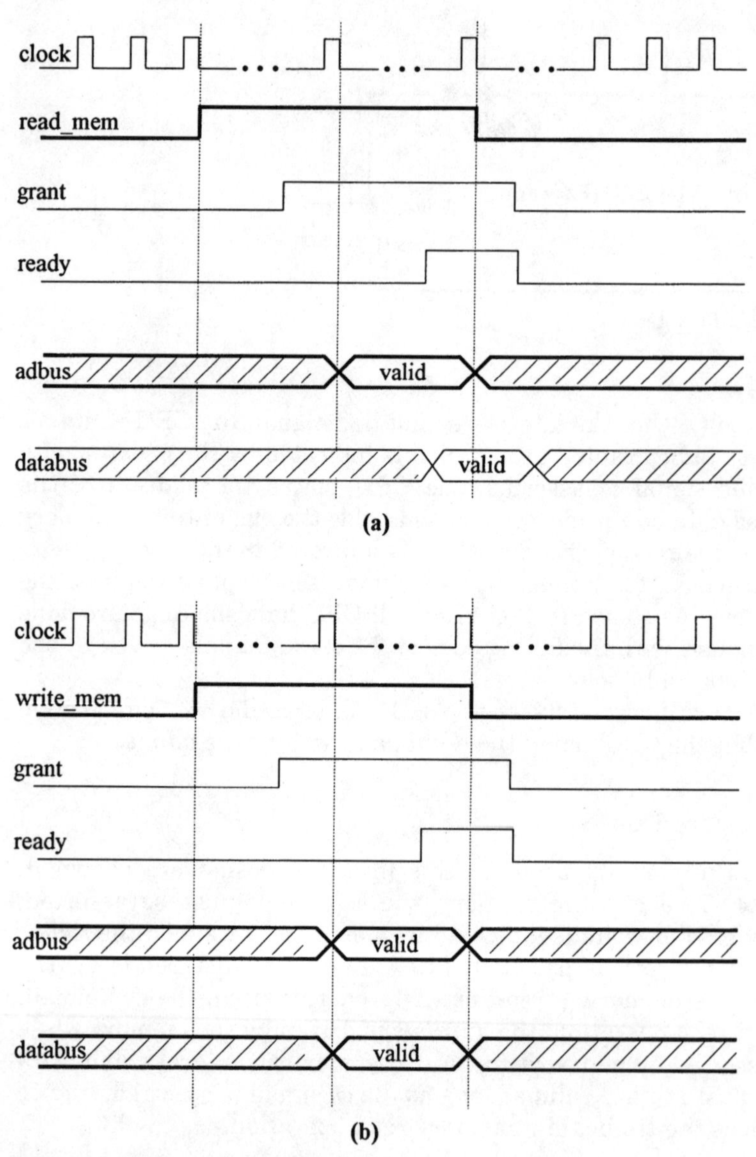

Figure 11.3 CPU read and write requests.

11.4.1 Arbitration operation

Figure 11.6 shows the block diagram of the bus arbiter used in our interface design. A device that requires bus access is assigned to one of the ports of this device. Port 3 has the highest priority, and port 0 has the lowest. Therefore, simultaneous bus requests from a device connected to port 0 will be granted only if no other device is requesting access to system busses.

Devices requesting bus access will raise their *read_request* or *write_request* lines. A request line stays asserted until the requesting device has been granted the bus and has completed its operation.

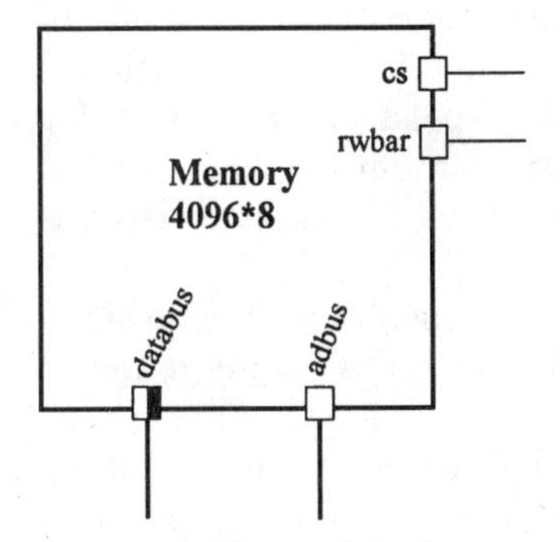

Figure 11.4 Memory interface.

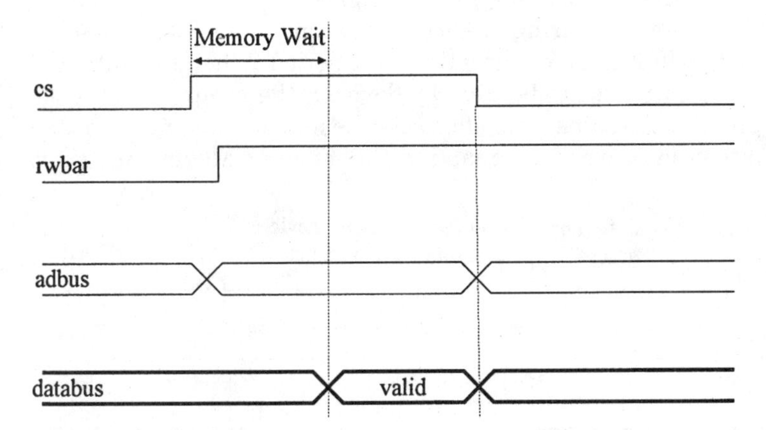

Figure 11.5 Memory read operation.

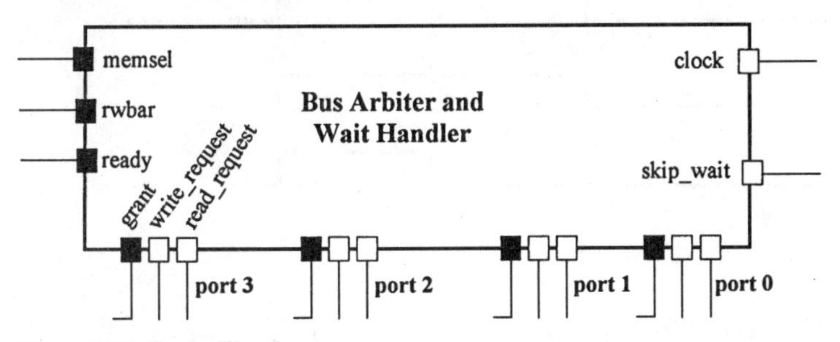

Figure 11.6 Controlling bus access.

The basic bus arbitration operation of this device is similar to that presented in Sec. 9.3. However, this component also handles wait states, which will be described next.

11.4.2 Wait operation

A device that is granted system busses requires access to these busses until its operation is completed. The time needed for the completion of such an operation depends on the relative speed of the requesting device and the memory. The amount of time required for a read or write operation after a device is granted access to system busses is referred to as the "wait state."

In our interface design, the component responsible for bus arbitration also handles wait states. Figure 11.7 shows the timing diagram for a read request being granted.

All operations of the bus arbiter are synchronized with the falling edge of its input clock. A device requesting a memory read operation asserts its need request line (shown as $read\_request_i$ in Fig. 11.7). The arbiter holds off on responding to this request until it is sure that all pending operations requiring bus access are completed. The first set of waveform dots in Fig. 11.7 shows this wait period. When the data and address busses are free to be used by device *i*, the arbiter asserts the $grant_i$ signal. At the same time, it selects the memory and provides the appropriate read or write command to the memory. Meanwhile, since

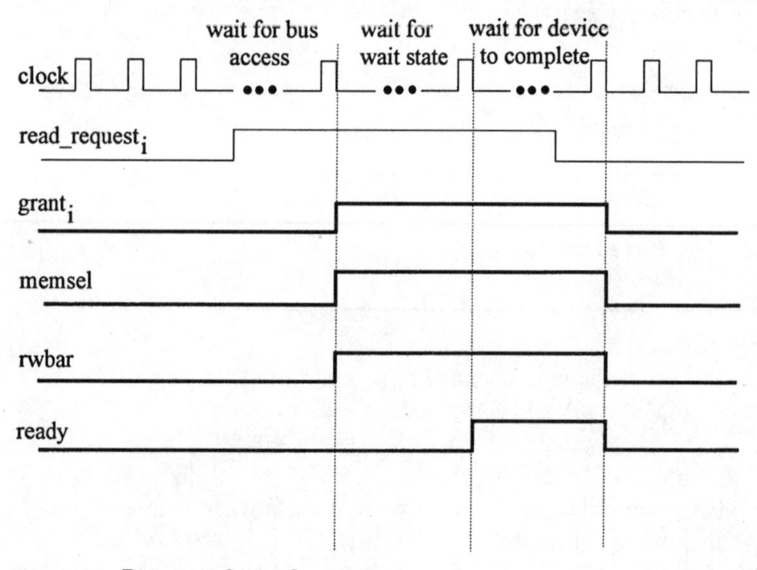

Figure 11.7 Bus grant for read operation.

grant$_i$ is asserted, device *i* will place its data or address on system busses (see, for example, Fig. 11.3*a* for a device performing a memory read operation).

When the memory is selected and is busy performing its specified operation, the arbiter of Fig. 11.6 starts counting the wait states. The *ready* signal will be issued to the requesting device to inform it of the completion of the operation after an appropriate number of wait states has elapsed. This wait period is shown by the second set of waveform dots in Fig. 11.7.

After a requesting device is informed of the completion of the bus operation, it removes its bus request by deasserting its request line. In the read cycle of Fig. 11.7, this is shown by *read_request*$_i$ going back to **0.** After the completion of an operation, the arbiter deasserts its bus grant to device *i* and removes the select memory line.

In addition to the signaling described above, the arbiter of Fig. 11.6 has a *skip_wait* input signal. When this is asserted while the arbiter is counting the wait states, it skips all remaining waiting clock periods and issues a *ready* signal on the next clock. This provides a mechanism for accessing fast I/O ports. We will use this feature in writing to DMA command registers.

11.4.3 Arbiter model

The Verilog description for a bus arbiter that satisfies the requirements stated above is shown in Fig. 11.8. Four *wait_state* parameters have been defined and initialized to **1.** The **integer** array *wait_states* is then set to these four parameters. For each requested device, the amount of waiting is looked up from this array. Also, all request and grant lines are 4-bit vectors, as shown in Fig 11.6. This enables our bus arbiter to handle four device requests for the memory.

An **always** block in Fig. 11.8 performs arbitration and handles wait states. After the falling edge of the *clock* signal, a loop searches for a read or write request. This search is done from bit 3 down to 0, giving highest priority to a device connected to port 3 of the arbiter. If a request is found, the corresponding grant signal, as well as memory select and read/write signals, will be asserted.

After asserting these signals, the process statement of Fig. 11.8 enters a loop for waiting for the appropriate number of wait states. As discussed before, assertion of the *skip_wait* signal must be able to cancel the waiting. Therefore, the loop performing the waiting executes a delay control statement one clock period at a time. The delay control statement is executed only if the *skip_wait* signal is not active. If this signal becomes **1** while waiting, the wait for the remaining clock periods will

```
`timescale 1ns/100ps

module arbitrator (read_request, write_request, grant,
     clock, skip_wait, memsel, rwbar, ready);

parameter wait_state1 = 1, wait_state2 = 1, wait_state3 = 1, wait_state4 = 1;
parameter clock_period = 1000;

input [3:0] read_request, write_request;
output [3:0] grant;
input clock, skip_wait;
output memsel, rwbar, ready;

integer wait_states [1:4];
initial begin
     wait_states [1] = wait_state1;
     wait_states [2] = wait_state2;
     wait_states [3] = wait_state3;
     wait_states [4] = wait_state4;
end
reg [3:0] grant;
reg memsel, rwbar, ready;
integer granted, i, j;

always @(negedge clock) begin : wait_cyclc
     granted = 0;
     #20;
     for (i = 0; i <= 3; i = i+1)  begin
          if (read_request[i] == 1 || write_request[i] == 1 && granted == 0)
          begin
               grant = 0; grant[i] = 1;
               memsel = 1;
               rwbar <= read_request[i];
               ready = 0;
               if (wait_states[i] != 0)
                    for (j = 1; j <= wait_states[i]; j = j+1)
                         if (skip_wait != 1) #clock_period;
               ready = 1;
               granted = 1;
          end else begin
               grant[i] = 0;
               memsel = 0;
          end
     end
end
endmodule
```

Figure 11.8 Arbiter Verilog code.

be skipped. After the wait loop has terminated, the *ready* signal is issued to inform the requesting device that the memory is ready and the read or write operation is completed.

11.5 DMA Device

The interfacing example in this chapter includes a serial device that is interfaced to our CPU system through a direct memory access channel. This interface, consisting of the serial interface, DMA controller, and a decoding logic, will be described in this section.

11.5.1 Serial connection

The serial-to-parallel adapter of Sec. 9.3 is to be interfaced to our CPU through a DMA. The Verilog code for *s2p* is shown in Fig. 9.50. This device (shown in Fig. 11.9) receives synchronous serial data, converts them to 8-bit parallel data, and makes them available on *parallel_out* outputs. Availability of parallel data is indicated by a **1** on *dataready.* This output stays asserted until the receiving device receives parallel data and asserts the *received* input of *s2p.* Two error indicator outputs of *s2p* flag overrun and framing errors in receiving serial data.

11.5.2 Interface through arbiter

The interface of the DMA controller is shown in Fig. 11.10. Ports of this component, shown on top of this figure, are device data and handshaking signals that connect to the corresponding *s2p* ports.

The DMA controller receives a byte of data from *s2p,* asks the bus arbiter of Sec. 11.4 for permission to use the memory, and, when permission is granted, completes its write operation. Prior to the transfer of data, the CPU programs the DMA controller for the memory location to write to.

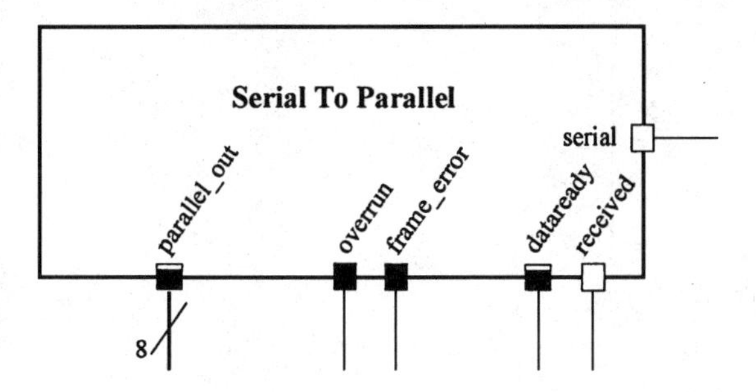

Figure 11.9 Interface of serial-to-parallel converter.

Interface of the DMA controller to the memory through the arbiter is similar to that of the CPU, as described in Sec. 11.2. The timing diagram of Fig. 11.3*b* applies to the DMA controller as well as the CPU. In our design, the arbiter port that the CPU is connected to has a higher priority than the port assigned to the DMA device. Busses shown on the left-hand side of the box in Fig. 11.10 connect to system data and address busses, and the handshaking signals shown here connect to port 0 of the arbiter.

11.5.3 Interface to CPU

Before the DMA controller can transfer data received from *s2p* to the memory, it has to be programmed by the CPU. The controller has four registers for communicating with the CPU. To initiate the transfer of a block of bytes from *s2p* to the memory, the CPU places the starting address of the block and the number of bytes in the block in the DMA control registers. The DMA controller will start moving data to the memory when it receives a start command from the CPU. As shown in Fig. 11.11, three 8-bit registers in the DMA controller are dedicated to the block starting address and the number of bytes in the block. A fourth register is used for commands from thc CPU to the DMA controller and for status reports back to the CPU. All these

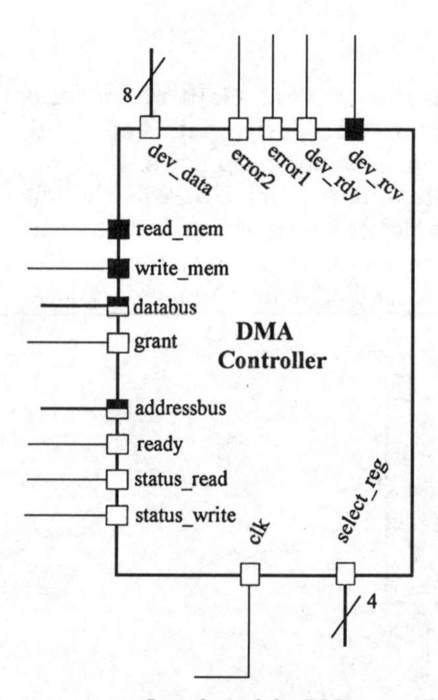

Figure 11.10 Interface of the DMA controller.

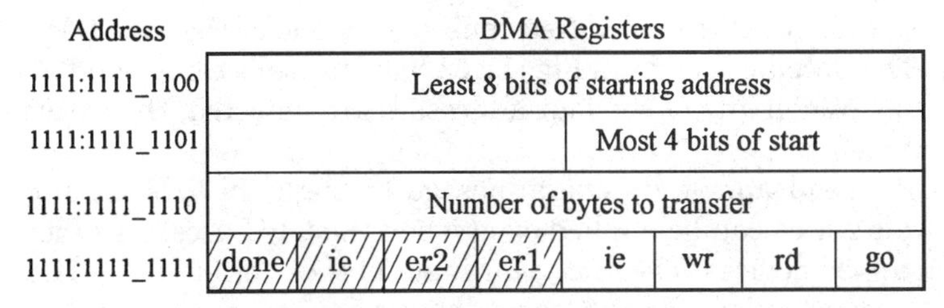

Figure 11.11 DMA registers.

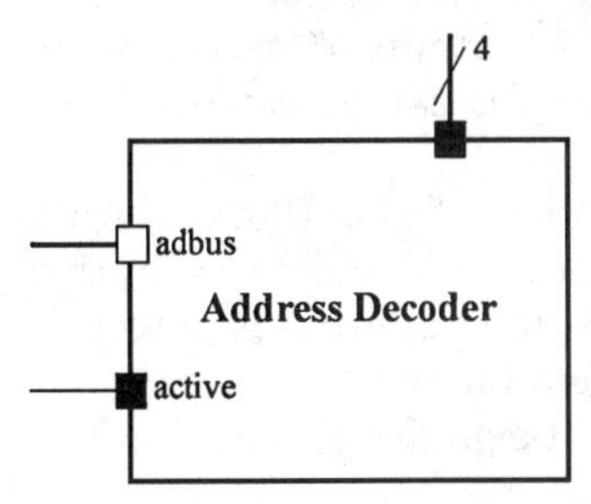

Figure 11.12 Decoder for selecting DMA registers.

registers are directly accessible by the CPU. Those bits that are read by the CPU are shaded in Fig. 11.11. The addresses shown in this figure correspond to the addressing schemes of Parwan as discussed in Chap. 10.

Store instructions from the CPU load the starting address in the registers that appear in the first and second DMA address locations, e.g., 1111:1111_1100 and 1111:1111_1101. This is followed by another store instruction for loading the number of bytes in location 1111:1111_1110 (the third DMA register). When the CPU is ready to start the transfer of data, it writes a byte to location 1111:1111_1111 to set the *go* and *wr* flags of the fourth DMA register to **1.** The *ie* flag may be set by the CPU to enable an interrupt when DMA completes its transfer. As soon as the *go* flag is set, the DMA controller begins to transfer data. When done, the controller writes its error status to the status register and sets the *done* flag to **1.** Like the command register, this register also appears at address 1111:1111_1111 in the CPU address space. The most-significant bits of the register in this location are used for status flags. The status interrupt flag may be used to interrupt the CPU if the *ie* flag of the command register is set. Otherwise, the CPU can monitor the status register to find out if the transfer of data has been completed. The status bits may be read by the CPU with a load instruction addressing this register.

A decoder connected to the address bus selects one of the four DMA registers. The decoder shown in Fig. 11.12 has four select lines and an *active* output. When any of the four addresses are detected, the active output becomes **1**.

Since CPU read and write operations go through the bus arbiter, memory wait states can be applied even when the CPU accesses other memory-mapped devices. The *skip_wait* input of the arbiter of Fig. 11.8 provides the mechanism for bypassing this unnecessary waiting. The *active* output of the decoder of Fig. 11.12 connects to the *skip_wait* output of the arbiter. When the DMA controller is addressed by the CPU, assertion of *active* causes the *grant* and *ready* outputs of the arbiter to be issued at the same time. Therefore, the CPU can access the DMA controller for a read or write in one cycle.

Figure 11.13 shows the Verilog description of the DMA address decoder. A 12-bit mask (*addresses*) masks out invalid addresses for this decoder and sets the *active* output to **1** if a valid address is detected. In an **always** statement, a **1** is assigned to a bit of the *selects* output of the decoder that is pointed to by the 2 least-significant bits of *adbus*.

11.5.4 DMA controller

Figure 11.14 shows the complete DMA device, consisting of the *s2p* adapter, an address decoder, and the controller. The functionality of the individual components of this figure and Verilog descriptions for the decoder and *s2p* have been presented. This section presents the Verilog description of the controller.

Figure 11.15 shows the module header and I/O declarations of the DMA controller. The inputs and outputs shown in this figure correspond

```
`timescale 1ns/100ps

module quad_adrdcd (adbus, active, selects);
input [11:0] adbus;
output active;
output [3:0] selects;
reg [3:0] selects;
parameter addresses = 12'b111111111100;
    assign active = (adbus & addresses) == addresses;
    initial selects = 0;
    always @(adbus) begin
        if ((adbus[1:0] >= 0) && (adbus[1:0] <= 3)) selects[adbus[1:0]] = 1;
        else selects = 0;
    end
endmodule
```

Figure 11.13 Verilog description of DMA register address decoder.

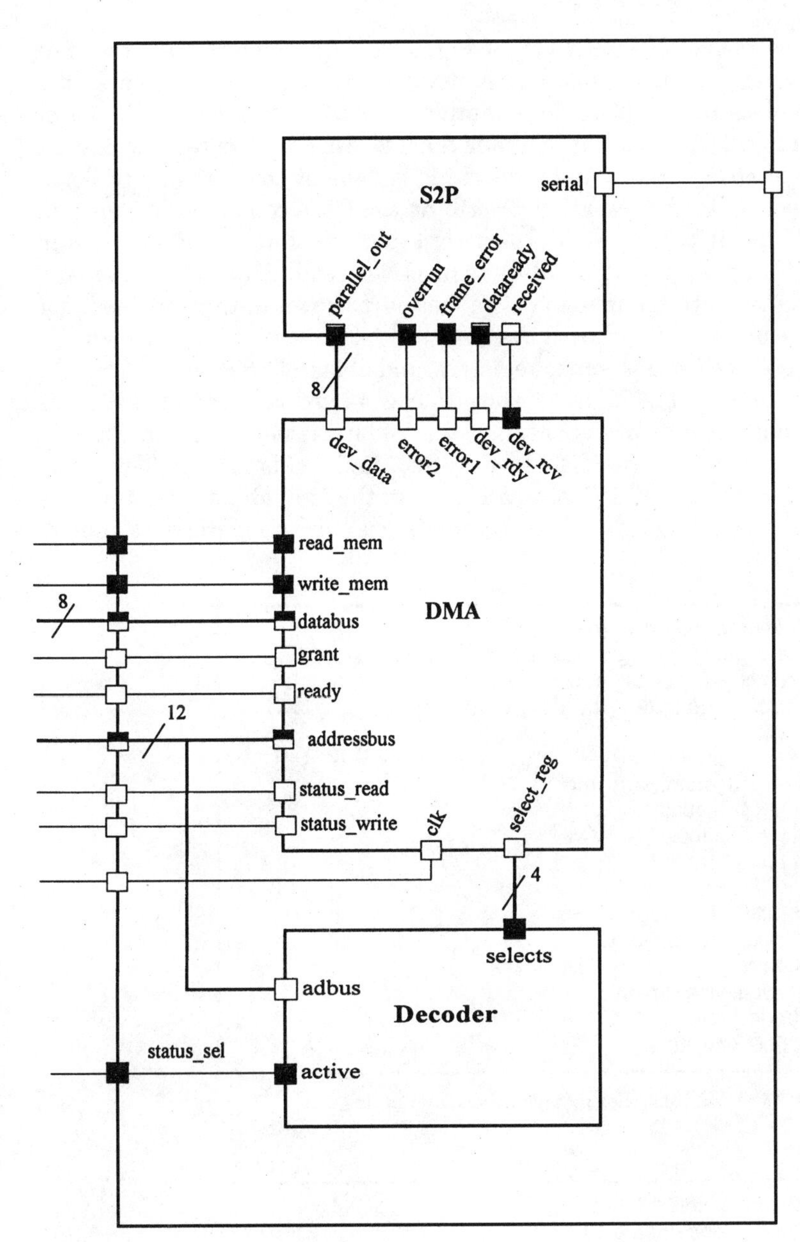

Figure 11.14 DMA device.

to those inside the DMA box of Fig. 11.14. Although the address bus is used only as input, for consistency among various components, it is declared as **inout.** Note, for example, that the DMA device (the outer box in Fig. 11.14) reads and writes the address bus. The reading is done by the decoder circuit, and the writing is done by the DMA controller.

The outline of the module describing the DMA controller is shown in Fig. 11.16. It begins with a part that declares internal registers and flags. The next part is an **always** block for reading data from *s2p* and writing them to the memory. The last part shown in this outline is for direct communication with the CPU.

In the declaration part, registers and flags of the *dma_controller,* shown in Fig. 11.17, are declared. The 4-byte register file includes DMA flags, a memory counter, and a 12-bit memory pointer. The use of these registers was discussed in Sec. 11.5.3 (Fig. 11.11). The register file, *rfile,* is initially set to all zeros in the **initial** block that follows it. Command flags *go, rd* (read), *wr* (write), and *ie* (interrupt enable)

```
module dma_controller (clk,
      read_mem, write_mem, databus, adbus, ready, grant,
      select_reg, status_rd, status_wr,
      error1, error2, dev_rdy, dev_rcv, dev_data);
input clk;
         // memory signals
output read_mem, write_mem;
inout [7:0] databus;
inout [11:0] adbus;
input ready, grant;
         // cpu signals
input [3:0] select_reg;
input status_rd, status_wr;
input error1, error2, dev_rdy;
         // device signals
output dev_rcv;
input [7:0] dev_data;
```

Figure 11.15 DMA controller module header and ports.

```
module dma_controller (. . .);
    Declarations (Figure 11.17)
//
    "get serial, put parallel" always statement (Figure 11.18)
//
    "direct CPU communications" blocks (Figure 11.19)
endmodule
```

Figure 11.16 Outline of DMA controller architecture.

```
//   Declarations (Figure 11.17)

reg dev_rcv, write_mem;

reg done; initial done = 0;
reg [7:0] rfile [0:3];
initial {rfile[3], rfile[2], rfile[1], rfile[0]} = 0;
wire [7:0] flags;
wire ie, wr, rd, go;
assign flags = rfile [3];
assign {ie, wr, rd, go} = flags [3:0];
```

Figure 11.17 DMA controller declarations.

are declared wires and are assigned to the least-significant bits of register 3 of *rfile*. Because *rfile* is a declared memory of 4 bytes, individual bits cannot be accessed. This is the reason that byte 3 of *rfile* is assigned to an 8-bit *flags* before its least-significant bits are accessed for the *ie, wr, rd,* and *go* flags.

Figure 11.18 shows the *get_put* sequential block, the flow into which is controlled by the positive edge of the *go* flag. When the *go* command is issued by the CPU, the process of data transfer begins. The *go* command is issued when the CPU writes a **1** to bit 0 of register 3 of the *rfile* register file. This is handled by the last part of the controller code, which is dedicated to direct communication between the CPU and the DMA controller. The *get_put* block and the blocks that handle direct CPU communication are concurrent.

To start the transfer of data, the number of bytes to transfer (*numb*) and the beginning pointer to the memory (*pntr*) will be looked up from the *rfile* register files. The 12 bits of the pointer are distributed in 2 bytes of the *rfile* memory. To read the most-significant 4 bits of the pointer from the memory file, the variable *tmp* is used to store byte 1 of this memory. Part-select on *tmp* concatenated with byte 0 of *rfile* constitutes a 12-bit address that is assigned to the *pntr* register.

For writing received data bytes into the memory, a while loop looping *numb* number of times receives a byte from *s2p* and puts it into the memory. In the part of this code indicated by "get data," the *dev_rdy* signal is waited for, and when this signal is asserted, a byte of data from the *s2p* output, *dev_data,* is put into the DMA controller internal buffer (*buff*). After a data byte is received, a synchronous positive pulse on *dev_rcv* provides the necessary handshaking to free the *s2p*. Two edge-triggered event control statements generate this positive pulse synchronous with *clk*.

In the second part of the *writing* while loop in the *get_put* sequential block of the DMA controller module, sending received data to the memory is done. The portion of the code beginning after the "put to mem" comment in Fig. 11.18 is responsible for this transfer. To transfer a byte to the memory, *write_mem* is issued to get permission from the arbiter to use system busses. When this permission is granted [@(**posedge** *grant)*], the byte received from the serial device (*buff*) will be **force**d onto the bidirectional data bus, and the memory location to write to (*pntr*) is **force**d onto the address bus. The variables *buff* and *pntr* will continue to drive *databus* and *adbus* until *ready* is issued. After the positive edge of *ready*, *databus* and *adbus* are **release**d.

As shown in Fig. 11.16, the last part of the Verilog description of the DMA controller is for communication with the CPU. Figure 11.19 shows this part of the code. Two **always** blocks respond to CPU read and write requests, and a third block loads byte 3 of *rfile* with error and status flags generated during the serial data reception.

The **always** block that responds to CPU read requests has a sequential block labeled *cpu_direct_read.* Flow into this block begins when a change occurs on the *select_reg* or *status_rd* input of the DMA controller. The

```
reg [7:0] buff;
integer numb;
reg [11:0] pntr; reg [7:0] tmp;
always @(posedge go) begin : get_put
    done = 0;
    numb = rfile [2];
    tmp = rfile [1];  pntr = {tmp [3:0], rfile [0]};
    if (wr == 1) begin
        while (numb) begin : writing
            numb = numb - 1;
            // get data
            wait (dev_rdy);
            buff = dev_data;
            @(posedge clk) dev_rcv = 1;
            @(negedge clk) dev_rcv = 0;
            // put to mem
            write_mem = 1;
            @(posedge grant);
            force databus = buff;
            force adbus = pntr;
            pntr = pntr + 1;
            @(posedge ready);
            release databus;
            release adbus;
            write_mem = 0;
        end
        done = 1;
    end
end
```

Figure 11.18 DMA controller "get data" and "put to mem" processes.

```
        integer jj;
        always @(select_reg or status_rd) begin : cpu_direct_read
            for (jj = 0; jj <= 3; jj = jj + 1)
                if (select_reg [jj] == 1 && status_rd == 1)
                    force databus = rfile [jj];
                else
                    release databus;
        end
        always @(negedge clk) begin : cpu_direct_write
            if (status_wr == 1) for (jj = 0; jj <= 3; jj = jj + 1)
                if (select_reg [jj] == 1) rfile [jj] = databus;
        end
        always @(negedge clk) begin : set_flags
            if (done == 1) rfile [3] = {1'b1, ie, error2, error1, flags[3:0]};
        end

endmodule
```

Figure 11.19 DMA controller "direct CPU communications" blocks.

4-bit *select_reg* input selects the *rfile* register to write to and is the output of the decoder circuit (Fig. 11.14) decoding DMA register addresses. The *cpu_direct_read* block uses the **force** statement to force a selected *rfile* register onto the *databus* when the *status_rd* read signal is active and the register is selected. When the read signal is deasserted or a selected register of *rfile* is deselected, a **release** statement is executed that releases (or frees) the *databus* of the DMA controller.

The second **always** block controls flow into the *cpu_direct_write* sequential block. This block is entered on the falling edge of the DMA *clk* signal. When it is entered, if writing is to take place—i.e., *status_wr* is **1** and a specific register of *rfile* is selected—*databus* will be written to the selected *rfile* register.

The Verilog description of *dma_serial_device,* consisting of the wiring of the DMA controller discussed above, a serial-to-parallel adapter, and the decoder of Sec. 11.5.3, is shown in Fig. 11.20. This description corresponds to the diagram of Fig. 11.14. Figure 11.1 shows how this unit is wired to other components of our interface design example. Details of this wiring will be discussed later in this chapter.

11.6 CPU Cache

As shown in Fig. 11.1, another component included in our interface design example is a CPU cache. The cache intercepts all CPU read and write operations, and if addressed data are found in the cache, they will be made available to the CPU. This eliminates lengthy memory wait states and speeds up CPU program execution.

```
module dma_serial_device (clk,
      read_mem, write_mem, databus, adbus, ready, grant,
      status_rd, status_wr,
      status_sel, serial_in);
input clk;
output read_mem, write_mem;
inout [7:0] databus;
inout [11:0] adbus;
input ready, grant;
input status_rd, status_wr;
output status_sel, serial_in;

wire s2p_er1, s2p_er2, s2p_rdy, s2p_rcv;
wire [7:0] s2p_par;
wire [7:0] cpu_mem_data;
wire [11:0] cpu_mem_addr;
wire [3:0] select_reg;

dma_controller dma (clk,
      read_mem, write_mem, databus, adbus, ready, grant,
      select_reg, status_rd, status_wr,
      s2p_er1, s2p_er2, s2p_rdy, s2p_rcv, s2p_par);

quad_adrdcd dcd (adbus, status_sel, select_reg);

serial2parallel s2p (serial_in,
      s2p_rcv, s2p_rdy, s2p_er1, s2p_er2, s2p_par);
endmodule
```

Figure 11.20 DMA serial device; description for diagram of Fig. 11.14.

11.6.1 Cache structure

Many cache schemes and methodologies exist. For this design example, we will implement a set-associative, two-way LRU cache with 32 cache lines, line size of 1 byte, and a write-through write policy. Figure 11.21 shows a block diagram of the cache design of this chapter. The working of this methodology will be discussed here for a CPU read operation.

When a read is to be performed, the CPU asserts the *read_mem* signal. Upon receiving *read_mem,* the cache unit issues *grant* so that the CPU places its address on the address bus. For the Parwan address bus, the seven most-significant address lines are treated as tag values and the 5 least-significant bits are the set value. The set field of an address points to a valid bit, a tag, and a cache line in each cache way. When a read operation is performed, a tag value addressed by the 5 least-significant address bits will be read from each cache way (one from way 0 and one from way 1). These two tag values will be compared against the seven most-significant lines of the address bus. If a match is found, a cache hit has occurred. In this case, the data in the cache way where the addressed tag matches the tag address lines are the data designated by the CPU and will be placed on the data bus. To inform the CPU of the availability of data, the cache controller issues *ready* to the CPU.

When a line of data is found in a cache way, it becomes the most recently used line in the cache. The line at the same location in the cache way where a line was not found becomes the least recently used line at this location. In a later read or write operation, if a cache miss occurs at this same cache location, the tag and line in the cache way

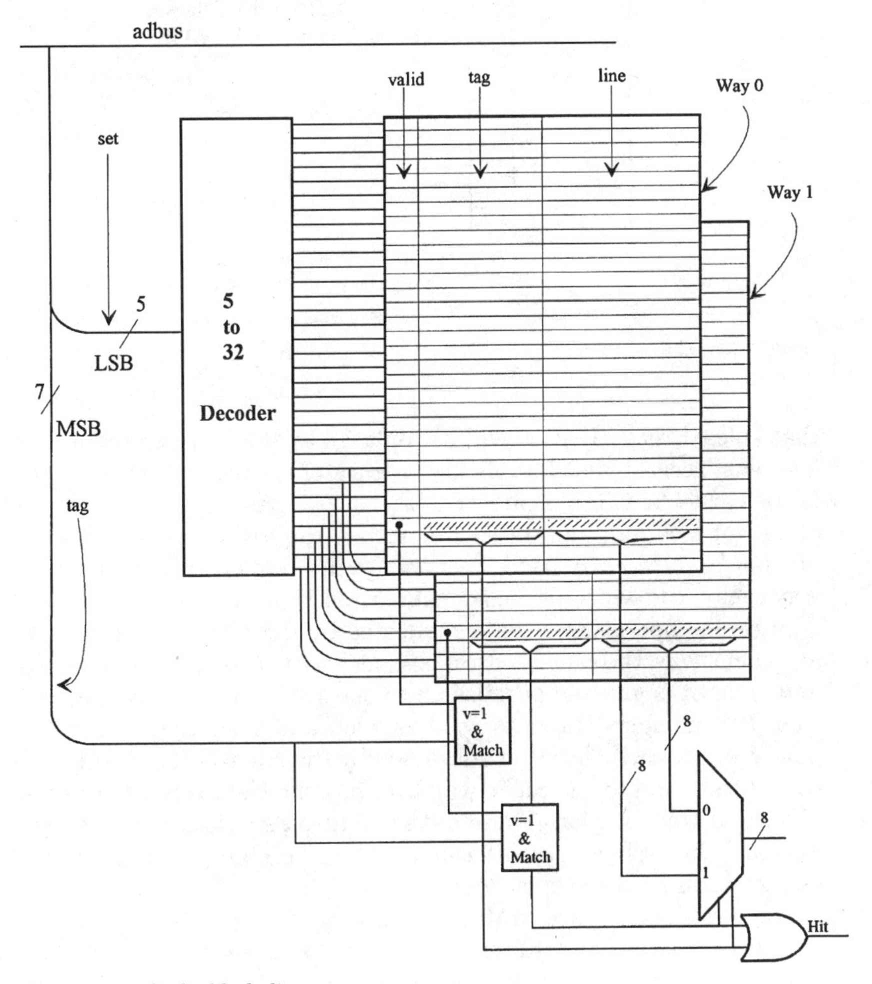

Figure 11.21 Cache block diagram.

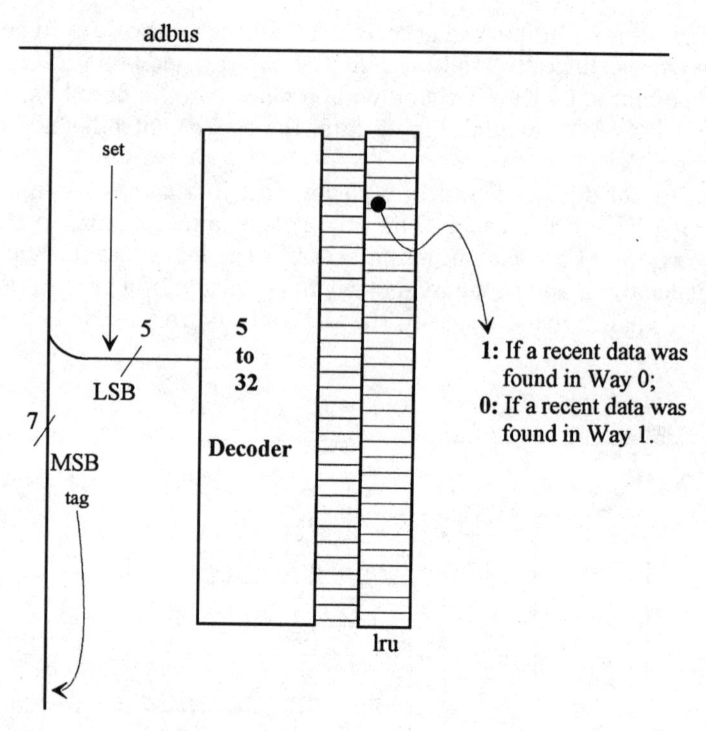

Figure 11.22 The *lru* table.

that is least recently used will be replaced by the new tag and line. An *lru* (least recently used) table keeps track of the usage of each location of the cache. This table, shown in Fig. 11.22, has one location for each cache set location, and its entries reflect the cache way number that has not been recently used. Because we are implementing only a two-way cache, the width of the *lru* table need be only 1 bit.

In a read operation, if data requested by the CPU are not found in any cache way, the cache controller issues a *read* request to the arbiter, and when it is granted permission to use system busses, it reads data from the memory. The data read as such and their tag value will be placed in the cache location addressed by the 5 least-significant bits of the address and in the cache way that has the least recently used line of data at this location. These data will also be put on the CPU *databus,* and the CPU will be informed of the availability of data by *ready* being issued by the cache controller.

The write policy used in this design is write-through. In this policy, in a write operation, data will be written to the memory as well as to the cache. Data written into the cache during a write operation will replace least recently used data, and these data will be marked as most recently used in the *lru* table. Obviously, CPU write operations are not affected by cache hardware using this write policy.

In the sections that follow, we will present the Verilog code for the cache presented here. Details of the cache controller and the interface of this unit with the CPU and the memory will be illustrated.

11.6.2 Cache interface

The interface aspect of the cache unit is shown in Fig. 11.23. This unit intercepts all CPU signals and busses that are used for memory read and write operations. To the functionality of the overall system, this cache is transparent. Therefore, removing it and connecting signals and busses of the same function together will have no effect on the functionality of the system.

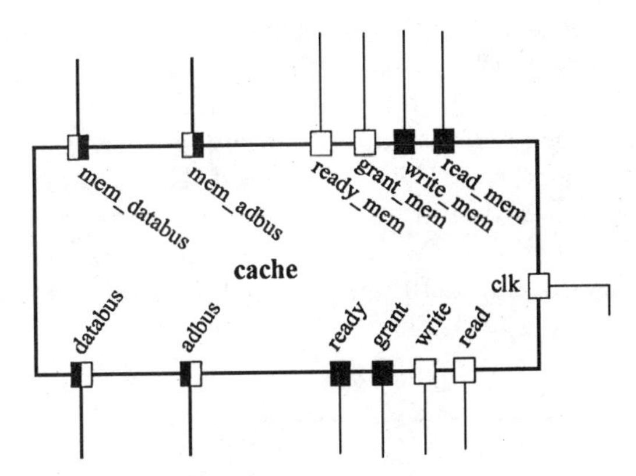

Figure 11.23 Cache interface.

```
`define ...

module cache_system (clk, read_mem, write_mem, grant_mem, ready_mem,
      mem_databus, mem_adbus, read, write, grant, ready, databus, adbus);
input clk;
output read_mem, write_mem;
input grant_mem, ready_mem;
inout [7:0] mem_databus;
inout [11:0] mem_adbus;
input read, write;
output grant, ready;
inout [7:0] databus;
inout [11:0] adbus;
```

Figure 11.24 Cache module header.

The module header for the Verilog code corresponding to this cache is shown in Fig. 11.24. Signal names in which "mem" appears connect to the memory, and the others connect to the CPU. The main system clock used by the CPU and the DMA controller connects to the *clk* input of this cache unit. A set of definitions before the module header defines the cache structure, which will be discussed shortly.

11.6.3 Cache structure modeling

The cache model consists of a memory structure, as shown in Figs. 11.21 and 11.22, and a controller. The complete Verilog description of our example cache consists of definitions (using the **`define** compiler directive) and declarations corresponding to the cache structure, and an **always** block that handles data in and out of this structure. Figure 11.25 shows the outline of this description.

Figure 11.26 shows cache structure definitions and declarations. The tag and set parts of the *adbus* are defined as *set_value* and *tag_value*. For each cache way, a cache entry addressed by *set_value* is read from the cache and is assigned to *entry* 16-bit **reg.** For cache way 0, *entry0* is used, and for cache way 1, *entry1* is used. Using **`define** directives, a valid bit, a 7-bit tag, and an 8-bit line are defined and aliased to fields of *entry0* and *entry1*. Two 16-bit memories of 32 words are declared for cache ways 0 and 1. In parallel with this memory, a 32-bit *lru* is declared for the *lru* table. This register is initialized to **0.**

```
stucture definitions
module cache_system . . .
structure declarations

miscellaneous declarations
    always @(posedge read or posedge write) begin
     initializations

     wait for request
     look for data in the cache
     For read, write
     If hit:
        For read, pass data to CPU
        For write, write data in cache and memory
     If miss:
        Find least recently used
        For write, write data in cache and memory
        For read, read from memory and pass on to CPU
endmodule
```

Figure 11.25 Outline of cache Verilog code.

The last declaration in Fig. 11.26 declares a temporary 32-bit register for reading cache way entries.

With this structure, *entry0* and *entry1* always contain cache entries pointed to by the *set_value* part of *adbus*. This also implies that *valid0, tag0, line0,* and *valid1, tag1, line1* always contain the cache way 0 and 1 valid bit, tag contents, and data that are being addressed by the *set* address part of the *adbus.*

Although *lru* is in parallel with cache memories, it is declared as a register and not a 1-bit memory array. The reason this is done is for ease of initializing it. As shown in Fig. 11.26, an initial statement is used for setting all bits of *lru* to 0.

11.6.4 Controller modeling

As shown in Fig. 11.25, cache controller operations consist of a search for data, a read or write when a cache hit occurs, and a read or write when a miss occurs. Figure 11.27 shows controller miscellaneous declarations for supporting these operations. The variable *hit* is a flag,

```
`define set_value adbus [4:0]
`define tag_value adbus [11:5]
`define valid0 entry0 [15]
`define tag0 entry0 [14:8]
`define line0 entry0 [7:0]
`define valid1 entry1 [15]
`define tag1 entry1 [14:8]
`define line1 entry1 [7:0]

. . . module header, Fig. 11.25

parameter sets = 32;
reg [15:0] way0 [0:sets-1], way1[0:sets-1];
reg [0:sets-1] lru; initial lru = 32'b0;

wire [15:0] entry0 = way0 [`set_value];
wire [15:0] entry1 = way1 [`set_value];
reg [15:0] entry;
```

Figure 11.26 Cache structure definitions and declarations.

```
reg grant, ready, write_mem, read_mem;
reg hit;
reg w;
integer free;
```

Figure 11.27 Controller miscellaneous local declarations.

w stores the cache way where a hit occurs, and *free* indicates a cache way that is free to be written to.

The controller operation begins when a read or write request is made by the CPU. This is done by an event control statement sensitive to edges of *read* and *write* inputs at the beginning of the **always** procedural block, as shown in Fig. 11.25. The code of Fig. 11.28 becomes active after such a request is made. As shown in this figure, *grant* is issued immediately after a CPU request. After the negative edge of the clock, the CPU drives the address bus with valid data. A cache hit occurs if in way 0 or way 1 of the cache, the tag part of the *adbus* is found to have a **1** for its valid bit value. If such occurs, *hit* is set to **1** and the way number is saved in *w*.

Figure 11.29 shows controller operations when a cache hit occurs. In this case the set location of *lru* receives the way number that does not contain the tag value and read and write operations continue. A read operation is completed by issuing *ready* to the CPU and driving the

```
grant = 1; ready = 0;
@(negedge clk);
hit = 0;
if (`valid0 == 1 && `tag0 == `tag_value) begin hit = 1; w = 0; end
if (`valid1 == 1 && `tag1 == `tag_value) begin hit = 1; w = 1; end
```

Figure 11.28 Controller search in cache.

```
if (hit == 1) begin : cache_hit
    lru [`set_value] = ~w;
    if (read == 1) begin
        ready = 1;
        force databus = w ? `line1 : `line0;
        @(negedge read);
        release databus;
    end else if (write == 1) begin
        if (w == 0) way0 [`set_value] = {1'b1, `tag0, databus};
        else way1 [`set_value] = {1'b1, `tag1, databus};
        write_mem = 1;
        @(posedge grant);
        force mem_databus = databus;
        force mem_adbus = adbus;
        @(posedge ready_mem);
        release mem_databus;
        release mem_adbus;
        write_mem = 0;
        ready = 1;
        @(negedge write);
    end
    ready = 0;
```

Figure 11.29 Controller code for a cache hit.

```
end else begin : cache_miss
    free = lru [`set_value];
    lru [`set_value] = ~free;
    if (write == 1) begin
        if (free == 0) way0 [`set_value] = {1'b1, `tag_value, databus};
        else way1 [`set_value] = {1'b1, `tag_value, databus};
        write_mem = 1;
        @(posedge grant_mem);
        force mem_databus = databus;
        force mem_adbus = adbus;
        @(posedge ready_mem);
        release mem_databus;
        release mem_adbus;
        write_mem = 0;
        ready = 1;
        @(negedge write);
        ready = 0;
    end else begin
        read_mem = 1;
        @(posedge grant_mem);
        force mem_adbus = adbus;
        @(posedge ready_mem);
        if (free == 0) way0 [`set_value] = {1'b1, `tag_value, mem_databus};
        else way1 [`set_value] = {1'b1, `tag_value, mem_databus};
        force databus = mem_databus;
        release mem_adbus; read_mem = 0;
        ready = 1;
        @(negedge read);
        ready = 0;
        release databus;
    end
end
```

Figure 11.30 Controller code for a cache miss.

data bus with data found in the cache line. For this purpose, data found in the cache at location *`set_value* are **force**d on *databus* while *read* is **1.** When *read* becomes **0,** a **release** statement releases the *databus.*

In the case of a write operation, the data on the data bus must be written to the cache and to the memory. The value of *w* selects the cache way to write to. A concatenation operation forms the 16-bit data to write to *way0* or *way1.* These data set the corresponding *valid* bit to **1,** write back the tag value to the cache, and write the contents of *databus* to the right-most bits of a cache entry. Write-through requires data written to cache to be written to the main memory. For a memory write, handshaking signals for getting permission from the arbiter, as discussed in Sec. 11.4, will be issued. For driving memory busses for writing to it, **force** statements are used. After the memory issues *ready_mem,* **release** statements release *mem_databus* and *mem_adbus* **inout net** type variables.

The last part of Fig. 11.25 shows read and write operations when a cache miss occurs. The details of this part of the controller code are shown in Fig. 11.30. In case of a cache miss, the cache way that is to become overwritten is looked up from the *lru* table. The variable *free* in Fig. 11.30 is used for this purpose. Following this variable assignment, the *lru* table at the referenced set location is updated to contain the cache way number opposite of the freed cache way. The code that follows these *lru* operations updates the cache at the location *`set_value* and the cache way *free.*

For a write operation, the tag value part of the new address and the new data that are on *databus* are written to the cache memory [in way *free* (0 or 1) and set location *`set_value*], and the valid bit is set to **1.** This is shown in the first part of Fig. 11.30 using an **if–else** statement. Writing to cache is followed by writing these same data into the memory at the specified memory location. For this purpose, *write_mem* is issued by the cache controller to get permission from the arbiter to use memory busses. After *grant_mem* becomes **1,** writing to the memory is done. The write cycle is completed by the cache controller releasing memory busses, issuing *ready* to the CPU, and waiting for the CPU to remove its *write* request.

For a read operation (the second part of Fig. 11.30), data are first read from the memory after the required handshaking with the arbiter. These data will be written in the cache and also sent to the CPU. The following paragraph describes Verilog details of these operations.

To read data from the memory, a *read_mem* is issued. After permission to use the busses is granted, a **force** statement is used to force *adbus* from the CPU onto the *mem_adbus* memory address bus. When memory is ready to drive the data bus with the requested data, the arbiter sets *ready_mem* to **1.** When this happens, a **1** for the *valid* bit, the *`tag_value* part of *adbus,* and the data from the memory contained on *mem_databus* will be concatenated and placed in the cache. The cache location is selected by *`set_value* and the cache way by the *free* variable. At the same time, data are written to the CPU *databus* and *ready* is issued. Data from this bus are released when *read* is deasserted by the CPU.

11.7 Complete System

Sections 11.1 through 11.6 presented components for a board-level interface design. This section shows the wiring of a board using the Parwan CPU of Chap. 10 with components from this chapter. The general outline of our design is shown in Fig. 11.1. Figure 11.31 shows the complete wiring, including bussing and handshaking and control signals.

The Verilog code for this part, shown in Fig. 11.32, consists of instantiation and interconnection of modules according to Fig. 11.31. Modules wired in this description are named *arb, dev, csh, cpu, mem,* and *ser.* Of these, *arb, dev,* and *csh* are those of Figs. 11.8, 11.20, and

11.25, respectively; *cpu* is the modified description of Parwan; *mem* is a memory model with a program for testing the interface board data transfer; and *ser* is a serial data generator. This module generates serial data on the *serial_in* input of *dev*.

Figure 11.33 shows the simulation of the interface design for several read cycles. All module activities are synchronized with the falling edge of the clock. A read cycle begins with the CPU issuing *cpu_read* at time 10. Since data at the requested address are not found in the cache, the cache unit issues *rd_req[1]* at time 11, requesting use of the memory. At this time, no other device is using the memory, therefore, the arbiter issues *grant_mem[1]* immediately after receiving *rd_req[1]*. Also at time 11, memory is selected (*cs* becomes **1**) and *ready* becomes **0.** After two wait states, at time 13, memory data become available and *ready* is issued by the arbiter. The cache controller receives data from the memory, writes them into the cache, and makes them available to the CPU starting at time 13. Availability of data is indicated to the CPU by *csh_ready* being set to **1.** At this same time, the cache controller removes its request for the use of memory busses by setting *rd_req[1]* back to **0.** A clock pulse later, at time 14, the arbiter frees memory by setting *cs* to **0.** At time 15, the CPU has received its requested data and sets *cpu_read* back to **0.** As a result, the *csh_ready* signal becomes **0.** The complete read cycle, which started at time 10, is completed at time 15. The five-clock-cycle period taken for this task includes two wait states for memory operations. The memory wait state is specified as an arbitrator parameter, which could be overwritten when *arb* is instantiated in Fig. 11.32.

11.8 Summary

In this chapter we presented a board-level design in Verilog. We illustrated the use of Verilog in a component-level design environment. Language constructs for behavioral descriptions and timing and event control were emphasized.

Several components with incompatible handshaking schemes were independently described. The interface of the memory component is nonresponsive, while other components such as the CPU and cache controller have two- or three-line fully responsive or partially responsive handshaking schemes. We have illustrated how such handshaking schemes can be described in Verilog and how Verilog constructs can be used for handing communication between various devices.

As opposed to Chap. 10, in which hardware details of a design were of concern, this chapter presented design at a higher level of abstraction. Verilog constructs used in this chapter were primarily at the behavioral level, as discussed in Chap. 9. The examples presented here

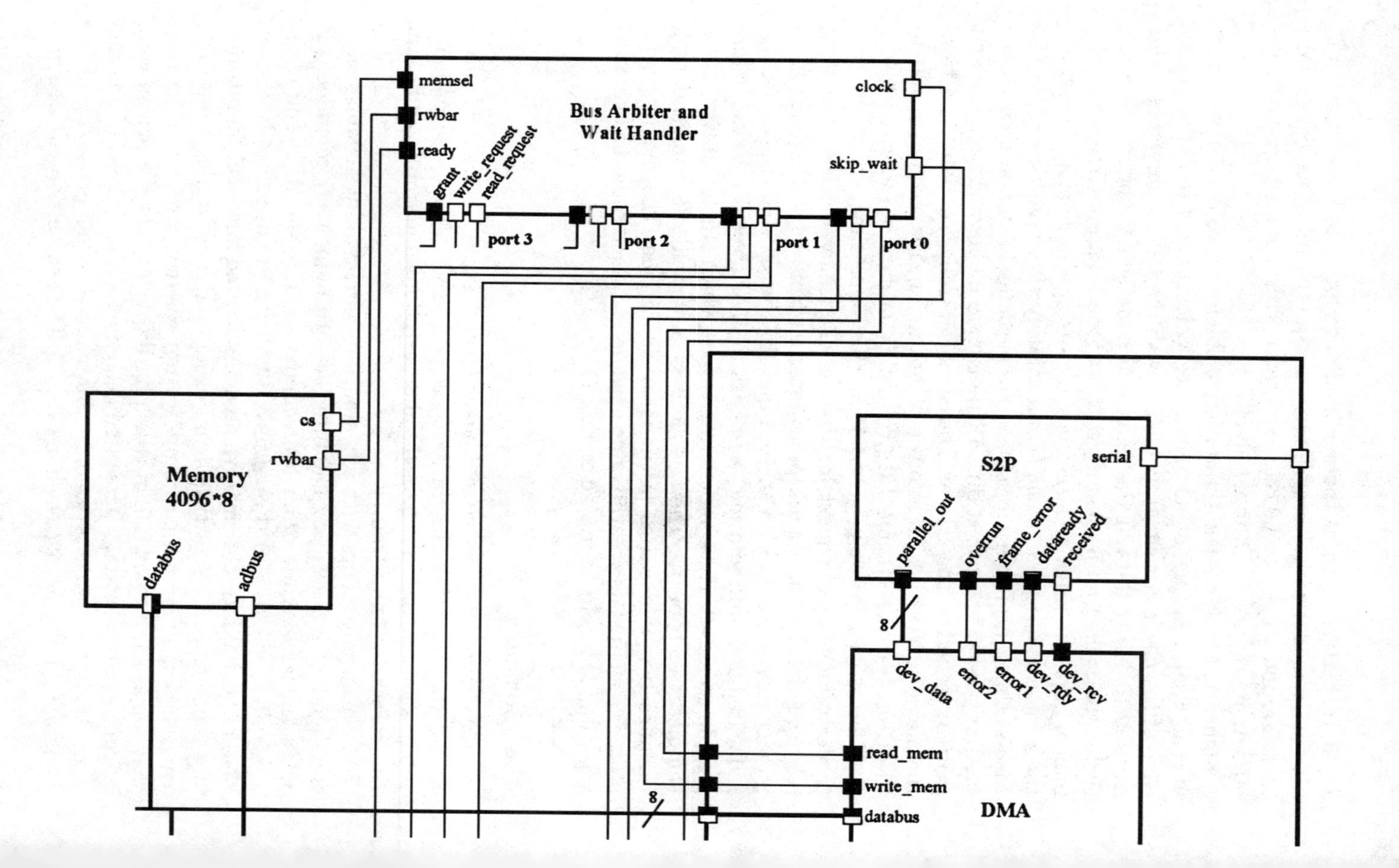

Bus Arbiter and
Wait Handler
memsel
rwbar
ready
grant
write_request
read_request
clock
skip_wait
port 3
port 2
port 1
port 0
Memory
4096*8
cs
rwbar
databus
adbus
S2P
serial
parallel_out
overrun
frame_error
dataready
received
8
dev_data
error2
error1
dev_rdy
dev_rcv
read_mem
write_mem
databus
DMA
8

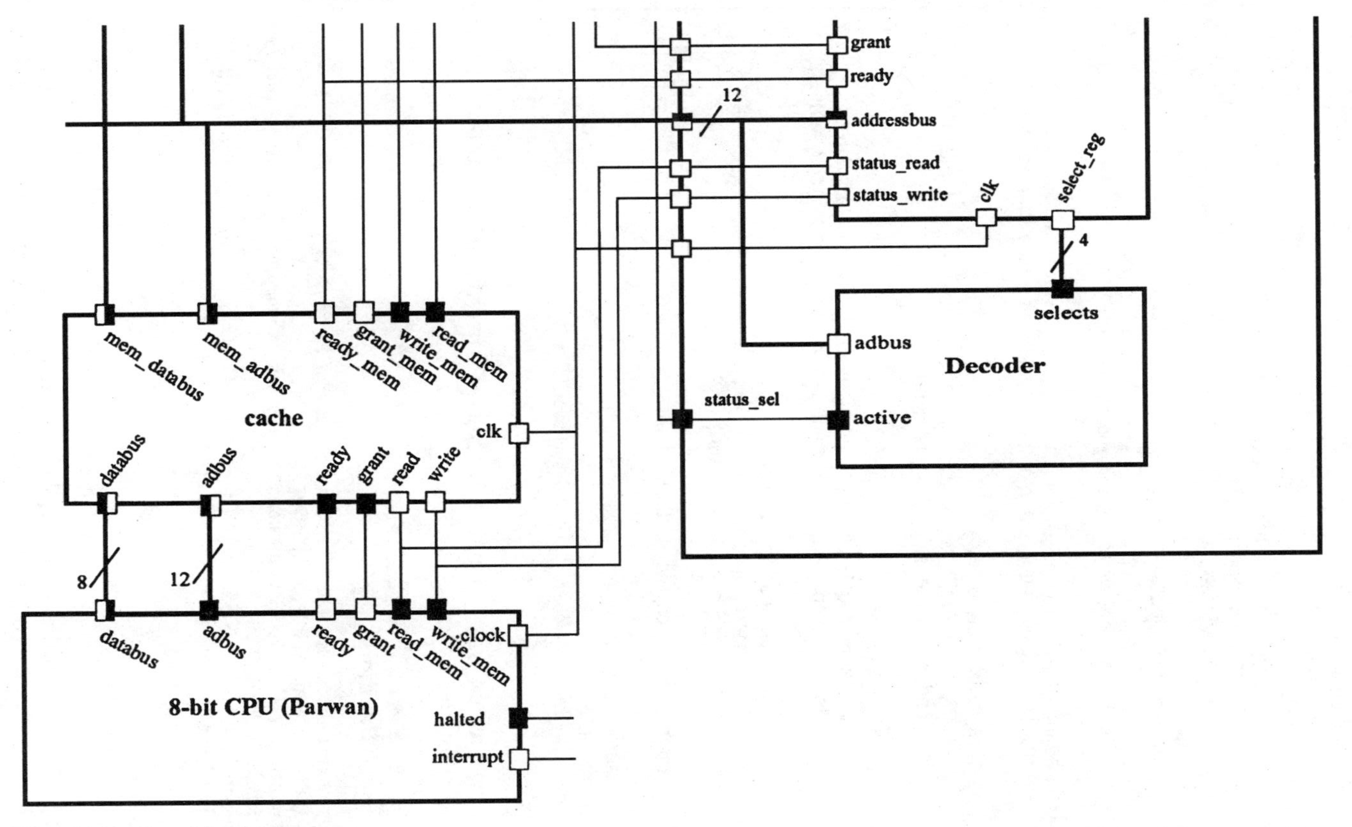

Figure 11.31 Board-level interface.

```
module parwan_tester;
   reg interrupt, clock;
   initial begin
      interrupt = 1'b1; clock = 1'b0;
      #4500 interrupt = 1'b0;
      #30000 $finish;
   end
   always #500 clock = ~clock;

   wire [3:0] rd_req, wr_req, grant_mem;

   arbitrator arb
      (rd_req, wr_req, grant_mem, clock, skip_wait, cs, rwbar, ready);

   dma_serial_device dev
      (clock, rd_req[0], wr_req[0], data, address, ready, grant_mem[0],
   cpu_read, cpu_write, skip_wait, serial_in)

   cache_system csh
      (clock, rd_req[1], wr_req[1], grant_mem[1], ready, data, address,
       cpu_read, cpu_write, csh_grant, csh_ready, cpu_data, cpu_address);

   par_central_processing_unit cpu
      (clock, interrupt, cpu_read, cpu_write, cpu_data, cpu_address,
       halted, csh_ready, csh_grant);

   memory mem (cs, rwbar, data, address);

   sergen ser (serial_in);
endmodule
```

Figure 11.32 Interface board Verilog description.

show various ways of using timing and event control statements in describing a design.

Further Reading

Armstrong, James R., and F. G. Gray, *Structured Logic Design with VHDL,* Prentice-Hall, Englewood Cliffs, N.J., 1993.

Hamacher, V. Carl, Z. G. Vranesic, and S. G. Zaky, *Computer Organization,* McGraw-Hill, New York, 1996.

Nelson, V. P., H. Troy Nagle, B. B. Carrol, and J. D. Irwin, *Digital Logic Circuit Analysis and Design,* Prentice-Hall, Englewood Cliffs, N.J., 1996.

Problems

11.1 Write a parallel-to-serial *p2s* interface adapter that receives 8-bit parallel data and converts them to serial data with the RS232 standard. Use the handshaking used for the *s2p* example in this chapter.

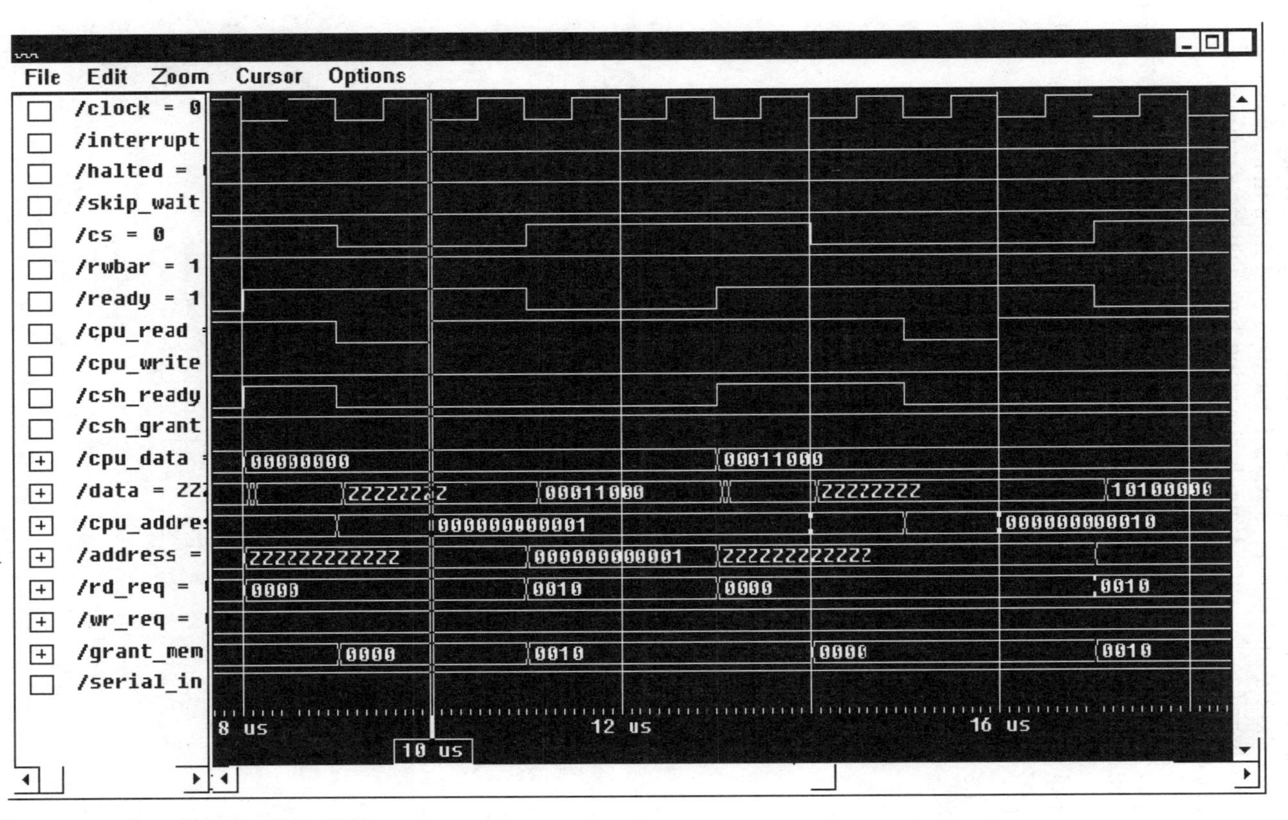

Figure 11.33 Interface board simulation run.

11.2 Write a DMA controller for outputting data. The CPU writes the size of the block of data and its starting address to the DMA controller registers. After the transfer of data is started, the data read from the memory will be transferred to the outputting device using handshaking signals similar to those used with the DMA controller of this chapter.

11.3 Interface the *p2s* of Prob. 11.1 to the Parwan CPU, using the DMA controller of Prob. 11.2. Use an arbiter and a decoder.

11.4 Modify the Verilog description of the cache controller presented in this chapter to a four-way cache. Use an LRU replacement scheme.

11.5 Redo Prob. 11.4 using a random replacement scheme.

Appendix

Frequently Used System Tasks and Functions

This appendix includes a list of frequently used Verilog system tasks and functions for reference. For each such utility, a brief description is provided. Examples for these and other system tasks are included in the chapters. The details of system tasks not discussed here and corresponding examples can be found in Chap. 7.

A.1 Display Tasks

$display

The **$display** task displays its arguments in the order in which they appear. Display will be done to the standard output device. When invoked, this task always inserts a newline character at the end of its output string. A string to be displayed and the format specification must appear in double quotes as an argument of this task. The task invocation

```
$display ( "Counter value is: %d", cnt);
```

prints the value of the *cnt* variable in decimal format. Decimal format is assumed if there is no format specification for a variable or expression.

$displayb

The **$displayb** task displays its arguments in the order in which they appear. Display will be done to the standard output device. When invoked, this task always inserts a newline character at the end of its output string. A string to be displayed and the format specifications must appear in double quotes as an argument of this task. The task invocation

$displayb ("Counter value is: %o", cnt);

prints the value of the *cnt* variable in octal format. Binary format is assumed if there is no format specification for a variable or expression.

$displayh

The **$displayh** task displays its arguments in the order in which they appear. Display will be done to the standard output device. When invoked, this task always inserts a newline character at the end of its output string. A string to be displayed and the format specifications must appear in double quotes as an argument of this task. The task invocation

$displayh ("Counter value is: %b", cnt);

prints the value of the *cnt* variable in binary format. Hexadecimal format is assumed if there is no format specification for a variable or expression.

$displayo

The **$displayo** task displays its arguments in the order in which they appear. Display will be done to the standard output device. When invoked, this task always inserts a newline character at the end of its output string. A string to be displayed and the format specifications must appear in double quotes as an argument of this task. The task invocation

$displayo ("Counter value is: %h", cnt);

prints the value of the *cnt* variable in hexadecimal format. Octal format is assumed if there is no format specification for a variable or expression.

$monitoron

This task turns on the monitor flag used by various forms of the **$monitor** system task. Monitoring will be enabled.

$monitoroff

This task turns off the monitor flag used by various forms of the **$monitor** system task. Monitoring will be disabled.

$monitor

While the monitor flag is on, when a variable or an expression on the argument list changes value, the entire argument list is displayed as in the **$display** system task.

$monitorb

While the monitor flag is on, when a variable or an expression on the argument list changes value, the entire argument list is displayed as in the **$displayb** system task.

$monitorh

While the monitor flag is on, when a variable or an expression on the argument list changes value, the entire argument list is displayed as in the **$displayh** system task.

$monitoro

While the monitor flag is on, when a variable or an expression on the argument list changes value, the entire argument list is displayed as in the **$displayo** system task.

$strobe

Using the same format as **$display,** the **$strobe** system task displays its arguments in a simulation cycle after all events have expired.

$strobeb

Using the same format as **$displayb,** the **$strobeb** system task displays its arguments in a simulation cycle after all events have expired.

$strobeh

Using the same format as **$displayh,** the **$strobeh** system task displays its arguments in a simulation cycle after all events have expired.

$strobeo

Using the same format as **$displayo,** the **$strobeo** system task displays its arguments in a simulation cycle after all events have expired.

$write

The **$write** task displays its arguments in the order in which they appear, using the same format as **$display.** Unlike the **$display** task, this task does not add a newline character to the end of its output, and so consecutive outputs continue on the same line.

$writeb

The **$writeb** task displays its arguments in the order in which they appear, using the same format as **$displayb.** Unlike the **$displayb** task, this task does not add a newline character to the end of its output, and so consecutive outputs continue on the same line.

$writeh

The **$writeh** task displays its arguments in the order in which they appear, using the same format as **$displayh.** Unlike the **$displayh** task, this task does not add a newline character to the end of its output, and so consecutive outputs continue on the same line.

$writeo

The **$writeo** task displays its arguments in the order in which they appear, using the same format as **$displayo.** Unlike the **$displayo** task, this task does not add a newline character to the end of its output, and so consecutive outputs continue on the same line.

A.2 File I/O Tasks

$fopen

The **$fopen** system function returns a file descriptor for the physical file specified as a string in the function argument. The following example makes *desc* a descriptor for the physical file *dataset.dat:*

```
integer desc=$fopen ("dataset.dat");
```

$fclose

The **$fclose** task closes an open file. The only argument of this task is a file descriptor for an open file.

$fdisplay

The **$fdisplay** task outputs its arguments in the order in which they appear. Writing will be done to a file specified by its descriptor. The file descriptor must appear first in the task argument list. This task uses the same format as the **$display** task. The task invocation

```
$fdisplay (desc, "Counter value is: %d", cnt);
```

prints the value of the *cnt* variable in decimal format. Decimal format is assumed if there is no format specification for a variable or expression.

$fdisplayb

The **$fdisplayb** task outputs its arguments in the order in which they appear. Writing will be done to a file specified by its descriptor. The file descriptor must appear first in the task argument list. This task uses the same format as the **$displayb** task. The task invocation

```
$fdisplayb (desc, "Counter value is: %d", cnt);
```

prints the value of the *cnt* variable in decimal format. Binary format is assumed if there is no format specification for a variable or expression.

$fdisplayh

The **$fdisplayh** task outputs its arguments in the order in which they appear. Writing will be done to a file specified by its descriptor. The file descriptor must appear first in the task argument list. This task uses the same format as the **$displayh** task. The task invocation

```
$fdisplayh (desc, "Counter value is: %d", cnt);
```

prints the value of the *cnt* variable in decimal format. Hexadecimal format is assumed if there is no format specification for a variable or expression.

$fdisplayo

The **$fdisplayo** task outputs its arguments in the order in which they appear. Writing will be done to a file specified by its descriptor. The file descriptor must appear first in the task argument list. This task uses the same format as the **$displayo** task. The task invocation

```
$fdisplayo (desc, "Counter value is: %d", cnt);
```

prints the value of the *cnt* variable in decimal format. Octal format is assumed if there is no format specification for a variable or expression.

$fmonitor

While the monitor flag is on, when a variable or an expression on the argument list changes value, the entire argument list is written into a file. The file is specified by its descriptor, which is the first argument in the task argument list. This task uses the same format as the **$display** task.

$fmonitorb

While the monitor flag is on, when a variable or an expression on the argument list changes value, the entire argument list is written into a file. The file is specified by its descriptor, which is the first argument in the task argument list. This task uses the same format as the **$displayb** task.

$fmonitorh

While the monitor flag is on, when a variable or an expression on the argument list changes value, the entire argument list is written into a file. The file is specified by its descriptor, which is the first argument in the task argument list. This task uses the same format as the **$displayh** task.

$fmonitoro

While the monitor flag is on, when a variable or an expression on the argument list changes value, the entire argument list is written into a file. The file is specified by its descriptor, which is the first argument in the task argument list. This task uses the same format as the **$displayo** task.

$fstrobe

Using the same format as **$display,** the **$fstrobe** system task writes its arguments into a file specified by its descriptor as the first argument of the task. Writing will be done in a simulation cycle after all events have expired.

$fstrobeb

Using the same format as **$displayb,** the **$fstrobeb** system task writes its arguments into a file specified by its descriptor as the first argument of the task. Writing will be done in a simulation cycle after all events have expired.

$fstrobeh

Using the same format as **$displayh,** the **$fstrobeh** system task writes its arguments into a file specified by its descriptor as the first argument of the task. Writing will be done in a simulation cycle after all events have expired.

$fstrobeo

Using the same format as **$displayo,** the **$fstrobeo** system task writes its arguments into a file specified by its descriptor as the first argument of the task. Writing will be done in a simulation cycle after all events have expired.

$fwrite

The **$fwrite** task is similar to the **$fdisplay** task except that it does not insert a newline character at the end of its output string. The descriptor for the file into which writing is done appears first in the argument list of this task.

$fwriteb

The **$fwriteb** task is similar to the **$fdisplayb** task except that it does not insert a newline character at the end of its output string. The descriptor for the file into which writing is done appears first in the argument list of this task.

$fwriteh

The **$fwriteh** task is similar to the **$fdisplayh** task except that it does not insert a newline character at the end of its output string. The descriptor for the file into which writing is done appears first in the argument list of this task.

$fwriteo

The **$fwriteo** task is similar to the **$fdisplayo** task except that it does not insert a newline character at the end of its output string. The descriptor for the file into which writing is done appears first in the argument list of this task.

$readmemb

A physical filename and a memory name are required arguments of the **$readmemb** task. When invoked, this task reads binary data from the file specified in its argument and loads these data into the memory specified as its second parameter. Optionally, invocation of this task may contain a range of memory words to fill.

If *mem* is declared as

```
reg [15:0] mem [0:511],
```

then the invocation shown below reads 16-bit words in binary from the *memdata.dat* file and loads these data into memory locations 12 to 412.

```
$readmemb("memdata.dat", mem, 12, 412);
```

$readmemh

A physical filename and a memory name are required arguments of the **$readmemh** task. When invoked, this task reads hexadecimal data from the file specified in its argument and loads these data into the memory specified as its second parameter. Optionally, invocation of this task may contain a range of memory words to fill.

If *mem* is declared as

```
reg [15:0] mem [0:511],
```

then the invocation shown below reads 16-bit words in hexadecimal from the *memdata.dat* file and loads these data into memory locations 12 to 412.

```
$readmemh("memdata.dat", mem, 12, 412);
```

A.3 Timescale Tasks

$printtimescale

The **$printtimescale** task prints the time unit and time precision specified by the **`timescale** directive. If this task is used without an argument, it considers the **`timescale** of the module within which it is invoked. If the task is used with an argument, the argument must be the hierarchical name of the module considered.

$timeformat

The **$timeformat** task specifies how the format specification reports time information for various forms of display tasks. Arguments of this task are unit number, precision number, suffix string, and field width. Unit number is an integer between 0 and –15 specifying time units 1 s to 1 fs, respectively. The precision number argument specifies the number of fractional digits of time reported. The suffix string argument is a string for textual representation of the time unit. The last argument specifies the width of the time information output string.

A.4 Simulation Control Tasks

$finish

When encountered in a procedural flow, the **$finish** system task terminates and exits the simulation. An integer between 0 and 2 passed to this task as an argument specifies the type of message printed when the task is invoked.

$stop

When invoked, the **$stop** system task suspends the simulation. An integer between 0 and 2 passed to this task as an argument specifies the type of message printed when the task is invoked.

A.5 Timing Check Tasks

$hold

The **$hold** system task reports a violation when a reference event occurs too close to a data event. A time limit specifying hold time is the allowed time distance between reference and data events. The first argument specifies the reference event, such as the clock edge. The second argument specifies the data signal. The third argument specifies the hold time. An example is

```
$hold (posedge clk, data, holdtime);
```

$period

The time distance between consecutive events of the same kind (positive or negative) is monitored by the **$period** task. The first argument is the reference event, and the second event is the time specifying the period.

$setup

The **$setup** system task reports a violation when a data event occurs too close to a reference event. A time limit, which is the setup time, specifies the allowed time distance between the data and reference events. The first argument specifies the name of the data signal. The second argument specifies the reference event, such as the clock edge. The third argument is the setup time. An example is

```
$setup (data, posedge clk, setuptime);
```

$skew

The **$skew** system task reports a violation when a reference event and a data event are too far apart in time. As in the **$hold** task, the reference event is the first argument, the data event is the second argument, and the skew time is the third argument of this task.

$nochange

The **$nochange** task reports a violation if during a level specified by the transition on the reference event of its first argument, its second argument changes value. Offset time values specified by the third and fourth arguments expand or shrink the time within which data events are monitored. The following statement reports a violation if *go* changes while *start* is 0:

```
$nochange (negedge start, go, 0, 0);
```

$recovery

The **$recovery** task is similar to the **$setup** task except that the **$recovery** task reports a violation if the data event and the reference event occur at the same simulation time.

$setuphold

The **$setuphold** task is invoked with arguments specifying a reference event, a data event, setup time, and hold time, in this order. This task performs both **$setup** and **$hold** tasks.

$width

The **$width** system task reports a violation when a reference event, specified by its first argument, occurs too close to an opposite event on this argument. The second argument of this task is the allowed pulse width.

Appendix

B

Compiler Directives

This appendix briefly describes Verilog HDL compiler directives. The use and examples of such a language utility were described in Chaps. 7 and 9.

`celldefine
`endcelldefine

Bracketing modules between **`celldefine** and **`endcelldefine** tags the modules as cells.

`default_nettype

The **`default_nettype** directive sets the type of implicit nets. The default is **wire.**

`define
`undef

The **`define** directive aliases an expression with a name. The **`undef** directive turns off aliases set by **`define.**

`ifdef
`else
`endif

Directives **`ifdef, `else** and **`endif** are if–then–else type bracketing for optional compilation of a Verilog code.

`include

The **`include** directive inserts text from an external file.

`unconnected_drive
`nounconnected_drive

The **`unconnected_drive** and **`nounconnected_drive** directives bracket a portion of code for which unconnected input ports will be treated as pulled up or pulled down instead of the normal default.

`resetall

The **`resetall** directive resets all directives to their default values.

`timescale

For setting time scale and time precision, **`timescale** is used.

Appendix

C

Verilog HDL Syntax

The syntax of the Verilog HDL as it appears in the IEEE Std. 1364-1995 document appears here. The language constructs are in alphabetical order. Language terminals are in bold typeface.

always_construct:: =
always
statement

binary_base:: =
`b | `B

binary_digit:: =
x | X | z | Z | 0 | 1

binary_number:: =
[size] binary_base binary-digit { _ | binary_digit }

binary_operator:: =
+ | - | * | / | %
| == | != | === | !== | && | || | < | <= | > | >=
| & | | | ^ | ^~ | ~^ | >> | <<

block_item_declaration:: =
parameter_declaration
| reg_ declaration
| integer_declaration
| real_declaration
| time_declaration
| realtime_declaration
| event_declaration

blocking_assignment:: =
reg_lvalue= [delay_or_event_control] expression

case_item:: =
expression { , expression } : statement_or_null
| **default** [:] statement_or_null

case_statement:: =
case (expression) case_item { case_item } **endcase**
| **casez** (expression) case_item { case_item } **endcase**
| **casex** (expression) case_item { case_item } **endcase**

charge_strength:: =
(**small**)
| (**medium**)
| (**large**)

cmos_switch_instance:: =
    [name_of_gate_instance] (output_terminal , input_terminal ,
    ncontrol_terminal , pcontrol_terminal)

cmos_switchtype:: =
    comos | rcmos

combinational_body::=
    table
      combinational_entry { combinational_entry }
    endtable

combinational_entry:: =
    level_input_list : output_symbol ;

comment:: =
      short_comment
    | long_comment

comment_text:: =
    { ANY_ASCII_CHARACTER }

concatenation:: =
    { expression { , expression } }

conditional_statement:: =
    if (expression) statement_or_null [**else** statement_or_null]

constant_expression:: =
      constant_primary
    | unary_operator constant_primary
    | constant_expression binary_operator constant_expression
    | constant_expression ? constant_expression : constant_expression
    | string

constant_mintypmax_expression:: =
      constant_expression
    | constant_expression : constant_expression : constant_expression

constant_primary:: =
      number
    | parameter_identifier
    | constant_concatenation
    | constant_multiple_concatenation

continuous_assign:: =
    assign [drive_strength] [delay3] list_of_net_assignments ;

controlled_timing_check_event:: =
    timing_check_event_control specify_terminal_descriptor
    [**&&&** timing_check_condition]

current_state:: =
    level_symbol

data_source_expression:: =
    expression

decimal_base:: =
    'd | 'D

decimal_digit:: =
  0 | 1 | 2 | 3 | 4 | 5 | 6 | 7 | 8 | 9

decimal_number:: =
  [sign] unsigned_number
  | [size] decimal_base unsigned_number

delay2:: =
  # daley_value
  | **#** (delay_value [, delay_value])

daley3:: =
  # delay_value
  | **#** (delay_value [, delay_value [, delay_value]])

delay_control:: =
  # delay_value
  | **#** (mintypmax_expression)

delay_or_event_control:: =
  delay_control
  | event_control
  | **repeat** (expression) event_control

delay_value:: =
  unsigned_number
  | parameter_identifier
  | constant_mintypmax_expression

description:: =
  module_declaration
  | udp_declaration

disable_statement:: =
  disable task_identifier ;
  | **disable** block_identifier ;

drive_strength:: =
  (strength0 , strength1)
  | (strength1 , strength0)
  | (strength0 , **highz1**)
  | (strength1 , **highz0**)
  | (**highz1** , strength0)
  | (**highz0** , strength1)

edge_control_specifier:: =
  edge [edge_descriptor [, edge_descriptor]]

edge_descriptor:: =
  01
  | **10**
  | **0x**
  | **x1**
  | **1x**
  | **0x**

edge_identifier:: =
  posedge | **negedge**

edge_indicator:: =
    (level_symbol level_symbol)
    | edge_symbol

edge_input_list:: =
    { level_symbol } edge_indicator { level_symbol }

edge_sensitive_path_declaration:: =
    parallel_edge_sensitive_path_description = path_delay_value
    | full_ edge_sensitive_path_description = path_delay_value

edge_symbol:: =
    r | R | f | F | p | P | n | N | *

enable_gate_instance:: =
    [name_of_gate_instance] (output_terminal , input_terminal ,
    enable_terminal)

enable_gate_type:: =
    bufif0 | bufif1 | notif0 | notif1

enable_terminal:: =
    scalar_expression

escaped_identifier:: =
    \ { ANY_ASCII_CHARACTER_EXCEPT_WHITE_SPACE } white_space

event_control:: =
    @ event_identifier
    | @ (event_expression)

event_declaration:: =
    event event_identifier { , event_identifer } ;

event_expression:: =
    expression
    | event_identifier
    | **posedge** expression
    | **negedge** expression
    | event_ expression **or** event_ expression

event_trigger:: =
    -> event_identifier ;

expression:: =
    primary
    | unary_operator primary
    | expression binary_operator expression
    | expression **?** expression **:** expression
    | string

full_edge_sensetive_path_description:: =
    ([edge_identifier] list_of_path_inputs *> list_of_path_outputs
    [polarity_operator] : data_source_ expression)

full_path_description:: =
    (list_of_path_inputs [polarity_operator] *> list_of_path_outputs)

function_call:: =
 function_identifier (expression { , expression })
 | name_of_system_function [(expression { , expression })]

function_declaration:: =
 function [range_or_type] function_identifier ;
 function_item_declaration { function_item_declaration }
 statement
 endfunction

function_item_declaration:: =
 block_item_declaration
 | input_declaration

gate_instantiation:: =
 n_input_gatetype [drive_strength] [delay2] n_input_gate_instance
 { , n_input_gate_instance } ;
 | n_output_gatetype [drive_strength] [delay2] n_output_gate_instance
 { , n_output_gate_instance } ;
 | enable_gatetype [drive_strength] [delay3] enable_gate_instance
 { , enable_gate_instance } ;
 | mos_switchtype [delay3] mos_switch_instance
 { , mos_switch_instance } ;
 | pass_switchtype pass_switch_instance { , pass_switch_instance } ;
 | pass_en_switchtype [delay3] pass_en_switch_instance
 { , pass_en_switch_instance } ;
 | comos_switchtype [delay3] cmos_switch_instance
 { , cmos_switch_instance } ;
 | **pullup** [pullup_strength] pull_gate_instance { , pull_gate_instance } ;
 | **pulldown** [pulldown_strength] pull_gate_instance
 { , pull_gate_instance } ;

hex_base:: =
 `h | `H

hex_digit:: =
 x | X | z | Z
 | **0 | 1 | 2 | 3 | 4 | 5 | 6 | 7 | 8 | 9**
 | **a | b | c | d | e | f | A | B | C | D | E | F**

hex_number:: =
 [size] hex_base hex_digit { _ | hex_digit }

identifier:: =
 IDENTIFIER [{ . IDENTIFIER }]
 /* The period may not be followed or proceeded by a space */

IDENTIFIER:: =
 Simple_identifier
 | escaped_identifier

init_val:: =
 1`b0 | 1`b1 | 1`bx | 1`bX | 1`B0 | 1`B1 | 1`Bx | 1`BX | 1 | 0

initial_construct:: =
 initial
 statement

inout_declaration:: =
 inout [range] list_of_port_identifiers ;

inout_terminal:: =
    terminal_identifier
    | terminal_identifier [constant_expression]

input_declaration:: =
    input [range] list_of_port_identifiers ;

input_terminal:: =
    input_port_identifier
    | inout_port_identifier

input_terminal:: =
    scalar_expression

integer_declaration:: =
    integer list_of_register_identifiers ;

level_input_list:: =
    level_symbol { level_symbol }

level_symbol: =
    0 | 1 | x | X | ? | b | B

limit_value:: =
    constant_mintypmax_expression

list_of_module_connections: =
    ordered_port_connection { , ordered_port_connection }
    | named_port_connection { , named_port_connection }

list_of_net_assignments:: =
    net_assignment { , net_assignment }

list_of_net_decl_assignments:: =
    net_decl_assignment { , net_decl_assignment }

list_of_ net_identifiers:: =
    net_identifier { , net_identifier }

list_of_param_assignments:: =
    param_assignment { , param_assignment }

list_of_path_delay_expressions:: =
    t_path_delay_expression
    | trise_path_delay_expression , tfall_path_delay_expression
    | trise_path_delay_expression , tfall_path_delay_expression ,
        tz_path_delay_expression
    | t01_path_delay_expression , t10_path_delay_expression ,
        t0z_path_delay_expression , tz1_path_delay_expression ,
        t1z_path_delay_expression , tz0_path_delay_expression
    | t01_path_delay_expression , t10_path_delay_expression ,
        t0z_path_delay_expression , tz1_path_delay_expression ,
        t1z_path_delay_expression , tz0_path_delay_expression ,
        t0z_path_delay_expression , tx1_path_delay_expression ,
        t1x_path_delay_expression , tx0_path_delay_expression ,
        txz_path_delay_expression , tzx_path_delay_expression

list_of_path_inputs:: =
    specify_input_terminal_descriptor { , specify_input_terminal_descriptor }

list_of_path_outputs:: =
    specify_output_terminal_descriptor { , specify_output_terminal_descriptor }

list_of_port_identifiers:: =
    port_identifier { , port_identifier }

list_of_ports:: =
    (port { , port })

list_of_real_identifiers:: =
    real_identifier { , real_identifiers }

list_of_register_identifiers: =
    register_name { , register_name }

list_of_specparam_assignments:: =
    specparam_assignment { , specparam_assignment }

long_comment:: =
    /* comment_text*/

loop_statement:: =
    forever statement
    | **repeat** (expression)statement
    | **while** (expression)statement
    for (reg_assignment ; expression ; reg_assignment)statement

mintypmax_ expression:: =
    expression
    | expression : expression : expression

module_declaration:: =
    module_keyword module_identifier [list_of_ports] ;
        { module_item }
    endmodule

module_instance:: =
    name_of_instance ([list_of_module_connections])

module_instantiation:: =
    modul_identifier [parameter_value_assignment] module_instance
        { , module_instance } ;

module_item:: =
    module_item_declaration
    | parameter_override
    | continuous_assign
    | gate_instantiation
    | udp_instantiation
    | module_instantiation
    | specify_block
    | initial_construct
    | always_construct

module_item_declaration:: =
    parameter_ declaration
    | input_ declaration
    | output_ declaration
    | inout_ declaration
    | net_ declaration

| reg_ declaration
| integer_ declaration
| real_ declaration
| time_ declaration
| realtime_ declaration
| event_ declaration
| task_ declaration
| function_ declaration

module_keyboard:: =
module | **macromodule**

mos_switch_instance:: =
[name_of_gate_instance] (output_terminal , input_terminal ,
enable_terminal)

mos_switchtype:: =
nmos | **pmos** | **rnmos** | **rpmos**

multiple_concatenation:: =
{ expression { expression { , expression } } }

n_input_gate_instance:: =
[name_of_gate_instance] (output_terminal , input_terminal
{ , input_terminal })

n_input_gatetype:: =
and | **nand** | **or** | **nor** | **xor** | **xnor**

n_output_gate_instance:: =
[name_of_gate_instance] (output_terminal { , output_terminal } ,
input_terminal)

n_output_gatetype:: =
buf | **not**

name_of_gate_instance:: =
gate_instance_identifier [range]

name_of_instance:: =
module_instance_identifier [range]

name_of_system_function:: =
$identifier

name_of_udp_instance:: =
udp_instance_identifier [range]

named_port_connection:: =
. port_identifier ([expression])

ncontrol_terminal:: =
scalar_ expression

net_assignment:: =
net_lvalue = expression

net_decl_assignment:: =
net_identifier = expression

net_declaration:: =
net_type [**vectored** | **scalared**] [range] delay3] list_of_net_identifires ;
| **trireg** [**vectored** || **scalared**] [charge_strength] [range] [delay3]
list_of_net_identifires ;
| net_type [**vectored** | **scalared**] [drive_strength] [range] [delay3]
list_of_net_decl_identifires ;

net_lvalue:: =
net_identifier
| net_identifier [expression]
| net_identifier [msb_constant_expression : lsb_constant_expression]
| net_concatenation

net_type:: =
wire | **tri** | **tri1** | **supply0** | **wand** | **triand** | **tri0** | **supply1** | **wor** | **trior**

next_state:: =
output_symbol | -

non_blocking_assignment:: =
reg_lvalue <= [delay_or_event_control] expression

notify_register:: =
register_identifier

number:: =
decimal_number
| octal_ number
| binary number
| hex_ number
| real_ number

octal_base:: =
`o** | **`O

octal_digit:: =
x | **X** | **z** | **Z** | **0** | **1** | **2** | **3** | **4** | **5** | **6** | **7**

octal_number:: =
[size] octal_base octal_digit { _ | octal_digit }

ordered_port_connection:: =
[expression]

output_declaration:: =
output [range] list_of_port_identifiers ;

output_identifier:: =
output_port_identifier
| inout_port_identifier

output_symbol:: =
0 | **1** | **x** | **X**

output_terminal:: =
terminal_identifier
| terminal_identifier [constant_expression]

par_block:: =
fork

[: block_identifier
{ block_item_declaration }]
{ statement }
join

parallel_edge_sensitive_path_description:: =
([edge_identifier] specify_input_terminal_descriptor =>
specify_output_terminal_descriptor [polarity_operator] :
data_source_expression)

parallel_path_description:: =
(specify_input_terminal_descriptor [polarity_operator] =>
specify_output_terminal_descriptor)

param_assignment:: =
parameter_identifier = constant_expression

parameter_declaration:: =
parameter list_of_param_assignments ;

parameter_override:: =
defparam list_of_param_assignments ;

parameter_value_assignment:: =
(expression { , expression })

pass_en_switchtype:: =
tranif0 | **tranif1** | **rtranif1** | **rtarnif0**

pass_en_switch_instance:: =
[name_of_gate_instance] (inout_terminal , inout_terminal ,
enable_terminal)

pass_switch_instance:: =
[name_of_gate_instance] (inout_terminal , inout_terminal)

pass_switchtype:: =
tran | **rtran**

path_declaration:: =
simple_ path_declaration ;
| edge_sensitive_ path_declaration ;
| state_dependent_ path_declaration ;

path_ delay_expression:: =
constant_mintypmax_expression

path_delay_value:: =
list_of_ path_delay_expressions
| (list_of_path_ delay_expressions)

pcontrol_terminal:: =
scalar_expression

polarity_operator:: =
+ | -

port:: =
[port_expression]
| . port_identifier ([port_expression])

port_expression:: =
    port_refrence
  | { port_refrence { , port_refrence } }

port_refrence:: =
    port_identifier
  | port_identifier [constant_expression]
  | port_identifier [msb_constant_expression : /sb_constant_expression]

primary:: =
    number
  | identifier
  | identifier [expression]
  | identifier [msb_constant_expression : /sb_constant_expression]
  | concatenation
  | multiple_concatenation
  | function_call
  | (mintypmax_expression)

procedural_continuous_assignment:: =
    assign reg_assignment ;
  | **deassign** reg_lvalue ;
  | **force** reg_assignment ;
  | **force** net_assignment ;
  | **release** reg_lvalue ;
  | **release** net_lvalue ;

procedural_timing_control_statement:: =
  delay_or_event_control statemenmt_or_null

pull_gate_instance:: =
  [name_of_gate_instance] (output_terminal)

pulldown_strength:: =
    (strength0 , strength1)
  | (strength1 , strength0)
  | (strength0)

pullup_ strength:: =
    (strength0 , strength1)
  | (strength1 , strength0)
  | (strength1)

pulse_control_specparam:: =
    PATHPULSE$ = (reject_limit_value [error)limit_value]) ;
  | **PATHPULSE$**specify_input_terminal_descriptor /*no space; continue*/
  $specify_output_terminal_descriptor = (reject_limit_value
  [, error_limit_value]) ;

range:: =
  [msb_constant_expression : lsb_constant_expression]

range_or_type:: =
  range | **integer** | **real** | **realtime** | **time**

real_declaration:: =
  real list_of_real_identifiers ;

real_number:: =
  [sign] unsigned_number . unsigned_number
  | [sign] unsigned_number [. unsigned_number] **e** [sign]
    unsigned_number
  | [sign] unsigned_number [. unsigned_number] **E** [sign]
    unsigned_number

realtime_declaration:: =
  realtime list_of_real_identifiers;

reg_assignment::=
  reg_lvalue = expression

reg_declaration:: =
  reg [range] list_of_register_identifiers ;

reg_lvalue:: =
  reg_identifier
  | reg_identifier **[** expression **]**
  | reg_identifier **[** msb_constant_expression **:** lsb_ constant_expression **]**
  | reg_concatenation

register_name:: =
  register_identifier
  | memory_identifier **[** upper_limit_constant_expression **:**
    lower_limit_constant_expression **]**

scalar_constant:: =
  1'b0 | 1'b1 | 1'B0 | 1'B1 | 'b0 | 'b1 | 'B0 | 'B1 | 1 | 0

scalar_timing_check_condition:: =
  expression
  | ~ expression
  | expression == scalar_constant
  | expression === scalar_constant
  | expression != scalar_constant
  | expression !== scalar_constant

seq_block:: =
  begin
    [: block_identifier
      { block_item_declaration }]
    { statement }
  end

seq_input_list:: =
  level_input_list | edge_input_list

sequential_body:: =
  [udp_initial_statement]
  table
    sequential_entry
    { sequential_entry }
  endtable

sequential_entry:: =
  seq_input_list **:** current_state **:** next_state **;**

short_comment:: =
  // comment_text **\n**

sign:: =
    + | -

simple_identifier:: =
    [a-zA-Z][a-zA-Z_$0-9]

simple_path_declaration:: =
    parallel_path_description = path_delay_value
    | full_path_description = path_delay_value

size:: =
    unsigned_number

source_text:: =
    { description }

specify_block:: =
    specify
    { specify_item }
    endspecify

specify_input_terminal_descriptor:: =
    input_identifier
    | input_identifier **[** constant_expression **]**
    | input_identifier **[** masb_constant_expression **:** lsb_constant_expression **]**

specify_item:: =
    specparam_declaration
    | path_declaration
    | system_timing_check

specify_output_terminal_descriptor:: =
    output_identifier
    | output_identifier [constant_expression]
    | output_identifier [msb_constant_expression : lsb_constant_expression]

specify_terminal_descriptor:: =
    specify_input_terminal_descriptor
    | specify_output_terminal_descriptor

specparam_assignment:: =
    specparam_identifier = constant_expression
        | pulse_control_specparam

specparam_declaration:: =
    specparam list_of_ specparam_assignments ;

state_dependent_path_declaration:: =
    if (conditional_expression) simple_path_declaration
    | **if** (conditional_expression)edge_sensitive_path_declaration
    | **ifnone** simple_path_declaration

statement:: =
    blocking_assignment ;
    | non_ blocking_assignment ;
    | procedural_continuous_ assignment ;
    | procedural_timing_control_statement
    | conditional_statement
    | case_statement
    | loop_ statement

| wait_ statement
| disable_ statement
| event_trigger
| seq_block
| par_block
| task_enable
| system_task_enable

statement_or_null:: =
statement | ;

strength0:: =
supply0 | **strong0** | **pull0** | **weak0**

strength1:: =
supply1 | **strong1** | **pull1** | **weak1**

string:: =
"{ ANY_ASCII_CHARACTERS_EXCEPT_NEWLINE }"

system_task_enable:: =
system_task_name [(expression { , expression })] ;

system_task_name:: =
$identifier
/* The $ cannot be followed by a space */

system_timing_check:: =
$setup (timing_check_event , timing_check_event , timing_ check_limit
[, notify_register]) ;
| **$hold** (timing_check_event , timing_check_event , timing_check_limit
[, notify_register]) ;
| **$period** (controlled_timing_check_event , timing_check_limit
[, notify_register]) ;
| **$width** (controlled_timing_check_event , timing_check_limit,
constant_expression [, notify_register]) ;
| **$skew** (timing_check_event , timing_check_event , timing_check_limit
[, notify_register]) ;
| **$recovery** (controlled_timing_check_event , timing_check_event,
timing_check_limit [, notify_register]) ;
| **$setuphold** (timing_check_event , timing_check_event ,
timing_check_limit , timing_check_limit [, notify_register]) ;

task_declaration:: =
task task_identifier ;
{ task_item_declaration }
statement_or_null
endtask

task_enable:: =
task_identifier [(expression { , expression }) ;

task_item_declaration:: =
block_item_ declaration
| input_ declaration
| output_ declaration
| inout_ declaration

time_ declaration:: =
time list_of_register_identifiers ;

timing_check_condition:: =
scalar_ timing_check_condition
| (scalar_ timing_check_condition)

timing_check_event:: =
[timing_check_event_control] specify_terminal_descriptor
[**&&&** timing_check_condition]

timing_check_event_control:: =
posedge
| **negedge**
| edge_control_specifier

timing_check_limit:: =
expression

udp_body:: =
combinational_body
| sequential_body

udp_declaration:: =
primitive udp_identifier (udp_port_list) ;
udp_port_declaration
{ udp_declaration }
udp_body
endprimitive

udp_initial_statement:: =
initial udp_output_port_identifier = init_val ;

udp_instance:: =
[name_of_udp_instance] (output_port_connection, input_port_connection
{ , input_port_connection })

udp_installation:: =
udp_identifier [drive_strength] delay2] udp_instance { , udp_instance } ;

udp_port_ declaration:: =
output_ declaration
| input_ declaration
| reg_ declaration

udp_port_list:: =
output_port_identifier , input_port_identifier { , input_port_identifier }

unary_operator:: =
+ | - | **!** | ~ | **&** | ~& | | | ~ | | ^ | ~^ | ^~

unsigned_number:: =
decimal_digit { _ | decimal_digit }

wait_statement:: =
wait (expression) statement_or_null

white_space:: =
space | tab | newline

Appendix

D

Parwan Verilog Descriptions

Two Verilog descriptions for the Parwan CPU were presented in Chap. 10. One was a behavioral description for presentation of the overall functionality of the CPU, and the other was a dataflow description that would be developed for a discrete manual design. This appendix presents these two descriptions in their complete forms, including modifications that are required for handling slow memories.

D.1 Complete Parwan Behavioral Description

The Verilog behavioral description of Parwan as presented in Chap. 10 is shown in this section.

```
//------------------------alu_operation_parameters
`define a_and_b 3'b000
`define b_compl 3'b001
`define a_input 3'b100
`define a_add_b 3'b101
`define b_input 3'b110
`define a_sub_b 3'b111
//------------------------par_control_parameters(state)
`define do_initials 4'd1
`define instr_fetch 4'd2
`define do_one_bytes 4'd3
`define opnd_fetch 4'd4
`define do_indirect 4'd5
`define do_two_bytes 4'd6
`define do_jsr 4'd7
`define continue_jsr 4'd8
`define do_branch 4'd9
//----------------------par_control_parameters(code)
`define single_byte_instructions 4'b1110
`define hlt 4'b0000
`define cla 4'b0001
`define cma 4'b0010
`define cmc 4'b0100
`define asl 4'b1000
`define asr 4'b1001
`define jsr 3'b110
`define bra 4'b1111
`define indirect 1'b1
`define jmp 3'b100
`define sta 3'b101
`define lda 3'b000
`define ann 3'b001
`define add 3'b010
`define sbb 3'b011
`define jsr_or_bra 2'b11
```

```
`include "par_para.v"
module par_central_processing_unit
    (clk, interrupt, read_mem, write_mem, databus, adbus, halted, ready, grant);
output read_mem, write_mem, halted;
reg read_mem, write_mem, halted;
output [11:0] adbus;
reg [11:0] adbus;
inout [7:0] databus;
input  clk, ready, grant, interrupt;
reg [2:0] present_state, next_state;
reg [8:0] ac, next_ac, ir, next_ir;
reg [4:0] sr, next_sr;
reg [11:0] pc, next_pc, mar, next_mar;
reg [9:0] ten_bit;
initial begin
    present_state <= `do_initials;
    next_state <= `do_initials;
```

```
end
always @(negedge clk or posedge interrupt) begin
    if (interrupt) begin
        present_state <= `do_initials;
         ac <= 8'b00000000;
        ir <= 8'b00000000;
        sr <= 4'b0000;
        pc <= 12'b000000000000;
        mar <= 12'b000000000000;
    end else begin
    if (~clk) begin
         ac <= next_ac;
         ir <= next_ir;
         sr <= next_sr;
         pc <= next_pc;
         mar <= next_mar;
         present_state <= next_state;
    end end
end

reg [7:0] dbus;
assign databus=dbus;
always @(present_state or interrupt or ready or grant or databus or ac or ir or sr or pc or mar)
begin
    next_pc <= pc;
    dbus <= 8'bZZZZZZZZ;
    adbus <= 12'bZZZZZZZZZZZZ;
    read_mem <= 1'b0;
    write_mem <= 1'b0;
    halted <= 1'b0;
    next_ir <= ir;
    next_ac <= ac;
    next_mar <= mar;
    next_sr <= sr;
    ten_bit = 10'b0000000000;
    case(present_state)
    `do_initials: begin          //------------------------------1
        if (interrupt) begin
            next_pc <= 12'b000000000000;
            next_state <= `do_initials;
        end else begin
            next_mar <= pc;
            next_state <= `instr_fetch;
        end end
    `instr_fetch: begin          //------------------------------2
        read_mem <= 1'b1;
        if (grant) begin
            adbus <= mar;
            if (ready) begin
                next_ir <= databus;
                next_pc <= pc + 1;
                next_state <= `do_one_bytes;
            end
            else next_state <= `instr_fetch;
        end else
            next_state <= `instr_fetch;
```

```
    end
`do_one_bytes: begin        //-----------------------------3
    next_mar <= pc;  // prepare for next memory read
    if (ir[7:4] != `single_byte_instructions)
        next_state <= `opnd_fetch;
    else begin
        case(ir[3:0])
        `cla: next_ac <= 8'b00000000;
        `cma: begin
        next_sr[0] <= ac[7];
            next_sr[1] <= ~|( ~ac );
            next_ac <= ~ ac;
     end
        `cmc: begin next_sr[2] <= ~ sr[2];
        `asl: begin
        next_sr[0] <= ac[6];
        next_sr[1] <= ~|ac[6:0];
        next_sr[2] <= ac[7];
        next_sr[3] <= ac[6] ^ ac[7];
        next_ac <=  {ac[6:0],1'b0};
        end
        `asr: begin
        next_sr[0] <= ac[6];
        next_sr[1] <= ~|ac[7:1];
        next_sr[2] <= ac[7];
        next_sr[3] <= ac[6] ^ ac[7];
        next_ac <=  {ac[6:0],1'b0};
        end
        `hlt: halted <= 1'b1;
        default: halted <= 1'b1;
        endcase
        next_state <= `instr_fetch;
    end
end
`opnd_fetch: begin          //-----------------------------4
    read_mem <= 1'b1;
    if (grant) begin
        adbus <= mar;
        if (ready) begin
            next_mar[7:0] <= databus;
            if (ir[7:6] != `jsr_or_bra) begin
                next_mar[11:8] <= ir[3:0];
                if (ir[4] == `indirect)
                    next_state <= `do_indirect;
                else
                    next_state <= `do_two_bytes;
            end    else begin
                if ( ~ir[5])
                    next_state <= `do_jsr;
                else
                    next_state <= `do_branch;
            end
            next_pc <= pc + 1;
        end else next_state <= `opnd_fetch;
    end else next_state <= `opnd_fetch;
end
```

```
`do_indirect: begin          //-----------------------------5
   read_mem <= 1'b1;
   if (grant) begin
      adbus <= mar;
      if (ready) begin
         next_mar[7:0] <= databus;
         next_state <= `do_two_bytes;
      end else
         next_state <= `do_indirect;
      end
   else next_state <= `do_indirect;
end
`do_two_bytes: begin          //-----------------------------6
   if (ir[7:5] == `jmp) begin
      next_pc <= mar;
      next_state <= `instr_fetch;
   end else begin
      if (ir[7:5] == `sta) begin
         write_mem <= 1'b1;
         if (grant) begin
            adbus <= mar;
            dbus <= ac;
            if (ready) next_state <= `do_initials;
            else next_state <= `do_two_bytes;
         end else
            next_state <= `do_two_bytes;
      end else begin
         if (~ir[7]) begin
            read_mem <= 1'b1;
            if (grant) begin
               adbus <= mar;
               if (ready) begin
                  if (~ir[6]) begin
                     if (~ir[5])
                        ten_bit = {sr [3:2], databus};
                     else
                        ten_bit = {sr [3:2], (ac & databus)};
                  end   else begin
                     if (~ir[5]) begin
                        ten_bit = ac + databus + sr[2];
                        if (ac[7] == databus[7] && ten_bit[7] != ac[7])
                           ten_bit[9]=1'b1;
                        else ten_bit[9]=1'b0;
                     end else begin
                        ten_bit = ac - databus - sr[2];
                        if (ac[7] == ~databus[7] && ten_bit[7] != ac[7])
                           ten_bit[9] = 1'b1;
                        else ten_bit[9]=1'b0;
                     end
                  end
                  next_sr <= {ten_bit [9], ten_bit [8], ~|ten_bit[7:0], ten_bit [7]};
                  next_ac <= ten_bit [7:0];
                  next_state <= `do_initials;
               end else next_state <= `do_two_bytes;
            end else next_state <= `do_two_bytes;
         end else next_state <= `do_initials;
```

```
                end
            end
        end
        `do_jsr: begin          //------------------------------7
            write_mem <= 1'b1;
            if (grant) begin
                adbus <= mar;
                dbus <= pc [7:0];
                if (ready) begin
                    next_pc <= mar;
                    next_state <= `continue_jsr;
                end else
                    next_state <= `do_jsr;
            end else next_state <= `do_jsr;
        end
        `continue_jsr: begin          //------------------------------8
            next_pc <= pc + 1;
            next_state <= `do_initials;
        end
        `do_branch: begin          //------------------------------9
            if (|{sr,ir[3:0]}) next_pc <= mar;
            next_state <= `do_initials;
        end
        default: next_state <= `do_initials;
        endcase
    end
endmodule
```

D.2 Complete Parwan Dataflow Description

This section shows the dataflow description of Parwan as presented in Chap. 10. Code fragments shown in the text are shown here in complete descriptions.

```
LIBRARY cmos;
USE cmos.basic_utilities.ALL;
--
PACKAGE alu_operations IS
  CONSTANT a_and_b   : qit_vector (5 DOWNTO 0) := "000001";
  CONSTANT b_compl   : qit_vector (5 DOWNTO 0) := "000010";
  CONSTANT a_input   : qit_vector (5 DOWNTO 0) := "000100";
  CONSTANT a_add_b   : qit_vector (5 DOWNTO 0) := "001000";
  CONSTANT b_input   : qit_vector (5 DOWNTO 0) := "010000";
  CONSTANT a_sub_b   : qit_vector (5 DOWNTO 0) := "100000";
END alu_operations;
```

```
`include "par_para.v"
module arithmetic_logic_unit(a_side,b_side,code,in_flags,z_out,out_flags);
 input  [7:0] a_side,b_side;
 input  [2:0] code;
 input  [3:0] in_flags;
 output  [3:0] out_flags;
 output [7:0] z_out;
 reg [7:0] z_out;
 reg [3:0] out_flags;
 reg [7:0] tl;
 reg v,c,z,n;
always @(a_side or b_side or code or in_flags)
 begin
  case (code)
  `a_add_b : begin
          {c,tl} = a_side+b_side+in_flags[2];
          if (a_side[7]==b_side[7] && tl[7]!=a_side[7])
                v=1'b1;
          else
           v=1'b0;
          end
  `a_sub_b : begin
          {c,tl} = a_side-b_side-in_flags[2];
          if (a_side[7]==~b_side[7] && tl[7]!=a_side[7])
                v=1'b1;
          else
           v=1'b0;
          end
  `a_and_b : begin
```

```
            tl = a_side & b_side;c=in_flags[2];v=in_flags[3];
                end
    `a_input : begin
            tl = a_side ;c=in_flags[2];v=in_flags[3];
                end
    `b_input : begin
            tl = b_side ;c=in_flags[2];v=in_flags[3];
                end
    `b_compl : begin
            tl = ~ b_side;c=in_flags[2];v=in_flags[3];
                end
    default: begin
           tl=8'b00000000;c=1'b0;v=1'b0;
               end
  endcase
  n = tl[7];
  z = ~|tl;
  z_out <= tl;
  out_flags <= {v,c,z,n};
 end
endmodule
//------------------------------------------------



module status_register_unit(in_flags,out_status,load,cm_carry,ck);
 input [3:0] in_flags;
 input load,cm_carry,ck;
 output [3:0] out_status;
 reg [3:0] status;
 reg [3:0] out_status;
initial
 begin
  out_status = 4'b0000;
 end
always @( ck )
 begin
  if ( ~ck )
   begin
    if(load)
         status <= in_flags;
    else
       begin
      if(cm_carry)
          status[2] <= ~status[2];
     end
     end
 end
always @( status )
 begin
  out_status <= status;
 end
endmodule


module shifter_unit(alu_side,arith_shift_left,arith_shift_right,in_flags,obus_side,out_flags);
```

```
 input [7:0] alu_side;
 input [3:0] in_flags;
 input arith_shift_left,arith_shift_right;
 output [7:0]obus_side;
 output [3:0]out_flags;
 reg [7:0] obus_side;
 reg [3:0] out_flags;
 reg [7:0] tl;
 reg c,v,z,n;
always @ (alu_side or arith_shift_left or arith_shift_right or in_flags)
 begin
  if (arith_shift_left)
   begin
    tl = {alu_side[6:0],1'b0};
      n = tl[7];
      z = ~|tl;
      c = alu_side[7];
      v = ^{alu_side[6],alu_side[7]};
     end
  else
   begin
    if (arith_shift_right)
     begin
      tl = {alu_side[7],alu_side[7:1]};
        n = tl[7];
        z = ~|tl;
        c = in_flags[2];
        v = in_flags[3];
       end
    else
       begin
      tl = alu_side[7:0];
        n = in_flags[0];
        z = in_flags[1];
        c = in_flags[2];
        v = in_flags[3];
     end
     end
  obus_side <= tl;
  out_flags <= {v,c,z,n};
 end
endmodule


module accumulatur_unit(i8,o8,load,zero,ck);
 input [7:0] i8;
 input load,zero,ck;
 output [7:0] o8;
 reg [7:0] o8;
always @ ( ck )
 begin
  if(~ ck )
   begin
    if(load)
     begin
      if(zero)
```

```
        o8 <= 8'b00000000;
     else
       o8 <= i8;
    end
     else
       ;
   end
 end
endmodule


module instruction_register_unit(i8,o8,load,ck);
 input [7:0] i8;
 input load,ck;
 output [7:0] o8;
 reg [7:0] o8;
always @ ( ck )
 begin
  if ( ~ck )
   begin
    if(load)
      o8 <= i8;
   end
 end
endmodule


module program_counter_unit(i12,o12,increment,load_page,load_offset,reset,ck);
 input [11:0] i12;
 input increment,load_page,load_offset,reset,ck;
 output [11:0] o12;
 reg [11:0] count ;
 reg [11:0] o12;
initial
 begin
    count = 12'b000000000000;
    o12 = 12'b000000000000;
 end
always @( ck )
 begin
  if ( ~ck )
   begin
    if(reset)
     count = 12'b000000000000;
    else
      begin
     if(increment)
        count <= count+1;
        else
         begin
       if(load_page)
           count[11:8] <= i12[11:8];
       if(load_offset)
           count[7:0] <= i12[7:0];
      end
    end
```

```
   end
  end
 always @( count )
  begin
   o12 <= count;
  end
 endmodule



 module memory_address_register_unit(i12,o12,load_page,load_offset,ck);
  input [11:0] i12;
  input load_page,load_offset,ck;
  output [11:0] o12;
  reg [11:0] o12;
 always @ ( ck )
  begin
   if ( ~ck )
    begin
     if( load_page )
       o12[11:8] <= i12[11:8];
     if( load_offset )
       o12[7:0] <= i12[7:0];
    end
  end
 endmodule

 module par_data_path(databus,adbus,clk,load_ac,zero_ac,load_ir,
           increment_pc,load_page_pc,load_offset_pc,reset_pc,
           load_page_mar,load_offset_mar,load_sr,cm_carry_sr,
           pc_on_mar_page_bus,ir_on_mar_page_bus,
           pc_on_mar_offset_bus,dbus_on_mar_offset_bus,
           pc_offset_on_dbus,obus_on_dbus,databus_on_dbus,
           mar_on_adbus,
           dbus_on_databus,
           arith_shift_left,arith_shift_right,alu_code,
           ir_lines,status);
  inout [7:0] databus;
  input clk,load_ac,zero_ac,load_ir,increment_pc,load_page_pc,load_offset_mar;
  input load_offset_pc,reset_pc,load_sr,cm_carry_sr,pc_on_mar_page_bus;
  input ir_on_mar_page_bus,pc_on_mar_offset_bus,dbus_on_mar_offset_bus;
  input pc_offset_on_dbus,obus_on_dbus,databus_on_dbus,mar_on_adbus;
  input dbus_on_databus,arith_shift_left,arith_shift_right,load_page_mar;
  input [2:0] alu_code;
  output [7:0] ir_lines;
  output [3:0] status;
  output [11:0] adbus;
  wire [7:0] ac_out,ir_out,alu_out,obus;
  wire [11:0] pc_out,mar_out;
  wire [7:0] dbus;
  wire [3:0] alu_flags,shu_flags,sr_out;
  wire [11:0] mar_bus;
 accumulatur_unit r1 (obus,ac_out,load_ac,zero_ac,clk);
 instruction_register_unit r2 (obus,ir_out,load_ir,clk);
 program_counter_unit r3 (mar_out,pc_out,increment_pc,load_page_pc,load_offset_pc,
 reset_pc,clk);
```

```
memory_address_register_unit r4 (mar_bus,mar_out,load_page_mar,load_offset_mar,clk);
status_register_unit r5 (shu_flags,sr_out,load_sr,cm_carry_sr,clk);
arithmetic_logic_unit l1 (dbus,ac_out,alu_code,sr_out,alu_out,alu_flags);
shifter_unit l2 (alu_out,arith_shift_left,arith_shift_right,alu_flags,obus,shu_flags);
//--------------------------------
assign mar_bus[7:0] = dbus_on_mar_offset_bus ?(dbus):(8'bZZZZZZZZ);
assign databus = dbus_on_databus ?(dbus):(8'bZZZZZZZZ);
assign dbus = obus_on_dbus ?(obus):(8'bZZZZZZZZ);
assign dbus = databus_on_dbus ?(databus):(8'bZZZZZZZZ);
assign mar_bus[11:8] = ir_on_mar_page_bus ?(ir_out[3:0]):(4'bZZZZ);
assign mar_bus[11:8] = pc_on_mar_page_bus ?(pc_out[11:8]):(4'bZZZZ);
assign mar_bus[7:0] = pc_on_mar_offset_bus ?(pc_out[7:0]):(8'bZZZZZZZZ);
assign dbus = pc_offset_on_dbus ?(pc_out[7:0]):(8'bZZZZZZZZ);
assign adbus = mar_on_adbus ?(mar_out):(12'bZZZZZZZZZZZZ);
assign status = sr_out;
assign ir_lines = ir_out;
endmodule


`include "par_para_1hot.v"
module par_cotrol_unit (clk, load_ac, zero_ac, load_ir,
        increment_pc, load_page_pc, load_offset_pc, reset_pc,
        load_page_mar, load_offset_mar, load_sr, cm_carry_sr,
        pc_on_mar_page_bus, ir_on_mar_page_bus,
        pc_on_mar_offset_bus, dbus_on_mar_offset_bus,
        pc_offset_on_dbus, obus_on_dbus, databus_on_dbus,
        mar_on_adbus,
        dbus_on_databus,
        arith_shift_left, arith_shift_right,
            alu_and, alu_not, alu_a, alu_add, alu_b, alu_sub,
        ir_lines, status,
        read_mem, write_mem, interrupt, halted, ready, grant);
output load_ac, zero_ac, load_ir, increment_pc, load_page_pc, load_offset_mar;
reg load_ac, zero_ac, load_ir, increment_pc, load_page_pc, load_offset_mar;
output load_offset_pc, reset_pc, load_sr, cm_carry_sr, pc_on_mar_page_bus;
reg load_offset_pc, reset_pc, load_sr, cm_carry_sr, pc_on_mar_page_bus;
output ir_on_mar_page_bus, pc_on_mar_offset_bus, dbus_on_mar_offset_bus;
reg ir_on_mar_page_bus, pc_on_mar_offset_bus, dbus_on_mar_offset_bus;
output pc_offset_on_dbus, obus_on_dbus, databus_on_dbus, mar_on_adbus;
reg pc_offset_on_dbus, obus_on_dbus, databus_on_dbus, mar_on_adbus;
output dbus_on_databus, arith_shift_left, arith_shift_right, load_page_mar;
reg dbus_on_databus, arith_shift_left, arith_shift_right, load_page_mar;
output read_mem, write_mem, halted;
reg read_mem, write_mem, halted;
output alu_and, alu_not, alu_a, alu_add, alu_b, alu_sub;
reg alu_and, alu_not, alu_a, alu_add, alu_b, alu_sub;
input [7:0] ir_lines;
input [3:0] status;
input clk, ready, grant, interrupt;
reg [9:1] present_state, next_state;

initial begin
   present_state <= `do_initials;
   next_state <= `do_initials;
end
```

```
always @(clk or interrupt) begin
   if (interrupt)
      present_state <= `do_initials;
   else begin
      if (~clk) present_state <= next_state;
   end
end

always @(present_state or ir_lines or status or interrupt or ready or grant)
begin
   {load_ac, zero_ac} <= 2'b00; load_ir <= 0;
   {increment_pc, load_page_pc, load_offset_pc, reset_pc} <= 4'b0000;
   {load_page_mar, load_offset_mar} <= 2'b00;
   {load_sr, cm_carry_sr} <= 2'b00;
   {pc_on_mar_page_bus, ir_on_mar_page_bus} <= 2'b00;
   {pc_on_mar_offset_bus, dbus_on_mar_offset_bus} <= 2'b00;
   {pc_offset_on_dbus, obus_on_dbus, databus_on_dbus} <= 3'b000;
   {mar_on_adbus, dbus_on_databus} <= 2'b00;
   {arith_shift_left, arith_shift_right} <= 2'b00;
   {alu_and, alu_not, alu_a, alu_add, alu_b, alu_sub} <= 6'b000000;
   {read_mem, write_mem} <= 2'b00; halted <= 0;
   case (present_state)

   `do_initials: begin : s1
      if (interrupt) begin
         reset_pc <= 1;
         next_state <= `do_initials;
      end else begin
         {pc_on_mar_page_bus, pc_on_mar_offset_bus} <= 2'b11;
         {load_page_mar, load_offset_mar} <= 2'b11;
         next_state <= `instr_fetch;
      end
   end

   `instr_fetch: begin : s2
      read_mem <= 1;
      if (grant) begin
         mar_on_adbus <= 1;
         if (ready) begin
            databus_on_dbus <= 1;
            alu_a <= 1;
            load_ir <= 1;
            increment_pc <= 1;
            next_state <= `do_one_bytes;
         end else
            next_state <= `instr_fetch;
      end else
            next_state <= `instr_fetch;
   end

   `do_one_bytes: begin : s3
      {pc_on_mar_page_bus, pc_on_mar_offset_bus} <= 2'b11;
      {load_page_mar, load_offset_mar} <= 2'b11;
      if (ir_lines[7:4] != `single_byte_instructions)
         next_state <= `opnd_fetch;
      else begin
```

```
        if (ir_lines[3:0] == `asl) arith_shift_left <= 1;
        if (ir_lines[3:0] == `asr) arith_shift_right <= 1;
        if (ir_lines[1] == 1) alu_not <= 1;
        else alu_b <= 1;
        if ((ir_lines[3] | ir_lines[1]) == 1) begin
            load_sr <= 1;
            load_ac <= 1;
        end
        if (ir_lines[2] == 1) cm_carry_sr <= 1;
        if ((ir_lines[3] == 0) & (ir_lines[0] == 1)) zero_ac <= 1;
        if (ir_lines[3:0] == `hlt) halted <= 1;
        next_state <= `instr_fetch;
    end
end

`opnd_fetch: begin : s4
    read_mem <= 1;
    if (grant) begin
        mar_on_adbus <= 1;
        if (ready) begin
            databus_on_dbus <= 1;
            dbus_on_mar_offset_bus <= 1;
            load_offset_mar <= 1;
            if (ir_lines[7:6] != `jsr_or_bra) begin
                ir_on_mar_page_bus <= 1;
                load_page_mar <= 1;
                if (ir_lines[4] == `indirect)
                    next_state <= `do_indirect;
                else
                    next_state <= `do_two_bytes;
            end else begin
                if (ir_lines[5]) next_state <= `do_jsr;
                else next_state <= `do_branch;
            end
            increment_pc <= 1;
        end else next_state <= `opnd_fetch;
    end else next_state <= `opnd_fetch;
end

`do_indirect: begin : s5
    read_mem <= 1;
    if (grant) begin
        mar_on_adbus <= 1;
        if (ready) begin
            databus_on_dbus <= 1;
            dbus_on_mar_offset_bus <= 1;
            load_offset_mar <= 1;
            next_state <= `do_two_bytes;
        end else next_state <= `do_indirect;
    end else next_state <= `do_indirect;
end

`do_two_bytes: begin : s6
    if (ir_lines[7:5] == `jmp) begin
        {load_page_pc, load_offset_pc} <= 2'b11;
        next_state <= `instr_fetch;
```

```
    end else begin
        if (ir_lines[7:5] == `sta) begin
            write_mem <= 1;
            if (grant) begin
                mar_on_adbus <= 1;
                alu_b <= 1;
                obus_on_dbus <= 1;
                dbus_on_databus <= 1;
                if (ready)
                    next_state <= `do_initials;
                else next_state <= `do_two_bytes;
            end else next_state <= `do_two_bytes;
        end else begin
            if (~ir_lines[7]) begin
                read_mem <= 1;
                if (grant) begin
                    mar_on_adbus <= 1;
                    if (ready) begin
                        databus_on_dbus <= 1;
                        load_ac <= 1;
                        load_sr <= 1;
                        if (~ir_lines[6]) begin
                            if (~ir_lines[5]) alu_a <= 1;
                            else alu_and <= 1;
                        end else begin
                            if (~ir_lines[5]) alu_add <= 1;
                            else alu_sub <= 1;
                        end
                            next_state <= `do_initials;
                    end else next_state <= `do_two_bytes;
                end else next_state <= `do_two_bytes;
            end else next_state <= `do_initials;
        end
    end
end

`do_jsr: begin : s7
    write_mem <= 1;
    if (grant) begin
        mar_on_adbus <= 1;
        pc_offset_on_dbus <= 1;
        dbus_on_databus <= 1;
        if (ready) begin
            load_offset_pc <= 1;
            next_state <= `continue_jsr;
        end else next_state <= `do_jsr;
    end else next_state <= `do_jsr;
end

`continue_jsr: begin : s8
    increment_pc <= 1;
    next_state <= `do_initials;
end

`do_branch: begin : s9
    if (|{status, ir_lines[3:0]}) load_offset_pc <= 1;
```

```
        next_state <= `do_initials;
    end

    default: begin : invalid_state
        next_state <= `do_initials;
    end
    endcase
end

endmodule



module par_central_processing_unit (clk, interrupt, read_mem, write_mem,
    databus, adbus, halted, ready, grant);
input clk, interrupt, ready, grant;
inout [7:0]databus;
output [11:0] adbus;
output read_mem, write_mem, halted;
wire load_ac, zero_ac;
wire load_ir;
wire increment_pc, load_page_pc, load_offset_pc, reset_pc;
wire load_page_mar, load_offset_mar;
wire load_sr, cm_carry_sr;
wire pc_on_mar_page_bus, ir_on_mar_page_bus;
wire pc_on_mar_offset_bus, dbus_on_mar_offset_bus;
wire pc_offset_on_dbus, obus_on_dbus, databus_on_dbus;
wire mar_on_adbus, dbus_on_databus;
wire arith_shift_left, arith_shift_right;
wire [2:0] alu_code;
wire [7:0] ir_lines;
wire [3:0] status;
par_cotrol_unit ctrl_u (clk,
        load_ac,zero_ac,
        load_ir,
        increment_pc,load_page_pc,load_offset_pc,reset_pc,
        load_page_mar,load_offset_mar,
        load_sr,cm_carry_sr,
        pc_on_mar_page_bus,ir_on_mar_page_bus,
        pc_on_mar_offset_bus,dbus_on_mar_offset_bus,
        pc_offset_on_dbus,obus_on_dbus,databus_on_dbus,
        mar_on_adbus,dbus_on_databus,
        arith_shift_left,arith_shift_right,
        alu_code,
        ir_lines,status,
        read_mem,write_mem,interrupt,
        halted,ready,grant);
par_data_path data_u(databus,adbus,
        clk,
        load_ac,zero_ac,
        load_ir,
        increment_pc,load_page_pc,load_offset_pc,reset_pc,
        load_page_mar,load_offset_mar,
        load_sr,cm_carry_sr,
        pc_on_mar_page_bus,ir_on_mar_page_bus,
```

```
        pc_on_mar_offset_bus,dbus_on_mar_offset_bus,
        pc_offset_on_dbus,obus_on_dbus,databus_on_dbus,
        mar_on_adbus,dbus_on_databus,
        arith_shift_left,arith_shift_right,
        alu_code,
        ir_lines,status);

endmodule
```

Appendix

E

Verilog Synthesis Examples

The examples of this appendix present Verilog code for synthesis of combinational and sequential circuits. The simpler examples are presented first. All examples are complete and can be synthesized using commercial synthesis tools. Each example is accompanied with a brief explanation of the key synthesis issues covered by the example. Section E.1 covers logic synthesis, E.2 covers sequential circuit synthesis, and E.3 covers state machine synthesizable code. The 21 examples presented here provide starting designers with most common tool-independent synthesizable styles.

E.1 Combinational Logic Synthesis

A Boolean expression used in a continuous assign statement will be synthesized to an equivalent logical circuit using library cells of the target hardware. The following may be synthesized to an array of complex gates, 2 or 3 NAND gates, or simply an AND and an OR gate array.

```
module and_or (a, b, c, z);
input [3:0] a, b, c;
output [3:0] z;
   assign z = a & b | c;
endmodule
```

Relational and arithmetic operators translate to logical circuits. The following description synthesizes to a comparator output selecting one of two inputs of a multiplexer to appear on the output of the circuit.

```
module bigger (a, b, z);
input [3:0] a, b;
output [3:0] z;
   assign z = (a > b) ? a : b;
endmodule
```

Condition operators are synthesizable. The following is a design of a 4-bit cascadable comparator. Multiple condition operators are cascaded to select one of three values for the output.

```
module cascadable_comparator (a, b, comp_in, comp_out);
input [3:0] a, b;
input [2:0] comp_in;
output [2:0] comp_out;
   assign comp_out =  (a > b) ? 3'b100 : ( (a < b) ? 3'b001 :
comp_in);
endmodule
```

For more complex decision making, multiple nestings of condition operators can be used. The description shown below synthesizes to a BCD to SSD decoder.

```
module seven_segment_decoder (bcd_in, ssd_out);
input [3:0] bcd_in;
output [6:0] ssd_out;
   assign  ssd_out =
      bcd_in == 4'b0000 ? 7'b1111110 :
      bcd_in == 4'b0001 ? 7'b0110000 :
      bcd_in == 4'b0010 ? 7'b1101101 :
      bcd_in == 4'b0011 ? 7'b1111001 :
      bcd_in == 4'b0100 ? 7'b0110011 :
      bcd_in == 4'b0101 ? 7'b1011011 :
      bcd_in == 4'b0110 ? 7'b1011111 :
      bcd_in == 4'b0111 ? 7'b1110000 :
      bcd_in == 4'b1000 ? 7'b1111111 :
      bcd_in == 4'b1001 ? 7'b1111011 :
                          7'b0000000;
endmodule
```

Output of one of two generated adders is selected by signal *s* to appear on the circuit output. Selection is done by a multiplexer. Outputs of the adders are not explicitly specified and may be combined with other logic.

```
module select_and_add (a, b, c, d, s, z);
input [7:0] a, b, c, d;
input s;
output [7:0] z;
   z = (s == 1) ? a + b : c + d;
endmodule
```

The following example explicitly specifies outputs of the adders. As in the previous example, signal *s* selects one of the two adder outputs. Add-first and multiplex outputs or multiplex inputs and add next can be enforced by use of explicit intermediate signals.

```
module select_and_add (a, b, c, d, s, z);
input [7:0] a, b, c, d;
input s;
```

```
output [7:0] z;
wire [7:0] ab, cd;
   assign cd = c + d;
   assign ab = a + b;
   assign z = (s == 1) ? ab : cd;
endmodule
```

Three-state bussing is used if *Z* values are assigned to **nets**. The following is a simple description that creates a three-state output bus with three 8-bit sources. The actual bus logic may be synthesized as three-state structures of basic logic gates depending on the library used.

```
module select_bus (a, b, c, sa, sb, sc, z);
input [7:0] a, b, c;
input sa, sb, sc;
output [7:0] z;
   assign z = (sa == 1) ? a : 8'bZZZZZZZZ;
   assign z = (sb == 1) ? b : 8'bZZZZZZZZ;
   assign z = (sc == 1) ? c : 8'bZZZZZZZZ;
endmodule
```

A general synthesizable style uses a procedural **always** block. For a combinational circuit a timing control statement includes all inputs of the combinational circuit. In a procedual block, if circuit outputs retain their values from one activation to another, latches will be synthesized on those output lines. In order to avoid latches, outputs of a combinational procedural block are set to their inactive values at the beginning of the block. In the example shown below, *flags, z_out* and *temp* are active high outputs and are initialized to 0.

```
module simple_alu (a, b, code_in, flags, z_out);
input [3:0] a, b;
input [2:0] code_in;
output [2:0] flags;
output [3:0] z_out;
reg [4:0] temp;
reg [3:0] z_out;
reg [2:0] flags;
   always @(a or b or code_in) begin
      flags <= 3'b000;  z_out <= 4'b0000;
       temp = 4'b0000;
      if (code_in == 3'b000) z_out <= a;
      else if (code_in == 3'b001) z_out <= b;
      else if (code_in == 3'b010) z_out <= a & b;
      else if (code_in == 3'b011) z_out <= a | b;
      else if (code_in == 3'b100) begin
         temp = a + b;
         {flags [2], z_out} <= temp;
      end else if (code_in == 3'b101) begin
         temp = a - b;
         {flags [2], z_out} <= temp;
      end else if (code_in [2:1] == 2'b11) begin
         if (a > b) begin
            if (code_in[0] == 0) begin
```

```
                z_out <= a;
                flags [1] <= 1;
             end else z_out <= b;
          end else if (a < b) begin
             if (code_in[0] == 0) z_out <= b;
             else begin
                z_out <= a;
                flags [0] <= 1;
             end
          end
       end
    end
endmodule
```

Loop, case, and other high-level procedural statements are synthesizable. As in the previous example all inputs are used in the event control statement at the beginning of the procedural block, and all outputs are set to their inactive values.

```
module priority_encoder (a, n, z, f);
input [7:0] a;
output [2:0] n;
output [7:0] z;
output f;
reg [2:0] n;
reg [7:0] z;
reg f;
reg [3:0] i;
   always @(a) begin
      n <= 3'b000; z <= 8'b00000000; f <= 0;
      for (i = 0; i <= 7; i = i + 1) begin
         if (f == 0 && a[i] == 1) begin
            n <= i;  z[i] <= 1; f <= 1;
         end
      end
   end
endmodule
```

The example below shows that sequential signal assignments provide another means of specifying logical expressions for synthesis.

```
module sequential_and_or (a, b, c, z);
input a, b, c;
output z;
reg z, t;
   always @ (a or b or c) begin
      t <= a & b;
      z <= t | c;
   end
endmodule
```

E.2 Sequential Circuit Synthesis

In the following description, signal z retains its value between activations of the procedural block. This description specifies a latching behavior and synthesizes as such. If the synthesis target library includes a latch to match or can be configured to match, this behavior, it will be used for generating output z; otherwise the latch will be built with discrete gates.

```
module basic_latch (s, r, z);
input s, r;
output z;
   always @( s or r)
      if (s == 1) z <= 1;
      else if (r == 1) z <= 0;
endmodule
```

The following describes a d-latch. As in the previous example, and unlike the combinational examples of Sec. E.1, the output of the circuit is not initialized in the procedural block. The output retains its value if d becomes active and c is 0.

```
module basic_d_latch (d, c, z);
input d, c;
output z;
reg z;
   always @ (d or c)
      if (c == 1) z <= d;
endmodule
```

Specifying a rising or a falling edge synthesizes to the appropriate type of flip-flop. For this purpose, an event control statement with **posedge** or **negedge** should be used.

```
module basic_d_flop (d, c, z);
input d, c;
output z;
reg z;
   always @ (posedge c) z <= d;
endmodule
```

A flip-flop can include asynchronous set and reset inputs. Enabling clock edge and active levels of clear and preset signals must be listed in a procedural block event control. The synthesis tool considers the clock as the signal for which an edge detection condition is not tested in the body of the procedural block.

```
module basic_srd_flop (d, c, s, r, z);
input d, c, s, r;
output z;
reg z;
   always @ (posedge c or posedge s or posedge r)
      if (s == 1 ) z <= 1;
      else if (r == 1) z <= 0;
      else z <= d;
endmodule
```

The description shown below uses an internal flag to hold the value of the *d* input on the rising edge of the clock. The flag value is then used in a combination block. The example shows using sequential and combination blocks.

```
module internal_flag (d, c, a, z);
input d, c, a;
output z;
reg flag, z;
   always @ (posedge c) flag <= d;
   always @ (a or d) z <= (flag == 1) ? a & d : 0;
endmodule
```

The styles presented above can be used for synthesis of most sequential circuits. Counters and special functional registers can be synthesized by assigning appropriate values to their outputs on the positive or negative edge of the clock.

```
module basic_up_counter (clock, count_out);
input clock;
output [7:0] count_out;
reg [7:0] count_out;
   always @ (posedge clock) count_out <= count_out + 1;
endmodule
```

As with the flip-flops, functional registers can also include asynchronous set and reset inputs. In the following description, the counter output is set to zero when the reset input becomes active.

```
module simple_up_counter (clock, reset, count_out);
input clock, reset;
output [7:0] count_out;
reg [7:0] count_out;
   always @ (posedge clock or posedge reset) begin
      if (reset == 1) count_out <= 0;
      else count_out <= count_out + 1;
   end
endmodule
```

For fully synchronous operation, all conditions of various operations of a sequential circuit must be moved inside the if-statement that detects the clock edge. The description shown below detects the edge of the clock and then zeros, increments, or decrements the counter based on *reset*, *updown*, *enable* inputs. This is a typical synthesizable description for a synchronous up-down counter.

```
module universal_synchronous_counter (clock, reset, enable,
updown, count_out);
input clock, reset, enable, updown;
output [7:0] count_out;
reg [7:0] count_out;
   always @ (posedge clock) begin
```

```
      if (reset == 1) count_out <= 0;
      else begin
         if (enable == 1 && updown == 1) count_out <= count + 1;
         else if (enable == 1 && updown == 0) count_out <= count
- 1;
      end
   end
endmodule
```

E.3 State Machine Synthesis

A state machine can be synthesized by use of a procedural block such as that used for the counter examples. States of the machine are defined by parameters. A variable—such as *current*, in the example below—holds the current active state. A case or if-statement tests this variable for every state of the machine and makes appropriate transitions based on the input values.

```
module moore_detector (x, clk, z);
input x, clk;
output z;
reg [2:0] current;
parameter [2:0]
            reset = 0,
            got1 = 1,
            got10 = 2,
            got101 =3,
            got1011 =4;
      initial current = reset;
      always @(posedge clk)
            case (current)
                  reset:
                        if (x==1) current = got1;  else
current = reset;
                  got1:
                        if (x==0) current = got10;  else
current = got1;
                  got10:
                        if (x==1) current = got101;  else
current = reset;
                  got101:
                        if (x==1) current = got1011;  else
current = got10;
                  got1011:
                        begin
                              if (x==1) current = got1;
else current = got10;
                        end
            endcase
      assign z = (current == got1011) ? 1 : 0;
endmodule
```

A synthesizable state machine can also be described by dedicating a procedural block for performing its logical and another for performing its sequential activities. In our example, three parameters define states of the machine, and *nxt* and *present* variables define inputs and outputs of state flip-flops respectively. This style allows flip-flop control in a block dedicated for the state machine registers. Coding style follows that described in Sec. E.1 for the combinational block and that described in Sec. E.2 for the sequential block.

```
module asynch_reset_detector (x, r, clk, z);
input x, r, clk;
output z;
reg [1:0] nxt, present;
reg z;
      parameter
              a = 0,
              b = 1,
              c = 2;
      initial nxt = a;
      always @(posedge clk or posedge r) begin : registering
              if (r==1) present = a;
              else present = nxt;
              end
      always @(present or x) begin : combinational
              z = 0;
              case (present)
                      a:
                              if (x==0) nxt = a;  else nxt = b;
                      b:
                              if (x==0) nxt = c;  else nxt = b;
                      c:
                              if (x==0) nxt = a;  else nxt = b;
                      default:
                              nxt = a;
              endcase
              if (present == c && x == 1) z = 1;
      end
endmodule
```

As in the previous example, only the *registering* block of the description of the state machine has to be modified for specification of preloading or resetting mechanism. The example shown below adds a synchronous reset to the out state machine example. Because only the clock edge is used for event control, inputs checked will only be considered synchronous with the clock.

```
module synch_reset_detector (x, r, clk, z);
input x, r, clk;
output z;
reg [1:0] nxt, present;
reg z;
      parameter
              a = 0,
              b = 1,
              c = 2;
```

```
    initial nxt = a;
    always @(posedge clk) begin : registering
            if (r==1) present = a;
            else present = nxt;
            end
    always @(present or x) begin : combinational
            z = 0;
            case (present)
                    a:
                            if (x==0) nxt = a;  else nxt = b;
                    b:
                            if (x==0) nxt = c;  else nxt = b;
                    c:
                            if (x==0) nxt = a;  else nxt = b;
                    default:
                            nxt = a;
            endcase
            if (present == c && x == 1) z = 1;
    end
endmodule
```

Appendix F

Software Accompanying this Book

This appendix describes software and programs that are included on the CD that accompanies this book. In the root directory of the CD five directories are *Examples*, *Navigator*, *Presentations*, *Simucad*, and *Veribest*. Contents of these directories will be presented here.

F.1 Examples

The *Examples* directory consists of *Silos3* and *MTI* subdirectories. Each subdirectory has Verilog code, test-benches, and simulation run results for the examples of the book. The examples are organized in directories according to the book chapters that they appear in.

The *Silos3* branch contains setup files and simulation run results of the of Simucad Silos3 Verilog simulation environment. For most of the examples, a project file (*.spj*) contains all files needed for simulation of the example and a test-bench for it. The test-bench is the top-level module in all projects and includes module instantiation and data generation for the module being tested.

The *MTI* branch has examples and simulation run results of Model Technology Verilog/VHDL simulator. List or postscript files containing simulation results for modules being tested are also included.

F.2 Navigator

Verilog Circuit Navigator is a Web-based electronic indexing tool for browsing examples of the book. Examples are organized according to their function and Verilog topic that they cover. Using the navigator, Verilog code for flip-flops, gates, state machines, and other hardware components can be accessed. Each code page has a brief description of the example, main Verilog constructs used in the example, and the section in the book that the example is described in. A Web page for an example also has pointers to a test-bench for the example and the simulation run results. Pointers to pages for module instances within a module are also provided.

Navigator can be used as an instruction tool or for looking up models and circuit templates. Circuits descriptions are available at various levels of abstraction. You can run the navigator by using D:\Navigator\index.htm (assuming that D is your CD drive) in the address field of your browser or by double clicking file *index.htm* in the Navigator subdirectory of the CD.

F.3 Presentations

Several VHDL and Verilog presentation slides are available in the *Presentations* directory of the CD. Presentations and short tutorial papers are in postscript or *.pdf* format.

F.4 Simucad Verilog Software

The *Silos3* subdirectory of the CD includes a self-extracting installation file for the Silos3 Verilog Simulation Environment. In addition, a serial adder example and its test-bench are included in this subdirectory. This example is used in the following description outlining steps for running the simulator.

To simulate the serial adder example in the Silos3 Simulation Environment, *seradd.v* circuit description file and *seraddTS.v* test-bench are needed. After starting the simulation environment, the following steps must be performed.

1. Select the **Project** menu to create a new project. A directory and a project filename must be entered. This example run uses *CD_demo* for the project name. The test-bench file, *seraddTS.v*, must appear last in the list of files.
2. In the main environment window click on the **GO** button to compile the Verilog files and run the simulation.
3. Open **Analyzer** and **Explorer** windows.
4. In the **Explorer** window click on the top-level module to expand and see module variables.

5. From the **Explorer** window drag variable names for which waveforms are to be displayed and drop them into the **Analyzer** window.
6. Zoom or scroll the **Analyzer** window to see the simulation run results.

F.5 Veribest VHDL Simulator

The *Veribest* subdirectory of the CD includes a self-extracting installation file for the Veribest VHDL Simulator. In addition, a serial adder example and its test-bench are included in this subdirectory. This example is used in the following description outlining steps for running the simulator.

To simulate the serial adder example using the Veribest VHDL Simulator, *seradd.vhd* circuit description file and *seraddTS.vhd* test-bench are needed. After starting the simulation environment, the following steps must be performed.

1. Open the **Create Workspace** window by pulling down the **Workspace** menu and selecting **New**. To create a new workspace, select a directory and enter a workspace name. This example run uses *CD_demo* for the workspace name.
2. In the window that opens after a workspace is created, add source file *seradd.vhd* and the test-bench file, *seraddTS.vhd*.
3. Pull down the **Workspace** menu and select **Compile All** to compile the serial adder and its test-bench.
4. Pull down the **Workspace** menu and select **Settings ...** to set the design root. For this purpose in the **Workspace Settings** window that opens, click on the **Simulate** tab, open the WORK library, select the top-level architecture (*INPUT_OUTPUT*) and click on **Set**.
5. Pull down the **Workspace** menu and select **Execute Simulator** to run the simulation.
6. Open **Waveform** and **Hierarchy Browser** windows.
7. In the **Hierarchy Browser** window, drag variable names for which waveforms are to be displayed and drop them in the **Waveform** window.
8. In the **Simulate** pull-down menu, select **Run Forever** to run the simulator until no more events occur in the simulation model.
9. Zoom or scroll waveforms in the **Waveform** window to see the simulation run results.

Index

Note: Bold page numbers indicate main references to indexed terms.

ABOUT THE AUTHOR

Zainalabedin Navabi, Ph.D., navabi@ece.neu.edu, is adjunct professor of electrical and computer engineering at Northeastern University and the author of both editions of *VDHL: Analysis and Modeling of Digital Systems*, published by McGraw-Hill. Since 1981, Dr. Navabi has worked in the design, definition, and implementation of hardware description languages and the synthesis and testing of digital systems. He has developed and supervised the development of many HDL-related software packages and tools, and has directed projects in VLSI design, test synthesis, simulation, synthesis, and other aspects of digital system automation. He has served as a consultant for several EDA companies developing HDL-based tools and environments. Dr. Navabi is a member of ACM, IEEE, IEEE Computer Society, and is an active participant in the IEEE DASC committee that sets standards related to hardware description languages.

SOFTWARE AND INFORMATION LICENSE

The software and information on this diskette (collectively referred to as the "Product") are the property of The McGraw-Hill Companies, Inc. ("McGraw-Hill") and are protected by both United States copyright law and international copyright treaty provision. You must treat this Product just like a book, except that you may copy it into a computer to be used and you may make archival copies of the Products for the sole purpose of backing up our software and protecting your investment from loss.

By saying "just like a book," McGraw-Hill means, for example, that the Product may be used by any number of people and may be freely moved from one computer location to another, so long as there is no possibility of the Product (or any part of the Product) being used at one location or on one computer while it is being used at another. Just as a book cannot be read by two different people in two different places at the same time, neither can the Product be used by two different people in two different places at the same time (unless, of course, McGraw-Hill's rights are being violated).

McGraw-Hill reserves the right to alter or modify the contents of the Product at any time.

This agreement is effective until terminated. The Agreement will terminate automatically without notice if you fail to comply with any provisions of this Agreement. In the event of termination by reason of your breach, you will destroy or erase all copies of the Product installed on any computer system or made for backup purposes and shall expunge the Product from your data storage facilities.

LIMITED WARRANTY

McGraw-Hill warrants the physical diskette(s) enclosed herein to be free of defects in materials and workmanship for a period of sixty days from the purchase date. If McGraw-Hill receives written notification within the warranty period of defects in materials or workmanship, and such notification is determined by McGraw-Hill to be correct, McGraw-Hill will replace the defective diskette(s). Send request to:

Customer Service
McGraw-Hill
Gahanna Industrial Park
860 Taylor Station Road
Blacklick, OH 43004-9615

The entire and exclusive liability and remedy for breach of this Limited Warranty shall be limited to replacement of defective diskette(s) and shall not include or extend to any claim for or right to cover any other damages, including but not limited to, loss of profit, data, or use of the software, or special, incidental, or consequential damages or other similar claims, even if McGraw-Hill has been specifically advised as to the possibility of such damages. In no event will McGraw-Hill's liability for any damages to you or any other person ever exceed the lower of suggested list price or actual price paid for the license to use the Product, regardless of any form of the claim.

THE McGRAW-HILL COMPANIES, INC. SPECIFICALLY DISCLAIMS ALL OTHER WARRANTIES, EXPRESS OR IMPLIED, INCLUDING BUT NOT LIMITED TO, ANY IMPLIED WARRANTY OF MERCHANTABILITY OR FITNESS FOR A PARTICULAR PURPOSE.

Specifically, McGraw-Hill makes no representation or warranty that the Product is fit for any particular purpose and any implied warranty of merchantability is limited to the sixty day duration of the Limited Warranty covering the physical diskette(s) only (and not the software or information) and is otherwise expressly and specifically disclaimed.

This Limited Warranty gives you specific legal rights; you may have others which may vary from state to state. Some states do not allow the exclusion of incidental or consequential damages, or the limitation on how long an implied warranty lasts, so some of the above may not apply to you.

This Agreement constitutes the entire agreement between the parties relating to use of the Product. The terms of any purchase order shall have no effect on the terms of this Agreement. Failure of McGraw-Hill to insist at any time on strict compliance with this Agreement shall not constitute a waiver of any rights under this Agreement. This Agreement shall be construed and governed in accordance with the laws of New York. If any provision of this Agreement is held to be contrary to law, that provision will be enforced to the maximum extent permissible and the remaining provisions will remain in force and effect.